Lecture Notes in Control and Information Sciences

Edited by M. Thoma

72

A. Isidori

Nonlinear Control Systems: An Introduction

Springer-Verlag
Berlin Heidelberg New York Tokyo

Author
Prof. Alberto Isidori
Dipartimento di Informatica e Sistemistica
Università di Roma »La Sapienza«
18 Via Eudossiana
00184 Rome (Italy)

ISBN 3-540-15595-3 Springer-Verlag Berlin Heidelberg New York Tokyo
ISBN 0-387-15595-3 Springer-Verlag New York Heidelberg Berlin Tokyo

Library of Congress Cataloging in Publication Data
Isidori, Alberto:
Nonlinear control systems: an introduction/A. Isidori.
Berlin; Heidelberg; New York; Tokyo: Springer 1985.
(Lecture notes in control and information sciences; Vol. 72)
NE: GT

Offsetprinting: Mercedes-Druck, Berlin
Binding: Lüderitz und Bauer, Berlin
2161/3020-543210

PREFACE

This volume was planned as a textbook for a graduate course on nonlinear multivariable feedback systems. Most of it was prepared while the author was teaching a similar course at the Department of Systems Sciences and Mathematics of the Washington University in St. Louis, in the 1983 fall semester. The purpose of the volume is to present a self-contained description of the fundamentals of the theory of nonlinear feedback control systems, with special emphasis on the differential-geometric approach.

In the last decade, differential geometry has proven to be as successful to the study of nonlinear systems as Laplace transform and complex functions theory were in the '50s to the study of single-input single-output linear systems and linear algebra in the '60s to the study of multivariable linear systems. Typical "synthesis" problems like disturbance isolation, noninteraction, shaping of the input-output response via feedback, can be dealt with relative ease, with tools that are well within the reach of a (mathematically oriented) control engineer. The purpose of this volume is to make the reader aquainted with major methods and results, and to make him able to explore the constantly growing literature.

The book is organized as follows. Chapter I introduces invariant distributions, a fundamental tool in the analysis of the internal structure of nonlinear systems. With the aid of this concept, it is shown that a nonlinear system locally exhibits Kalman-like decompositions into "reachable/unreachable" parts and/or "observable/unobservable" parts. Chapter II explains to what extent there may exist global decompositions, corresponding to a partition of the whole state space into "lower dimensional" reachability and/or indistinguishability subsets. Chapter III describes various "formats" in which the input-output map of a nonlinear system may be represented, and provides a short description of the fundamentals of realization theory. Chapters IV and V deal with the synthesis of feedback control laws. In the first of these, disturbance decoupling and noninteracting control are dealt with, along the so-called "geometric approach", that proved to be quite successful for the solution of similar synthesis problems in linear multivariable systems. In Chapter V it is shown that nonlinear state-feedback may be used in order to make a given system to behave, internally and/or externally, like a linear one. In particular, feedback may be used in order to shape the input-output behavior in some prescribed way.

The reader is supposed to be familiar with the basic concepts of linear systems theory. Moreover, some knowledge of the fundamentals of differential geometry is required. There are several excellent textbooks available to this end, and some of them are quoted among the references. However, in order to make the volume as much as possible self-contained, and particularly to unify the notations, the most important notions and results of frequent usage are collected - without proof - in the Appendix.

The author of this book is particularly grateful to Professor A. Ruberti, for his constant encouragement, to Professors J. Zaborszky and T.J. Tarn for their interest and generous support, to Professor A.J. Krener who, especially in the course a joint research venture, was a source of inspiration for many of the ideas developed in this volume. The author would also like to thank Professor M. Thoma for his encouragement during the preparation of this work and Professors C. Byrnes, M. Fliess, P. Kokotovic and S. Monaco for many stimulating discussions.

Rome, March 1985

TABLE OF CONTENTS

CHAPTER I

LOCAL DECOMPOSITIONS OF CONTROL SYSTEMS

1. Introduction

In this section we review some basic results from the theory of linear systems, with the purpose of describing some fundamental properties which find close analogues in the theory of nonlinear systems.

Usually, a linear control system is described by equations of the form

$$\dot{x} = Ax + Bu$$

$$y = Cx$$

in which the state x belongs to X, an n-dimensional vector space and the input u and the output y belong respectively to an m-dimensional vector space U and ℓ-dimensional vector space Y. The mappings $A : X \to X$, $B : U \to X$, $C : X \to Y$ are linear mappings.

Suppose that there exists a d-dimensional subspace V of X with the following property:

(i) V is invariant under the mapping A, i.e. is such that $Ax \in V$ for all $x \in V$;

then, it is known from linear algebra that there exists a basis for X (namely, any basis $(v_1,\ldots,v_n)$ with the property that $(v_1,\ldots,v_d)$ is also a basis for V) in which A is represented by means of a block-triangular matrix

$$\begin{pmatrix} A_{11} & A_{12} \\ 0 & A_{22} \end{pmatrix}$$

whose elements on the lower $(n-d)$ rows and left d columns are vanishing.

Moreover, if this subspace V is such that:

(ii) V contains the image of the mapping B, i.e. is such that $Bu \in V$ for all $u \in U$;

then, choosing again the same basis as before for X, regardless of the choice of basis in U, the mapping B is represented by a matrix

$$\begin{pmatrix} B_1 \\ 0 \end{pmatrix}$$

whose last n-d rows are vanishing.

Thus, if there exists a subspace V which satisfies (i) and (ii), then there exists a choice of coordinates for X in which the control system is described by a set of differential equations of the form

$$\dot{x}_1 = A_{11}x_1 + A_{12}x_2 + B_1u$$

$$\dot{x}_2 = A_{22}x_2$$

By x_1 and x_2 we denote the d-vector and, respectively, the n-d vector formed by taking the first d and, respectively, the last n-d coordinates of a point x of X in the selected basis.

The representation thus obtained is particularly interesting when studying the behavior of the system under the action of the control u. At time T, the coordinates of x(T) are

$$x_1(T) = \exp(A_{11}T)x_1(0) + \int_0^T \exp(A_{11}(T-\tau))A_{12}\exp(A_{22}\tau)d\tau x_2(0) + $$
$$+ \int_0^T \exp(A_{11}(T-\tau))B_1u(\tau)d\tau$$

$$x_2(T) = \exp(A_{22}T)x_2(0)$$

From this we see that the set of coordinates denoted with x_2 does not depend on the input u but only on the time T. The set of points that can be reached at time T, starting from x(0), under the action of the input lies inside the set of points of X whose x_2 coordinate is equal to $\exp(A_{22}T)x_2(0)$. In other words, if we let $x^o(T)$ denote the point of X reached at time T when $u(t) = 0$ for all $t \in [0,T]$, we observe that the state x(T) may be expressed as

$$x(T) = x^o(T) + v$$

where v is a vector in V. Therefore, the set of points that can be reached at time T, starting from x(0), lies inside the set

$$S_T = x^o(T) + V$$

Let us now make the additional assumption that the subspace V, which is the starting point of our considerations, is such that:

(iii) V is the smallest subspace which satisfies (i) and (ii) (i.e. is contained in any other subspace of X which satisfies both (i) and (ii)).

It is known from the linear theory that this happens if and only if

$$V = \sum_{i=0}^{n-1} \operatorname{Im}(A^i B)$$

and, moreover, that in this case the pair (A_{11}, B_1) is a reachable pair, i.e. satisfies the condition

$$\operatorname{rank}(B_1 \; A_{11}B_1 \; \dots \; A_{11}^{d-1}B_1) = d$$

or, in other words, for each $x_1 \in \mathbf{R}^d$ there exists an input u, defined on $[0,T]$, such that

$$x_1 = \int_0^T \exp(A_{11}(T-\tau))B_1 u(\tau)d\tau$$

Then, if V is such that the condition (iii) is also satisfied, starting from $x(0)$ we can reach at time T any state of the form $x^o(T) + v$ with $v \in V$ or, in other words, any state belonging to the set S_T. This set is therefore exactly the set of the states reachable at time T starting from $x(0)$.

This result suggests the following considerations. Given a linear control system, let V be the smallest subspace of X satisfying (i) and (ii). Associated with V there is a *partition* of X into subsets of the form

$$x + V$$

with the property that each one of these subsets coincides with the set of points reachable at some time T starting from a suitable point of X. Moreover, these subsets have the structure of a d-dimensional *flat submanifold* of X.

An analysis similar to the one developed so far can be carried out by examining the interaction between state and output. In this case we consider a d-dimensional subspace W of X such that

(i) W is invariant under the mapping A
(ii) W is contained into the kernel of the mapping C (i.e. is such that $Cx = 0$ for all $x \in W$)
(iii) W is the largest subspace which satisfies (i) and (ii) (i.e. contains any other subspace of X which satisfies both (i) and (ii)).

Then, there is a choice of coordinates for X in which the control system is described by equations of the form

$$\dot{x}_1 = A_{11}x_1 + A_{12}x_2 + B_1u$$

$$\dot{x}_2 = A_{22}x_2 + B_2u$$

$$y = C_2x_2$$

From this we see that the set of coordinates denoted with x_1 has no influence on the output y. Thus any two initial states whose last n-d coordinates coincide produce two identical outputs under any input, i.e. are indistinguishable. Actually, any two states whose last n-d coordinates coincide are such that their difference is an element of W and, then, we may conclude that any two states belonging to a set of the form x+W are indistinguishable.

Moreover, we know that the condition (iii) is satisfied if and only if

$$W = \bigcap_{i=0}^{n-1} \ker(CA^i)$$

and, if this is the case, the pair (C_2, A_{22}) is observable, i.e. satisfies the condition

$$\text{rank}(C_2' \quad A_{22}'C_2' \ \ldots (A_{22}')^{d-1}C_2') = d$$

or, in other words,

$$C_{22}\exp(A_{22}t)x_2 \equiv 0 \Rightarrow x_2 = 0$$

Then, if two initial states are such that their difference does not belong to W, they may be distinguished from each other by the output produced under zero input.

Again we may synthesize the above discussion with the following considerations. Given a linear control system, let W be the largest subspace of X satisfying (i) and (ii). Associated with W there is a *partition* of X into subsets of the form

$$x + W$$

with the property that each one of these subsets coincides with the set of points that are indistinguishable from a fixed point of X. Moreover, these subsets have the structure of a d-dimensional *flat submanifold* of X.

In the following sections of this chapter and in the following chapter we shall deduce similar decompositions for nonlinear control systems.

2. Distributions on a Manifold

The easiest way to introduce the notion of distribution Δ on a manifold N is to consider a mapping assigning to each point p of N a *subspace* $\Delta(p)$ of the tangent space T_pN to N at p. This is not a rigorous definition, in the sense that we have only defined the domain N of Δ without giving a precise characterization of its codomain. Deferring for a moment the need for a more rigorous definition, we proceed by adding some conditions of regularity. This is imposed by assuming that for each point p of N there exist a neighborhood U of p and a set of smooth vector fields defined on U, denoted $\{\tau_i : i \in I\}$, with the property that,

$$\Delta(q) = \text{span}\{\tau_i(q) : i \in I\}$$

for all $q \in U$. Such an object will be called a smooth *distribution* on N. Unless otherwise noted, in the following sections we will use the term "distribution" to mean a smooth distribution.

Pointwise, a distribution is a linear object. Based on this property, it is possible to extend a number of elementary concepts related to the notion of subspace. Thus, if $\{\tau_i : i \in I\}$ is a set of vector fields defined on N, their *span*, written $\text{sp}\{\tau_i : i \in I\}$, is the distribution defined by the rule(*)

$$\text{sp}\{\tau_i : i \in I\} : p \mapsto \text{span}\{\tau_i(p) : i \in I\}$$

If Δ_1 and Δ_2 are two distributions, their *sum* $\Delta_1 + \Delta_2$ is defined by taking

$$\Delta_1 + \Delta_2 : p \mapsto \Delta_1(p) + \Delta_2(p)$$

and their *intersection* $\Delta_1 \cap \Delta_2$ by taking

$$\Delta_1 \cap \Delta_2 : p \mapsto \Delta_1(p) \cap \Delta_2(p)$$

(*) In order to avoid confusions, we use the symbol $\text{span}\{\cdot\}$ to denote any $\mathbb{R}$-linear combination of elements of some $\mathbb{R}$-vector space (in particular, tangent vectors *at a point*). The symbol $\text{sp}\{\cdot\}$ is used to denote a *distribution* (or a *codistribution*, see later).

A distribution Δ_1 *is contained* in the distribution Δ_2 and is written $\Delta_1 \subset \Delta_2$ if $\Delta_1(p) \subset \Delta_2(p)$ for all $p \in N$. A vector field τ *belongs* to a distribution Δ and is written $\tau \in \Delta$ if $\tau(p) \in \Delta(p)$ for all $p \in N$.

The *dimension* of a distribution Δ at $p \in N$ is the dimension of the subspace $\Delta(p)$ of T_pN.

Note that the span of a given set of smooth vector fields is a smooth distribution. Likewise, the sum of two smooth distributions is smooth. However, the intersection of two such distributions may fail to be smooth. This may be seen in the following example.

(2.1) *Example*. Let $M = \mathbb{R}^2$, and

$$\Delta_1 = sp\{\frac{\partial}{\partial x_1} + \frac{\partial}{\partial x_2}\}$$

$$\Delta_2 = sp\{(1+x_1)\frac{\partial}{\partial x_1} + \frac{\partial}{\partial x_2}\}$$

Then we have

$$(\Delta_1 \cap \Delta_2)(x) = \{0\} \quad \text{if} \quad x_1 \neq 0$$

$$(\Delta_1 \cap \Delta_2)(x) = \Delta_1(x) = \Delta_2(x) \quad \text{if} \quad x_1 = 0$$

This distribution is not smooth because it is not possible to find a smooth vector field on $\mathbb{R}^2$ which is zero everywhere but on the line $x_1 = 0$. □

Since sometimes it is useful to take the intersection of smooth distributions Δ_1 and Δ_2 , one may overcome the problem that $\Delta_1 \cap \Delta_2$ is possibly non-smooth with the aid of the following concepts. Suppose Δ is a mapping which assigns to each point $p \in N$ a subspace $\Delta(p)$ of T_pN and let $M(\Delta)$ be the set of all smooth vector fields defined on N which at p take values in $\Delta(p)$, i.e.

$$M(\Delta) = \{\tau \in V(N) : \tau(p) \in \Delta(p) \text{ for all } p \in N\}$$

Then, it is not difficult to see that the span of $M(\Delta)$, in the sense defined before, is a smooth distribution contained in Δ.

(2.2) *Remark*. Recall that the set $V(N)$ of all smooth vector fields defined on N may be given the structure of a vector space over $\mathbb{R}$ and, also, the structure of a module over $C^\infty(N)$, the ring of all smooth real-valued functions defined on N. The set $M(\Delta)$ defined before (which is non-empty because the zero element of $V(N)$ belongs to $M(\Delta)$ for any

Δ) is a subspace of the vector space V(N) and a submodule of the module V(N). From this is it easily seen that the span of $M(\Delta)$ is contained in Δ. □

Note that if Δ' is any smooth distribution contained in Δ, then Δ' is contained in the span of $M(\Delta)$, so the span of $M(\Delta)$ is actually the *largest* smooth distribution contained in Δ. To identify this distribution we shall henceforth use the notation

$$\text{smt}(\Delta) \overset{\Delta}{=} \text{sp}\ M(\Delta)$$

i.e. we look at the span of $M(\Delta)$ as the "smoothing" of Δ. Note also that if Δ is smooth, then smt$(\Delta) = \Delta$.

Thus, if $\Delta_1 \cap \Delta_2$ is non-smooth, we shall rather consider the distribution smt$(\Delta_1 \cap \Delta_2)$.

(2.3) *Remark*. Note that $M(\Delta)$ may not be the unique subspace of V(N), or submodule of V(N), whose span coincides with smt(Δ). But if M' is any other subspace of V(N), or submodule of V(N), with the property that sp M' = smt(Δ), then $M' \subset M(\Delta)$.

(2.4) *Example*. Let $N = \mathbb{R}$, and

$$\Delta = \text{sp}\{x \frac{\partial}{\partial x}\}$$

Then $M(\Delta)$ is the set of all vector fields of the form $c(x)\frac{\partial}{\partial x}$ where $c(x)$ is a smooth function defined on $\mathbb{R}$ which vanishes at $x = 0$. Clearly Δ is smooth and coincides with smt(Δ). There are many submodules of $V(\mathbb{R})$ which span Δ, for instance

$$M_1 = \{\tau \in V(\mathbb{R}) : \tau(x) = c(x)x \frac{\partial}{\partial x} \text{ and } c \in C^\infty(\mathbb{R})\}$$

$$M_2 = \{\tau \in V(\mathbb{R}) : \tau(x) = c(x)x^2 \frac{\partial}{\partial x} \text{ and } c \in C^\infty(\mathbb{R})\}$$

Both are submodules of $M(\Delta)$, M_2 is a submodule of M_1 but M_1 is not a submodule of M_2 because is not possible to express every function $c(x)x$ as $\hat{c}(x)x^2$ with $\hat{c} \in C^\infty(\mathbb{R})$. □

(2.5) *Remark*. The previous considerations enable us to give a rigorous definition of a smooth distribution in the following way. A smooth distribution is a submodule M of V(N) with the following property: if θ is a smooth vector field such that for all $p \in N$

$$\theta(p) \in \text{span}\{\tau(p) : \tau \in M\}$$

then θ belongs to M. □

Other important concepts associated with the notion of distribution are the ones related to the "behavior" of a given Δ as a "function" of p. We have already seen how it is possible to characterize the quality of being smooth, but there are other properties to be considered.

A distribution Δ is *nonsingular* if there exists an integer d such that

(2.6) $$\dim \Delta(p) = d$$

for all $p \in N$. A singular distribution, i.e. a distribution for which the above condition is not satisfied, is sometimes called a distribution of variable dimension. If a distribution Δ is such that the condition (2.6) is satisfied for all p belonging to an open subset U of N, then we say that Δ is nonsingular on U. A point p is a *regular* point of a distribution Δ if there exists a neighborhood U of p with the property that Δ is nonsingular on U.

There are some interesting properties related to these notions, whose proof is left to the reader.

(2.7) *Lemma*. Let Δ be a smooth distribution and p a regular point of Δ. Suppose $\dim \Delta(p) = d$. Then there exist an open neighborhood U of p and a set $\{\tau_1,\ldots,\tau_d\}$ of smooth vector fields defined on U with the property that every smooth vector field τ belonging to Δ admits on U a representation of the form

(2.8) $$\tau = \sum_{i=1}^{d} c_i \tau_i$$

where each c_i is a real-valued smooth function defined on U. □

A set of d vector fields which makes (2.8) satisfied will be called *a set of local generators* for Δ at p.

(2.9) *Lemma*. The set of all regular points of a distribution Δ is an open and dense submanifold of N.

(2.10) *Lemma*. Let Δ_1 and Δ_2 be two smooth distributions with the property that Δ_2 is nonsingular and $\Delta_1(p) \subset \Delta_2(p)$ at each point p of a dense submanifold of N. Then $\Delta_1 \subset \Delta_2$.

(2.11) *Lemma*. Let Δ_1 and Δ_2 be two smooth distributions with the property that Δ_1 is nonsingular, $\Delta_1 \subset \Delta_2$ and $\Delta_1(p) = \Delta_2(p)$ at each point p of a dense submanifold of N. Then $\Delta_1 = \Delta_2$. □

We have seen before that the intersection of two smooth distributions may fail to be smooth. However, around a regular point this cannot happen, as we see from the following result.

(2.12) *Lemma*. Let p be a regular point of Δ_1, Δ_2 and $\Delta_1 \cap \Delta_2$. Then there exists a neighborhood U of p with the property that $\Delta_1 \cap \Delta_2$ restricted to U is smooth. □

A distribution is *involutive* if the Lie bracket $[\tau_1,\tau_2]$ of any pair of vector fields τ_1 and τ_2 belonging to Δ is a vector field which belongs to Δ, i.e. if

$$\tau_1 \in \Delta,\ \tau_2 \in \Delta \Rightarrow [\tau_1,\tau_2] \in \Delta$$

(2.13) *Remark*. It is easy to see that a nonsingular distribution of dimension d is involutive if and only if, at each point p, any set of local generators $\tau_1,\ldots,\tau_d$ defined on a neighborhood U of p is such that

$$[\tau_i,\tau_j] = \sum_{\ell=1}^{d} c_{ij}^{\ell}\tau_\ell$$

where each c_{ij}^{ℓ} is a real-valued smooth function defined on U. □

If f is a vector field and Δ a distribution on N we denote by $[f,\Delta]$ the distribution

$$[f,\Delta] = \mathrm{sp}\{[f,\tau] \in V(N) : \tau \in \Delta\} \qquad (2.14)$$

Note that $[f,\Delta]$ is a smooth distribution, even if Δ is not. Using this notation, one can say that a distribution is involutive if and only if $[f,\Delta] \subset \Delta$ for all $f \in \Delta$.

Sometimes, it is useful to work with objects that are dual to the ones defined above. In the same spirit of the definition given at the beginning of this section, we say that a *codistribution* Ω on N is a mapping assigning to each point p of N a subspace $\Omega(p)$ of the cotangent space $T_p^*(N)$. A smooth codistribution is a codistribution Ω on N with the property that for each point p of N there exist a neighborhood U of p and a set of smooth covector fields (smooth one-forms) defined on U, denoted $\{\omega_i : i \in I\}$, such that

$$\Omega(q) = \mathrm{span}\{\omega_i(q) : i \in I\}$$

for all $q \in U$.

In the same manner as we did for distributions we may define the dimension of a codistribution at p, and construct codistributions by taking the span of a given set of covector fields, or else by adding or intersecting two given codistributions, etc. always looking at a pointwise characterization of the objects we are dealing with.

Sometimes, one can construct codistributions starting from given distributions and conversely. The natural way to do this is the following: given a distribution Δ on N, the *annihilator* of Δ, denoted $\Delta^{\perp}$, is the codistribution on N defined by the rule

$$\Delta^{\perp} : p \mapsto \{ v^* \in T_p^*N : \langle v^*, v \rangle = 0 \text{ for all } v \in \Delta(p) \}$$

Conversely, the *annihilator* of Ω, denoted $\Omega^{\perp}$, is the distribution defined by the rule

$$\Omega^{\perp} : p \mapsto \{ v \in T_pN : \langle v^*, v \rangle = 0 \text{ for all } v^* \in \Omega(p) \}$$

Distributions and codistributions thus related possess a number of interesting properties. In particular, the sum of the dimensions of Δ and of $\Delta^{\perp}$ is equal to the dimension of N. The inclusion $\Delta_1 \subset \Delta_2$ is satisfied if and only if the inclusion $\Delta_1^{\perp} \supset \Delta_2^{\perp}$ is satisfied. The annihilator $(\Delta_1 \cap \Delta_2)^{\perp}$ of an intersection of distributions is equal to the sum $\Delta_1^{\perp} + \Delta_2^{\perp}$.

Like in the case of the distributions, some care is required when dealing with the quality of being smooth for codistributions constructed in some of the ways we described before. Thus it is easily seen that the span of a given set of smooth covector fields, as well as sum of two smooth codistributions is again smooth. But the intersection of two such codistributions may not need to be smooth.

Moreover, the annihilator of a smooth distribution may fail to be smooth, as it is shown in the following example.

(2.15) *Example*. Let $N = \mathbb{R}$

$$\Delta = \mathrm{sp}\{x \frac{\partial}{\partial x}\}$$

Then

$$\Delta^{\perp}(x) = \{0\} \quad \text{if } x \neq 0$$

$$\Delta^{\perp}(x) = T_x^*N \quad \text{if } x = 0$$

and we see that $\Delta^{\perp}$ is not smooth because it is not possible to find a smooth covector field on $\mathbb{R}$ which is zero everywhere but on the point

$x = 0$. □

Or, else, the annihilator of a smooth codistribution may not be smooth, as in the following example.

(2.16) *Example*. Consider again the two distributions Δ_1 and Δ_2 described in the Example (2.1). One may easily check that

$$\Delta_1^{\perp} = \mathrm{sp}\{dx_1 - dx_2\}$$

$$\Delta_2^{\perp} = \mathrm{sp}\{dx_1 - (1+x_1)dx_2\}$$

The intersection $\Delta_1 \cap \Delta_2$ is not smooth but its annihilator $\Delta_1^{\perp} + \Delta_2^{\perp}$ is smooth, because both $\Delta_1^{\perp}$ and $\Delta_2^{\perp}$ are smooth. □

One may easily extend Lemmas (2.7) to (2.12). In particular, if p is a regular point of a codistribution Ω and $\dim \Omega(p) = d$, then it is possible to find an open neighborhood U of p and a set $\{\omega_1,\ldots,\omega_d\}$ of smooth covector fields defined on U, such that every smooth covector field ω belonging to Ω can be expressed on U as

$$\omega = \sum_{i=1}^{d} c_i \omega_i$$

where each c_i is a real-valued smooth function defined on U. The set $\{\omega_1,\ldots,\omega_d\}$ is called a set of local generators for Ω at p.

We have seen before that the annihilator of a smooth distribution Δ may fail to be smooth. However, around a regular point of Δ this cannot happen, as we see from the following result.

(2.17) *Lemma*. Let p be a regular point of Δ. Then p is a regular point of $\Delta^{\perp}$ and there exists a neighborhood U of p with the property that $\Delta^{\perp}$ restricted to U is smooth. □

We conclude this section with some notations that are frequently used. If f is a vector field and Ω a codistribution on N we denote by $L_f\Omega$ the smooth codistribution

$$L_f\Omega = \mathrm{sp}\{L_f\omega \in V^*(N) : \omega \in \Omega\} \tag{2.18}$$

If h is a real-valued smooth function defined on N, one may associate with h a distribution, written $\ker(h_*)$, defined by

$$\ker(h_*) : p \mapsto \{v \in T_pN : h_*v = 0\}$$

One may also associate with h a codistribution, taking the span of the covector field dh. It is easy to verify that the two objects thus defined are one the annihilator of the other, i.e. that

$$(sp(dh))^{\perp} = ker(h_*).$$

3. Frobenius Theorem

In this section we shall establish a correspondence between the notion of distribution on a manifold N and the existence of partitions of N into lower dimensional submanifolds. As we have seen at the beginning of this chapter, partitions of the state space into lower dimensional submanifolds are often encountered when dealing with reachability and/or observability of control systems.

We begin our analysis with the following definition. A *nonsingular* d-dimensional distribution Δ on N is *completely integrable* if at each $p \in N$ there exists a cubic coordinate chart (V,ξ) with coordinate functions $\xi_1,\ldots,\xi_n$, such that

$$\Delta(q) = \operatorname{span}\{(\frac{\partial}{\partial\xi_1})_q,\ldots,(\frac{\partial}{\partial\xi_d})_q\} \tag{3.1}$$

for all $q \in V$.

There are two important consequences related to the notion of completely integrable distribution. First of all, observe that if there exists a cubic coordinate chart (V,ξ), with coordinate functions $\xi_1,\ldots,\xi_n$, such that (3.1) is satisfied, then any *slice* of V passing through any point p of V and defined by

$$S_p = \{q \in V : \xi_i(q) = \xi_i(p);\ i = d+1,\ldots,n\} \tag{3.2}$$

(which is a d-dimensional imbedded submanifold of N), has a tangent space which, at any point q, coincides with the subspace $\Delta(q)$ of T_qN.

Since the set of all such slices is a *partition* of V, we may see that a completely integrable distribution Δ induces, locally around each point $p \in N$, a partition into lower dimensional submanifolds, and each of these submanifolds is such that its tangent space, at each point, agrees with the distribution Δ at that point.

The second consequence is that a completely integrable distribution is *involutive*. In order to see this we use the definition of involutivity and compute the Lie bracket of any pair of vector fields belonging to Δ. For, recall that in the ξ coordinates, any vector field τ defined on N is represented by a vector of the form

$$\tau(\xi) = (\tau_1(\xi) \ldots \tau_n(\xi))'$$

The components of this vector are related to the value of the vector field τ at a point q by the expression

$$\tau(q) = \tau_1(\xi(q))\left(\frac{\partial}{\partial \xi_1}\right)_q + \ldots + \tau_n(\xi(q))\left(\frac{\partial}{\partial \xi_n}\right)_q$$

If τ is a vector field of Δ and (3.1) is satisfied, the last n-d components $\tau_{d+1}(\xi), \ldots, \tau_n(\xi)$ must vanish. Moreover, if θ is any other vector field of Δ, also the last n-d components of its local representation

$$\theta(\xi) = (\theta_1(\xi) \ldots \theta_n(\xi))'$$

must vanish. From this one deduces immediately that also the last n-d components of the vector

$$\frac{\partial \theta}{\partial \xi}\tau(\xi) - \frac{\partial \tau}{\partial \xi}\theta(\xi)$$

are vanishing. Since this vector represents locally the vector field $[\tau,\theta]$ one may conclude that $[\tau,\theta]$ belongs to Δ, i.e. that Δ is involutive.

We have seen that involutivity is a *necessary* condition for the complete integrability of a distribution. However, it can be proved that this condition is also sufficient, as it is stated below

(3.3) *Theorem* (Frobenius). A nonsingular distribution is completely integrable if and only if it is involutive

Proof. Let d denote the dimension of Δ. Since Δ is nonsingular, given any point $p \in N$ it is possible to find d vector fields $\tau_1, \ldots, \tau_d \in \Delta$ with the property that $\tau_1(q), \ldots, \tau_d(q)$ are linearly independent for all q in a suitable neighborhood U of p. In other words, these vector fields are such that

$$\Delta(q) = \text{span}\{\tau_1(q), \ldots, \tau_d(q)\}$$

for all $q \in U$.

Moreover, let $\tau_{d+1}, \ldots, \tau_n$ be any other set of vector fields with the property that $\text{span}\{\tau_i(p) : i = 1, \ldots, n\} = T_pN$. With each vector field τ_i, $i = 1, \ldots, n$, we associate its flow $\Phi_t^{\tau_i}$ and we consider the mapping

$$F : C_\varepsilon(0) \longrightarrow N$$

$$: (\xi_1, \dots, \xi_n) \longmapsto \Phi_{\xi_1}^{\tau_1} \circ \Phi_{\xi_2}^{\tau_2} \circ \dots \circ \Phi_{\xi_n}^{\tau_n}(p)$$

where $C_\varepsilon(0) = \{\xi \in \mathbb{R}^n : |\xi_i| < \varepsilon,\ 1 \le i \le n\}$.
If ε is sufficiently small, this mapping:

(i) is defined for all $\xi \in C_\varepsilon(0)$ and is a diffeomorphism onto its image

(ii) is such that for all $\xi \in C_\varepsilon(0)$

$$F_*\left(\frac{\partial}{\partial \xi_i}\right)_\xi \in \Delta(F(\xi)) \qquad i = 1,\dots,d \quad (*)$$

We show now that (i) and (ii) are true and, later, that both imply the thesis.

Proof of (i). We know that for each $p \in N$ and sufficiently small $|t|$ the flow $\Phi_t^\tau(p)$ of a vector field τ is defined and this makes the function F defined for all $(\xi_1,\dots,\xi_n)$ with sufficiently small $|\xi_i|$. Moreover, since a flow is smooth, so is F. We prove that F is a local diffeomorphism by showing that the rank of F at 0 is equal to n.

To this purpose, we first compute the image under F_* of the tangent vector $\left(\frac{\partial}{\partial \xi_i}\right)_\xi$ at a point $\xi \in C_\varepsilon(0)$. Suppose F is expressed in local coordinates. Then, it is known that the coordinates of $F_*\left(\frac{\partial}{\partial \xi_i}\right)_\xi$ in the basis $\left\{\left(\frac{\partial}{\partial \varphi_1}\right)_q, \dots, \left(\frac{\partial}{\partial \varphi_n}\right)_q\right\}$ of the tangent space to N at the point $q = F(\xi)$ coincide with the elements of the i-th column of the jacobian matrix

$$\frac{\partial F}{\partial \xi}$$

By taking the partial derivative of F with respect to ξ_i we obtain

$$\frac{\partial F}{\partial \xi_i} = (\Phi_{\xi_1}^{\tau_1})_* \dots (\Phi_{\xi_{i-1}}^{\tau_{i-1}})_* \frac{\partial}{\partial \xi_i}(\Phi_{\xi_i}^{\tau_i} \circ \dots \circ \Phi_{\xi_n}^{\tau_n}(p)) =$$

$$= (\Phi_{\xi_1}^{\tau_1})_* \dots (\Phi_{\xi_{i-1}}^{\tau_{i-1}})_* \tau_i \circ \Phi_{\xi_i}^{\tau_i} \circ \dots \circ \Phi_{\xi_n}^{\tau_n}(p) =$$

$$= (\Phi_{\xi_1}^{\tau_1})_* \dots (\Phi_{\xi_{i-1}}^{\tau_{i-1}})_* \tau_i \circ \Phi_{-\xi_{i-1}}^{\tau_{i-1}} \circ \dots \circ \Phi_{-\xi_1}^{\tau_1}(F(\xi))$$

(*) Note that $\left(\frac{\partial}{\partial \xi_i}\right)_\xi$ is a tangent vector at the point ξ of $C_\varepsilon(0)$.

In particular, at $\xi = 0$, since $F(0) = p$,

$$F_*(\frac{\partial}{\partial \xi_i})_0 = \tau_i(p)$$

The tangent vectors $\tau_1(p),\ldots,\tau_n(p)$ are by assumption linearly independent and this proves that F_* has rank n at p.

Proof of (ii). From the previous computations, we deduce also that, at any $\xi \in C_\varepsilon(0)$,

$$F_*(\frac{\partial}{\partial \xi_i})_\xi = (\Phi^{\tau_1}_{\xi_1})_* \ldots (\Phi^{\tau_{i-1}}_{\xi_{i-1}})_* \tau_i \circ \Phi^{\tau_{i-1}}_{-\xi_{i-1}} \circ \ldots \circ \Phi^{\tau_1}_{-\xi_1}(q)$$

where $q = F(\xi)$.

If we are able to prove that for all q in a neighborhood of p, for $|t|$ small, and for any two vector fields τ and θ belonging to Δ,

$$(\Phi^\theta_t)_* \tau \circ \Phi^\theta_{-t}(q) \in \Delta(q)$$

i.e. that $(\Phi^\theta_t)_* \tau \circ \Phi^\theta_{-t}$ is a (locally defined) vector field of Δ, then we easily see that (ii) is true.

To prove the above, one proceeds as follows. Let θ be a vector field of Δ and set

$$V_i(t) = (\Phi^\theta_{-t})_* \tau_i \circ \Phi^\theta_t(q)$$

for $i = 1,\ldots,d$.

Then, from a well known property of the Lie bracket we have

$$\frac{dV_i}{dt} = (\Phi^\theta_{-t})_*[\theta,\tau_i] \circ \Phi^\theta_t(q)$$

Since both τ_i and θ belong to Δ and Δ is involutive, there exist functions λ_{ij} defined locally around p such that

$$[\theta,\tau_i] = \sum_{j=1}^{d} \lambda_{ij}\tau_j$$

and, therefore,

$$\frac{dV_i}{dt} = (\Phi^\theta_{-t})_*[\sum_{j=1}^{d} \lambda_{ij}(\Phi^\theta_t(q))]\tau_j \circ \Phi^\theta_t(q) = \sum_{j=1}^{d} \lambda_{ij}(\Phi^\theta_t(q))V_j(t)$$

The functions $V_i(t)$ are seen as solutions of a linear differential equation and, therefore, it is possible to set

$$[V_1(t)\dots V_d(t)] = [V_1(0)\dots V_d(0)]X(t)$$

where $X(t)$ is a $d\times d$ fundamental matrix of solutions. By multiplying on the left both sides of this equality by $(\Phi_t^\theta)_*$ we get

$$[\tau_1\circ\Phi_t^\theta(q)\dots\tau_d\circ\Phi_t^\theta(q)] = [(\Phi_t^\theta)_*\tau_1(q)\dots(\Phi_t^\theta)_*\tau_d(q)]X(t)$$

and also, by replacing q with $\Phi_{-t}^\theta(q)$

$$[\tau_1(q)\dots\tau_d(q)] = [(\Phi_t^\theta)_*\tau_1\circ\Phi_{-t}^\theta(q)\dots(\Phi_t^\theta)_*\tau_d\circ\Phi_{-t}^\theta(q)]X(t)$$

Since $X(t)$ is nonsingular for all t we have that, for $i=1,\dots,d$,

$$(\Phi_t^\theta)_*\tau_i\circ\Phi_{-t}^\theta(q) \in \text{span}\{\tau_1(q),\dots,\tau_p(q)\}$$

i.e.

$$(\Phi_t^\theta)_*\tau_i\circ\Phi_{-t}^\theta(q) \in \Delta(q)$$

This result, bearing in mind the possibility of expressing any vector τ of Δ in the form

$$\tau = \sum_{i=1}^{d} c_i\tau_i$$

completes the proof of (ii).

From (i) and (ii) the thesis follows easily. Actually, (i) makes it possible to consider on the neighborhood $V = F(C_\varepsilon(0))$ of p the coordinate chart (V,F^{-1}). By definition, the tangent vector $(\frac{\partial}{\partial\xi_i})_q$ at a point $q\in V$ coincides with the image under F_* of the tangent vector $(\frac{\partial}{\partial\xi_i})_\xi$ at the point $\xi = F^{-1}(q)\in C_\varepsilon(0)$. From (ii) we see that the tangent vectors

$$(\frac{\partial}{\partial\xi_1})_q, \dots, (\frac{\partial}{\partial\xi_d})_q$$

are elements of $\Delta(q)$. Since these vectors are linearly independent, they span $\Delta(q)$ and (3.1) is satisfied. □

There are several interesting system-theoretic consequences of Frobenius' Theorem. The most important one is found in the correspondence, established by this Theorem, between involutive distributions and local partitions of a manifold into lower dimensional submanifolds. As we have seen, given a nonsingular and completely integrable, i.e. involutive, d-dimensional distribution Δ on a manifold N, around each $p \in N$ it is possible to find a coordinate neighborhood V on which Δ induces a partition into submanifolds of dimension d, which are slices (and, therefore, imbedded submanifolds) of V. Conversely, given any coordinate neighborhood V, a partition of V into d-dimensional slices defines on V a nonsingular completely integral distribution of dimension d.

We examine some examples in order to further clarify these concepts

(3.4) *Example*. Let $N = \mathbb{R}^n$ and let $x = (x_1,\dots,x_n)$ be a point on $\mathbb{R}^n$. Suppose V is a subspace of $\mathbb{R}^n$, of dimension d, spanned by the vectors

$$v_i = (v_{i1},\dots,v_{in}) \qquad 1 \le i \le d$$

We may associate with V a distribution, denoted Δ_V , in the following way. At each $x \in \mathbb{R}^n$, $\Delta_V(x)$ is the subspace of $T_x\mathbb{R}^n$ spanned by the tangent vectors

$$\sum_{j=1}^{n} v_{ij} \left(\frac{\partial}{\partial x_j}\right)_x \qquad 1 \le i \le d$$

It is easily seen that this distribution is nonsingular and involutive, thus completely integrable.

Now, suppose we perform a (linear) change of coordinates in $\mathbb{R}^n$, $\xi = \xi(x)$ such that

$$\xi_i(v_j) = \delta_{ij}$$

In the ξ coordinates, the subspace V will be spanned by vectors of the form $(1,0,\dots,0)$, $(0,1,\dots,0)$, etc., while the subspace $\Delta_V(x)$ by the tangent vectors $\left(\frac{\partial}{\partial \xi_1}\right)_x,\dots,\left(\frac{\partial}{\partial \xi_d}\right)_x$. Thus, we see that the condition (3.1) is satisfied globally on $\mathbb{R}^n$ in the ξ coordinates.

The slices

$$S = \{x \in \mathbb{R}^n : \xi_i(x) = c_i \ , \qquad i = d+1,\dots,n\}$$

characterize a global partition of $\mathbb{R}^n$ and each of these is such that its tangent space, at each point x, is exactly $\Delta_V(x)$. It is worth noting that each of these slices corresponds to a set of the form

$$x + V$$

Thus, the partitions of the state space X discussed in section 1 may be thought of as global partitions induced by a distribution associated with a given subspace of X.

(3.5) *Example*. Let $N = \mathbb{R}^2$ and let $x = (x_1, x_2)$ be a point on $\mathbb{R}^2$. Consider the one-dimensional nonsingular distribution

$$\Delta = \mathrm{sp}\{(\exp x_2)\frac{\partial}{\partial x_1} + \frac{\partial}{\partial x_2}\}$$

If we want to find a change of coordinates that makes (3.1) satisfied, we may proceed as follows. Recall that, given a coordinate chart with coordinate functions ξ_1, ξ_2, a tangent vector v at x may be represented as

$$v = v_1(\frac{\partial}{\partial \xi_1})_x + v_2(\frac{\partial}{\partial \xi_2})_x$$

where the coefficients v_1 and v_2 are such that $v_1 = L_v\xi_1$ and $v_2 = L_v\xi_2$.

Since the tangent vector

$$\tau(x) = (\exp x_2)(\frac{\partial}{\partial x_1})_x + (\frac{\partial}{\partial x_2})_x$$

spans $\Delta(x)$ at each $x \in \mathbb{R}^2$, if we want that (3.1) is satisfied we have to have

$$\tau(x) = (L_\tau\xi_1)(\frac{\partial}{\partial \xi_1})_x + (L_\tau\xi_2)(\frac{\partial}{\partial \xi_2})_x = (\frac{\partial}{\partial \xi_1})_x$$

for all $x \in U$, or

$$1 = (L_\tau\xi_1) = (\exp x_2)\frac{\partial \xi_1}{\partial x_1} + \frac{\partial \xi_1}{\partial x_2}$$

$$0 = (L_\tau\xi_2) = (\exp x_2)\frac{\partial \xi_2}{\partial x_1} + \frac{\partial \xi_2}{\partial x_2}$$

A solution of this set of partial differential equations is given by

$$\xi_1 = \xi_1(x) = x_2$$

$$\xi_2 = \xi_2(x) = x_1 - \exp(x_2)$$

The mapping $\xi = \xi(x)$ is a diffeomorphism $\xi : \mathbb{R}^2 \to \mathbb{R}^2$ and solves the problem of finding the change of coordinates that makes (3.1) satisfied. Note that $\mathbb{R}^2$ is globally partitioned into one-dimensional slices, each one being the locus where the function $\xi_2(x)$ is constant, i.e. the locus of points (x_1, x_2) such that

$$x_1 = \exp(x_2) + \text{constant} \quad \square$$

The procedure described in the Example (3.5) may easily be extended. For, let Δ be a nonsingular involutive distribution of dimension d. Let (U,φ) be a coordinate chart with coordinate functions $\varphi_1,\ldots,\varphi_n$. Given any point $p \in U$ it is possible to find d vector fields $\tau_1,\ldots,\tau_d \in \Delta$ with the property that $\tau_1(q),\ldots,\tau_d(q)$ are linearly independent for all q in a suitable neighborhood $U' \subset U$ of p. In other words, these vectors are such that

$$\Delta(q) = \text{span}\{\tau_1(q),\ldots,\tau_d(q)\}$$

for all $q \in U'$.

In the coordinates $\varphi_1,\ldots,\varphi_n$, each of these vector fields is locally expressed in the form

$$\tau_i = \sum_{j=1}^{n} (L_{\tau_i}\varphi_j)\left(\frac{\partial}{\partial\varphi_j}\right)$$

If (V,ξ) is another coordinate chart around p with coordinate functions $\xi_1,\ldots,\xi_n$ the corresponding expressions for τ_i has the form

$$\tau_i = \sum_{j=1}^{n} (L_{\tau_i}\xi_j)\left(\frac{\partial}{\partial\xi_j}\right)$$

For (3.1) to be satisfied, i.e. for

$$\text{sp}\{\tau_1,\ldots,\tau_d\} = \text{sp}\left\{\frac{\partial}{\partial\xi_1},\ldots,\frac{\partial}{\partial\xi_d}\right\}$$

on V, we must have

$$L_{\tau_i}\xi_j = 0$$

on V, for $i = 1,\ldots,d$ and $j = d+1,\ldots,n$ and, moreover

$$\text{rank}\begin{pmatrix} L_{\tau_1}\xi_1 & \cdots & L_{\tau_1}\xi_d \\ & \cdots & \\ L_{\tau_d}\xi_1 & \cdots & L_{\tau_d}\xi_d \end{pmatrix} = d$$

on V. These conditions characterize a set of partial differential equations on V, which has to be satisfied by the new coordinate functions $\xi_1,\ldots,\xi_n$.

Setting, as usual

$$\tau_{ij}(x) = (L_{\tau_i}\varphi_j)\circ\varphi^{-1}(x)$$

where $x = (x_1,\ldots,x_n) \in \mathbb{R}^n$, it is possible to express the functions $L_{\tau_i}\xi_j$ involved in the previous conditions as follows

$$L_{\tau_i}\xi_j(x) = \sum_{k=1}^{n} \tau_{ik}(x)\frac{\partial(\xi_j\circ\varphi^{-1})}{\partial x_k} \quad .$$

Therefore, using just $\xi_j(x)$ to denote the composite function $\xi_j\circ\varphi^{-1}(x)$, one has

$$L_{\tau_i}\xi_j(x) = \sum_{k=1}^{n} \tau_{ik}(x)\frac{\partial \xi_j}{\partial x_k}$$

Setting

$$T(x) = \begin{pmatrix} \tau_{11}(x)\ldots\tau_{d1}(x) \\ \cdot \quad \ldots \quad \cdot \\ \tau_{1n}(x)\ldots\tau_{dn}(x) \end{pmatrix}$$

the previous equations for $L_{\tau_i}\xi_j$ become

$$\frac{\partial\xi}{\partial x}T(x) = \begin{pmatrix} K(x) \\ \\ 0_{(n-d)\times d} \end{pmatrix} \tag{3.6}$$

in which K(x) is some $d\times d$ matrix of real valued functions, nonsingular for all $x \in \varphi(V)$.

Thus, we may conclude that finding a coordinate transformation $\xi = \xi(x)$ that makes (3.1) satisfied corresponds to solving a partial differential equation of the form (3.6).

Note that the matrix T(x) is a matrix of rank d at $x = \varphi(p)$ because the tangent vectors $\tau_1(p),\ldots,\tau_d(p)$ are linearly independent. Therefore the matrix $\frac{\partial\xi}{\partial x}$ can be nonsingular at $x = \varphi(p)$ and this, according to the rank Theorem, guarantees that $\xi = \xi(x)$ is a local diffeomorphism.

(3.7) *Remark*. There are alternative ways to describe the equation (3.6).

For instance, one may easily check that solving these equations corresponds to find n-d functions $\lambda_1,\dots,\lambda_{n-d}$ defined on a neighborhood V of p with values in $\mathbb{R}$ with the following properties

(i) the tangent covectors $d\lambda_1(p),\dots,d\lambda_{n-d}(p)$ are linearly independent

(ii) $\langle d\lambda_i(q),\tau_j(q)\rangle = 0$ for all $q \in V$, $i = 1,\dots,n-d$ and $j=1,\dots,d$.

In fact, if (V,ξ) is a coordinate chart that makes (3.6) satisfied, then the functions

$$\lambda_i = \xi_{i+d}$$

will satisfy (i) and (ii). Conversely, if $\lambda_1,\dots,\lambda_{n-d}$ is a set of functions that satisfies (i) and (ii), then it is always possible to find d functions $\xi_1,\dots,\xi_d$ defined on V and with values in $\mathbb{R}$ which, together with the functions $\xi_{d+1} = \lambda_1,\dots,\xi_n = \lambda_{n-d}$, define a coordinate chart (V,ξ) with ξ solving the equations (3.6).

From (ii) we deduce also that there is a set of covector fields $\{d\lambda_1,\dots,d\lambda_{n-d}\}$ with the property that at each $q \in V$, $\langle d\lambda_i(q),v\rangle = 0$ for all $v \in \Delta(q)$. Thus

$$d\lambda_i(q) \in \Delta^{\perp}(q) \qquad i = 1,\dots,n-d$$

Moreover, the tangent covectors $d\lambda_1(q),\dots,d\lambda_{n-d}(q)$ are linearly independent for all q in a neighborhood of p and $\Delta^{\perp}(q)$ has exactly dimension n-d. Therefore, we may conclude that the set of covector fields $\{d\lambda_1,\dots,d\lambda_{n-d}\}$ spans $\Delta^{\perp}$ locally around p.

In short, we may state this result by saying that a nonsingular distribution of dimension d is integrable if and only if its annihilator is locally spanned by n-d *exact* one-forms.

(3.8) *Remark*. We note that the involutivity of Δ corresponds to the property that any two columns $\tau_i(x)$ and $\tau_j(x)$ of the matrix $T(x)$ are such that

$$\left(\frac{\partial\tau_i}{\partial x}\tau_j(x) - \frac{\partial\tau_j}{\partial x}\tau_i(x)\right) \in \mathrm{Im}(T(x))$$

for all $x \in \varphi(V)$.

(3.9) *Remark*. We know that, given a set of functions $\{\lambda_i : i \in I\}$, defined on N and with values in $\mathbb{R}$, we can define a codistribution $\Omega = \mathrm{sp}\{d\lambda_i : i \in I\}$. It is easily seen that if Ω is nonsingular then $\Omega^{\perp}$ is completely integrable. For, let d denote the dimension of Ω, take a point $p \in N$ and a set of functions $\lambda_1,\dots,\lambda_d$ with the property that

$$\Omega(p) = \text{span}\{d\lambda_1(p),\ldots,d\lambda_d(p)\}$$

If U is a neighborhood of p with the property that $d\lambda_1(q),\ldots,d\lambda_d(q)$ are linearly independent at all $q \in U$, it is seen that Ω is spanned on U by the exact forms $d\lambda_1,\ldots,d\lambda_d$. As a consequence of our earlier discussions, $\Omega^{\perp}$ is completely integrable. □

The notion of complete integrability can be extended to a given collection of distributions. There are two cases of special importance in the applications.

Let $\Delta_1,\Delta_2,\ldots,\Delta_r$ be a collection of *nested* distributions, i.e. a set of distributions with the property

$$\Delta_1 \subset \Delta_2 \subset \ldots \subset \Delta_r .$$

A collection of nested nonsingular distributions on N is completely integrable if at each point $p \in N$ there exists a coordinate chart (V,ξ) with coordinate functions $\xi_1,\ldots,\xi_n$ such that

$$\Delta_i(q) = \text{span}\{(\frac{\partial}{\partial\xi_1})_q,\ldots,(\frac{\partial}{\partial\xi_{d_i}})_q\}$$

for all $q \in V$, where d_i denotes the dimension of Δ_i.

The following results extends Frobenius Theorem

(3.10) *Theorem*. A collection $\Delta_1 \subset \Delta_2 \subset \ldots \subset \Delta_r$ of nested nonsingular distributions is completely integrable if and only if each distribution of the collection is involutive.

Proof. The same construction described in the proof of Theorem (3.3) can be used. □

A collection $\Delta_1,\ldots,\Delta_r$ of distributions on N is said to be *independent* if

(i) Δ_i is nonsingular, for all $i = 1,\ldots,r$

(ii) $\Delta_i \cap (\sum_{j\neq i} \Delta_j) = 0$, for all $i = 1,\ldots,r$

A collection of distributions $\Delta_1,\ldots,\Delta_r$ is said to *span the tangent space* if for all $q \in N$

$$\Delta_1(q) + \Delta_2(q) + \ldots + \Delta_r(q) = T_qN.$$

An independent collection of distributions $\Delta_1,\ldots,\Delta_r$ which spans the tangent space is said to be *simultaneously integrable* if at each point $p \in N$ there exists a coordinate chart (V,ξ), with coordinate functions $\xi_1,\ldots,\xi_n$ such that

$$\Delta_i(q) = \text{span}\{(\frac{\partial}{\partial \xi_{s_i+1}})_q,\ldots,(\frac{\partial}{\partial \xi_{s_{i+1}}})_q\} \tag{3.11}$$

for all $q \in V$, where $s_1 = 0$ and

$$s_i = \dim(\Delta_1 + \ldots + \Delta_{i-1})$$

for $i = 2,\ldots,r+1$.

The following result is an additional extension of Frobenius Theorem

(3.12) *Theorem.* An independent collection of distributions $\Delta_1,\ldots,\Delta_r$ which spans the tangent space is simultaneously integrable if and only if, for all $1 \le i \le r$, the distribution

$$D_i = \sum_{\substack{j=1 \\ j\neq i}}^{r} \Delta_j \tag{3.13}$$

is involutive.

Proof. Sufficiency. Let $n_i = \dim(\Delta_i)$. Using Theorem(3.3), at each point p one may find a neighborhood V of p and, for each $1 \le i \le r$, a set of coordinate functions ξ_j^i, $1 \le j \le n$, defined on V with the property that

$$D_i = \text{sp}\{\frac{\partial}{\partial \xi_j^i} : 1 \le j \le n-n_i\}$$

An easy computation shows that the covector fields

$$d\xi^1_{n-n_1+1},\ldots,d\xi^1_n,\ldots,d\xi^r_{n-n_r+1},\ldots,d\xi^r_n$$

are linearly independent at p. Thus, the set of functions $\{\xi_j^i : n-n_i+1 \le j \le n;\ 1 \le i \le r\}$ defines on V a set of coordinate functions.

Since D_i is tangent to the slice of V where all the coordinate functions $\xi^i_{n-n_i+1},\ldots,\xi^i_n$ are held constant, one deduces that Δ_i is tangent to the slice of V where all the coordinate functions $\xi^k_{n-n_k+1},\ldots,\xi^k_n$, for all $k \neq i$, are held constant. This yields (3.11).

The necessity is a straightforward consequence of the definition.

4. Invariant Distributions

The notion of distribution invariant under a vector field plays, in the theory of nonlinear control systems, a role similar to the one played in the theory of linear systems by the notion of subspace invariant under a linear mapping.

A distribution Δ on N is *invariant* under a vector field f if the Lie bracket $[f,\tau]$ of f with every vector field $\tau \in \Delta$ is a vector field which belongs to Δ, i.e. if

$$(4.1) \qquad [f,\Delta] \subset \Delta$$

(4.2) *Remark*. There is a natural way to see that the previous definition generalizes the notion of invariant subspace. Let $N = \mathbb{R}^n$, A a linear mapping $A : \mathbb{R}^n \to \mathbb{R}^n$ and V a subspace of $\mathbb{R}^n$ invariant under A, i.e. such that $AV \subset V$. Suppose V is spanned by the vectors

$$v_i = (v_{i1},\ldots,v_{in}) \qquad 1 \leq i \leq d$$

and consider, as in the Example (3.4), the flat distribution Δ_V spanned by the vector fields

$$\tau_i = \sum_{j=1}^{n} v_{ij} \frac{\partial}{\partial x_j} \qquad 1 \leq i \leq d$$

With the mapping A we associate a vector field f_A represented, in the canonical basis $(\frac{\partial}{\partial x_1})_x,\ldots,(\frac{\partial}{\partial x_n})_x$ of $T_x\mathbb{R}^n$ by the vector

$$f_A(x) = Ax$$

(note that the right-hand-side of this expression represent of coordinates of an element of the tangent space at x to $\mathbb{R}^n$ and not a vector of coordinates of a point in $\mathbb{R}^n$).

It is easily seen that the distribution Δ_V is invariant under the vector field f_A in the sense of our previous definition. For, observe that any vector field τ in Δ_V can be represented in the form (2.8) where $c_1,\ldots,c_d$ is any set of real-valued functions defined locally around x.

Computing the Lie bracket of f_A and τ we have

$$[f_A,\tau] = \sum_{i=1}^{d} [f_A,c_i\tau_i] = \sum_{i=1}^{d} c_i[f_A,\tau_i] + \sum_{i=1}^{d} (L_{f_A}c_i)\tau_i$$

Moreover,
$$[f_A,\tau_i](x) = \frac{\partial \tau_i}{\partial x} f_A(x) - \frac{\partial f_A(x)}{\partial x}\tau_i(x) = -A\tau_i(x)$$

Note that $\tau_i(x)$, regarded as a point of $\mathbb{R}^n$, is an element of V, so also $A\tau_i(x) \in V$. Then, for each x, $[f_A, \tau_i](x) \in \Delta_V(x)$ and

$$[f_A, \tau](x) \in \Delta_V(x)$$

that proves the assertion. □

The notion of invariance under a vector field is particularly useful when referred to completely integrable distributions, because it provides a way of simplifying the local representation of the given vector field.

(4.3) *Lemma.* Let Δ be a nonsingular involutive distribution of dimension d and assume that Δ is invariant under the vector field f. Then, at each point $p \in N$ there exists a coordinate chart (U, ξ) with coordinate functions $\xi_1, \ldots, \xi_n$, in which the vector field f is represented by a vector of the form

$$f(\xi) = \begin{pmatrix} f_1(\xi_1, \ldots, \xi_d, \xi_{d+1}, \ldots, \xi_n) \\ \cdot \quad \cdot \quad \cdot \\ f_d(\xi_1, \ldots, \xi_d, \xi_{d+1}, \ldots, \xi_n) \\ f_{d+1}(\xi_{d+1}, \ldots, \xi_n) \\ \cdot \quad \cdot \quad \cdot \\ f_n(\xi_{d+1}, \ldots, \xi_n) \end{pmatrix} \tag{4.4}$$

Proof. The distribution Δ, being nonsingular and involutive, is integrable and, therefore, at each point $p \in N$ there exists a coordinate chart (U, ξ) that makes (3.1) satisfied for all $q \in U$. Now, let $f_1(\xi), \ldots, f_n(\xi)$ denote the coordinates of f(q) in the canonical basis of T_qN associated with (U, ξ), and recall that

$$f(q) = \sum_{i=1}^{n} f_i(\xi(q)) \left(\frac{\partial}{\partial \xi_i}\right)_q$$

The invariance condition (4.1) implies, in particular, that

$$[f, \frac{\partial}{\partial \xi_j}](q) \in \operatorname{span}\{(\frac{\partial}{\partial \xi_1})_q, \ldots, (\frac{\partial}{\partial \xi_d})_q\}$$

for all $q \in U$ and $j = 1, \ldots, d$. Therefore we must have that

$$[f, \frac{\partial}{\partial \xi_j}] = \sum_{i=1}^{n} [f_i \frac{\partial}{\partial \xi_i}, \frac{\partial}{\partial \xi_j}] = -\sum_{i=1}^{n} (\frac{\partial f_i}{\partial \xi_j}) \frac{\partial}{\partial \xi_i} \in \operatorname{sp}\{\frac{\partial}{\partial \xi_1}, \ldots, \frac{\partial}{\partial \xi_d}\}$$

From this we see that the coefficients $\frac{\partial f_i}{\partial \xi_j}$ are such that

$$\frac{\partial f_i}{\partial \xi_j} = 0$$

for all $i = d+1,\ldots,n$ and $j = 1,\ldots,d$ and all $\xi \in \xi(U)$. The components $f_{d+1},\ldots,f_n$ are thus independent of the coordinates $\xi_1,\ldots,\xi_d$, and the (4.4) are proved. □

The following properties of invariant distributions will be also used later on.

(4.5) *Lemma*. Let Δ be a distribution invariant under the vector fields f_1 and f_2. Then Δ is also invariant under the vector field $[f_1,f_2]$.

Proof. Suppose τ is a vector field in Δ. Then, from the Jacobi identity we get

$$[[f_1,f_2],\tau] = [f_1,[f_2,\tau]]-[f_2,[f_1,\tau]]$$

By assumption $[f_2,\tau] \in \Delta$ and so is $[f_1,[f_2,\tau]]$. For the very same reasons $[f_2,[f_1,\tau]] \in \Delta$ and thus from the above equality we conclude that $[[f_1,f_2],\tau] \in \Delta$. □

(4.6) *Remark*. Note that the notion of invariance under a given vector field f is still meaningful in the case of a distribution Δ which is not smooth. In this case, it is simply required that the Lie bracket $[f,\tau]$ of f with every smooth vector field in Δ be a vector field in Δ. Since $[f,\tau]$ is a smooth vector field, it follows that if Δ is a (possibly) non-smooth distribution invariant under the vector field f, then also $\mathrm{smt}(\Delta)$ is invariant under f. □

When dealing with codistributions, one can as well introduce the notion of invariance under a vector field in the following way.

A codistribution Ω on M is *invariant* under a vector field f if the Lie derivative along f of any covector field $\omega \in \Omega$ is a covector field which belongs to Ω, i.e. if

$$L_f\Omega \subset \Omega \tag{4.7}$$

It is easily seen that this is the dual version of the notion of invariance of a distribution.

(4.8) *Lemma*. If a smooth distribution Δ is invariant under the vector field f, then the codistribution $\Omega = \Delta^{\perp}$ is invariant under f. If a smooth codistribution Ω is invariant under the vector field f, then the distribution $\Delta = \Omega^{\perp}$ is invariant under f.

Proof. We shall make use of the identity

$$\langle L_f\omega, \tau \rangle = L_f\langle \omega, \tau \rangle - \langle \omega, [f,\tau] \rangle$$

Suppose Δ is invariant under f and let τ be any vector field of Δ. Then $[f,\tau] \in \Delta$. Let ω be any covector field in Ω. Then, by definition

$$\langle \omega, \tau \rangle (p) = 0$$

for all $p \in N$, and also

$$\langle \omega, [f,\tau] \rangle (p) = 0$$

This yields

$$\langle L_f\omega, \tau \rangle (p) = 0$$

Since Δ is a smooth distribution, given any vector v in $\Delta(p)$ we may find a vector field τ in Δ with the property that $\tau(p) = v$ and, then, the previous result shows that

$$\langle L_f\omega(p), v \rangle = 0$$

for all $v \in \Delta(p)$, i.e. that $L_f\omega(p) \in \Omega(p)$. From this it is concluded that $L_f\omega$ is a covector field in Ω.

The second part of the statement is proved in the same way. □

(4.9) *Remark.* Note that in the previous Lemma, first part, we don't need to assume that the annihilator $\Delta^{\perp}$ of Δ is smooth, nor, in the second part, that the annihilator $\Omega^{\perp}$ of Ω is smooth. However, if both Δ and $\Delta^{\perp}$ are smooth, we conclude from the Lemma that the invariance of Δ under f implies and is implied by the invariance of $\Delta^{\perp}$ under the same vector field. In view of Lemma (2.17) this is true, in particular, whenever Δ is nonsingular. □

By making use of these notions one may give a dual formulation of Lemma (4.3). Instead of a nonsingular and involutive distribution Δ, we have to consider (see Remark (3.7)) a nonsingular codistribution Ω of dimension n-d with the property that for each $p \in N$ there exist a neighborhood U of p and n-d functions $\xi_{d+1},\dots,\xi_n$ defined on U with values in $\mathbb{R}$ such that

$$\Omega(q) = \text{span}\{d\xi_{d+1}(q),\ldots,d\xi_n(q)\}$$

for all $q \in U$.

If Ω satisfies these assumptions and if Ω is also invariant under f, then it is possible to find d more real-valued functions $\xi_1,\ldots,\xi_d$ defined on U with the property that, choosing as local coordinates on U the functions ξ_i , $1 \leq i \leq n$, for each $q \in U$ the vector field f is represented by a vector $f(\xi)$ of the form (4.4).

5. Local Decompositions of Control Systems

Throughout these notes we deal with nonlinear control systems described by equations of the form

$$\dot{x} = f(x) + \sum_{i=1}^{m} g_i(x)u_i \tag{5.1a}$$

$$y_i = h_i(x) \qquad (i = 1,\ldots,\ell) \tag{5.1b}$$

The state x of this system belongs to an open subset N of $\mathbb{R}^n$, while the m components $u_1,\ldots,u_m$ of the input and, respectively, the ℓ components $y_1,\ldots,y_\ell$ of the output are real-valued functions of time. We shall make later on some further assumptions on the class of admissible input functions to be considered. The vector fields $f,g_1,\ldots,g_m$ are smooth vector fields defined on N and assumed to be complete. The output maps $h_1,\ldots,h_\ell$ are real-valued smooth functions defined on N.

(5.2) *Remark*. One may define systems with the same structure as (5.1), with the state evolving on some abstract manifold N (not necessarily diffeomorphic to an open subset of $\mathbb{R}^n$). In this case, instead of (5.1), which is an ordinary differential equation defined on an open subset of $\mathbb{R}^n$, one should consider a description based upon an ordinary differential equation defined on the abstract manifold N. The vector fields $f,g_1,\ldots,g_m$ will be defined on N and so the output functions $h_1,\ldots,h_\ell$. If we let p denote a point in N then, instead of (5.1), we may use a description of the form

$$\dot{p} = f(p) + \sum_{i=1}^{m} g_i(p)u_i \tag{5.3a}$$

$$y_i = h_i(p) \qquad (i = 1,\ldots,\ell) \tag{5.3b}$$

with the understanding that $\dot{p}$ stands for the tangent vector at the point p to the smooth curve which characterizes the solution of (5.3a) for some fixed initial condition.

If this is the case, then (5.1) may be regarded as a local representation of (5.3) in some coordinate chart (U,φ) with the understanding that $x = \varphi(p)$. □

The theory developed so far enables us to obtain for this class of systems decompositions similar to those described at the beginning of the Chapter. The relevant results may be formalized in the following way.

(5.4) *Proposition.* Let Δ be a nonsingular involutive distribution of dimension d and assume that Δ is invariant under the vector fields $f,g_1,\ldots,g_m$. Moreover, suppose that the distribution $\mathrm{sp}\{g_1,\ldots,g_m\}$ is contained in Δ. Then, for each point $\bar{x} \in N$ it is possible to find an open subset U of $\bar{x}$ and a local coordinates transformation $\xi = \xi(x)$ defined on U, such that, in the new coordinates, the control system (5.1a) is represented by equations of the form

$$\dot{\xi}_1 = f_1(\xi_1,\xi_2) + \sum_{i=1}^{m} g_{i1}(\xi_1,\xi_2)u_i \tag{5.5a}$$

$$\dot{\xi}_2 = f_2(\xi_2) \tag{5.5b}$$

where (ξ_1,ξ_2) is a partition of ξ and $\dim(\xi_1) = d$.

Proof. From Lemma (4.3) it is known that there exists, around each $\bar{x} \in N$, a coordinate chart (U,ξ) with coordinate functions $\xi_1,\ldots,\xi_n$ with the property that the vector fields $f,g_1,\ldots,g_m$ are represented in form (4.4). Moreover, since by assumption $g_i \in \Delta$ for all $i=1,\ldots,m$, then the vector fields $g_1,\ldots,g_m$ in the same coordinate chart are represented by vectors whose last (n-d)-components are vanishing. This coordinate chart (U,ξ) may obviously be considered as a local change of coordinates around $\bar{x}$ and therefore the Proposition is proved. □

(5.6) *Proposition.* Let Δ be a nonsingular involutive distribution of dimension d and assume that Δ is invariant under the vector fields $f,g_1,\ldots,g_m$. Moreover, assume that the codistribution $\mathrm{sp}\{dh_1,\ldots,dh_\ell\}$ is contained in the codistribution $\Delta^\perp$. Then, for each $\bar{x} \in N$ it is possible to find an open subset U of $\bar{x}$ and a local coordinates transformation $\xi = \xi(x)$ defined on U, such that, in the new coordinates, the control system (5.1) is represented by equations of the form

$$\dot{\xi}_1 = f_1(\xi_1,\xi_2) + \sum_{i=1}^{m} g_{i1}(\xi_1,\xi_2)u_i \tag{5.7a}$$

(5.7b) $$\dot{\xi}_2 = f_2(\xi_2) + \sum_{i=1}^{m} g_{i2}(\xi_2)u_i$$

(5.7c) $$y_i = h_i(\xi_2)$$

where (ξ_1,ξ_2) is a partition of ξ and $\dim(\xi_1) = d$.

Proof. As before, we know that there exists, around each $\bar{x} \in N$, a coordinate chart (U,ξ), with coordinate functions $\xi_1,\ldots,\xi_n$, with the property that the vector fields $f,g_1,\ldots,g_m$ are represented in the form (4.4). Moreover, we have assumed that

$$\Delta \subset [\mathrm{sp}\{dh_1,\ldots,dh_\ell\}]^\perp$$

For each point x of the selected coordinate chart we have in particular, for $j = 1,\ldots,d$,

$$\left(\frac{\partial}{\partial \xi_j}\right)_x \in \Delta(x) \subset [\sum_{i=1}^{\ell} \mathrm{span}\{dh_i(x)\}]^\perp = \bigcap_{i=1}^{\ell} [\mathrm{span}\{dh_i(x)\}]^\perp$$

As a consequence, for $j = 1,\ldots,d$ and $i = 1,\ldots,\ell$ and for all $x \in U$

$$\langle dh_i(x), \left(\frac{\partial}{\partial \xi_j}\right)_x \rangle = 0$$

or, in other words, we see that the local representation of h_i in the selected coordinate chart is such that

$$\frac{\partial h_i}{\partial \xi_j} = 0$$

for all $j = 1,\ldots,d$ and $i = 1,\ldots,\ell$ and for all $\xi \in \xi(U)$. We conclude that h_i depends only on the local coordinates $\xi_{d+1},\ldots,\xi_n$ on U and this completes the proof. □

The two local decompositions thus obtained are very useful in understanding the input-state and state-output behavior of the control system (5.1).

Suppose that the inputs u_i are piecewise constant functions of time, i.e. that there exist real numbers $T_o = 0 < T_1 < T_2 < \ldots$ such that

$$u_i(t) = \bar{u}_i^k \text{ for } T_k \leq t < T_{k+1}$$

Then, on the time interval $[T_k,T_{k+1})$, the state of the system evolves along the integral curve of the vector field

$$f + g_1\bar{u}_1^k + \ldots + g_m\bar{u}_m^k$$

passing through the point $x(T_k)$. In particular, if the initial state x^o at time $t = 0$ is contained in some neighborhood U of N, then for small t the state $x(t)$ evolves in U.

Suppose now that the assumptions of the Proposition (5.4) are satisfied and that x^o belongs to the domain U of the coordinate transformation $\xi(x)$. If the input u is such that the $x(t)$ evolves in U, we may use the equations (5.5) to describe the behavior of the system. From these we see that the local coordinates $(\xi_1(t),\xi_2(t))$ of $x(t)$ are such that $\xi_2(t)$ is not affected by the input. In particular, let $x^o(T)$ denote the point of U reached at time T when $u(t) = 0$ for all $t\in[0,T]$, i.e. the point

$$x^o(T) = \Phi_T^f(x^o)$$

Φ_T^f being the flow of the vector field f, and let $(\xi_1^o(T),\xi_2^o(T))$ denote the local coordinates of $x^o(T)$. We see that the set of points that can be reached at time T, starting from x^o, lies inside the set of points whose local coordinates ξ_2 are equal to $\xi_2^o(T)$. This set is actually a *slice* of U passing through the point $x^o(T)$.

Thus, we see that locally the system displays a behavior strictly analogous to the one described in section 1. Locally, the state space may be partitioned into submanifolds (the slices of U), all of dimension d, and the points reachable at time T, along trajectories that stay in U for all $t \in [0,T]$, lie inside the slice passing through the point $x^o(T)$ reached under zero input.

The Proposition (5.6) is useful in studying state-output interactions. Suppose we take two initial states x^a and x^b belonging to U with local coordinates (ξ_1^a,ξ_2^a) and (ξ_1^b,ξ_2^b) such that

$$\xi_2^a = \xi_2^b$$

i.e. two initial states belonging to the same *slice* of U. Let $x_u^a(t)$ and $x_u^b(t)$ denote the values of the states reached at time t, starting from x^a and x^b, under the action of the same input u. From the equation (5.7b) we see immediately that, if the input u is such that $x_u^a(t)$ and $x_u^b(t)$ both evolve in U, the ξ_2 coordinates of $x_u^a(t)$ and of $x_u^b(t)$ are the same, no matter which input u we consider. Actually these coordinates $\xi_2^a(t)$ and $\xi_2^b(t)$ are solutions of the same differential equation (the equation (5.7b)) with the same initial condition. If we take into account also the (5.7c) we have the equality

$$h_i(\xi_2^a(t)) = h_i(\xi_2^b(t))$$

which holds for every input u. We may conclude that x^a and x^b are indistinguishable.

Again, we find that locally the state space may be partitioned into submanifolds (the slices of U), all of dimension d, and pair of points of each slice both produce the same output (i.e. are indistinguishable) under any input u which keeps the state trajectory evolving on U.

In the next sections we shall reach stronger conclusions, showing that if we add to the hypotheses contained in the Propositions (5.4) and (5.6) the further assumption that the distribution Δ is "minimal" (in the case of Proposition (5.4)) or "maximal" (in the case of Proposition (5.6)), then from the decompositions (5.5) and (5.7) one may obtain more informations about the set of states reachable from x^o and, respectively, indistinguishable from x^o.

We conclude this section with a remark about a dual version of Proposition (5.6).

(5.8) *Remark.* Suppose that Ω is a nonsingular codistribution of dimension n-d with the property that for each $x \in M$ there exist a neighborhood U of x and n-d real-valued functions $\xi_{d+1},\dots,\xi_n$ defined on U such that

$$\Omega(x) = \mathrm{span}\{d\xi_{d+1}(x),\dots,d\xi_n(x)\}$$

for all $x \in U$. Let $\xi_1,\dots,\xi_d$ be other functions defining, together with $\xi_{d+1},\dots,\xi_n$, a coordinate transformation on U. In these coordinates, the one-form dh_i will be represented by a row vector

$$dh_i(\xi) = (\gamma_{i1}(\xi)\dots\gamma_{in}(\xi))$$

whose components are related to the value of dh_i at x by the expression

$$dh_i(x) = \gamma_{i1}(\xi(x))(d\xi_1)_x + \dots + \gamma_{in}(\xi(x))(d\xi_n)_x$$

If we assume that the covector fields $dh_1,\dots,dh_\ell$ belong to Ω, then, since Ω is spanned by $d\xi_{d+1},\dots,d\xi_n$ on U, we must have

$$\gamma_{ij}(\xi) = 0$$

for all $1 \leq i \leq \ell$, $1 \leq j \leq d$ and all ξ in $\xi(U)$. But since

$$\gamma_{ij}(\xi) = \frac{\partial h_j}{\partial \xi_j}$$

one concludes that $h_1,\ldots,h_\ell$ are independent of $\xi_1,\ldots,\xi_d$ on U, like in (5.7c). □

6. Local Reachability

In the previous section we have seen that if there is a nonsingular distribution Δ of dimension d with the properties that:

(i) Δ is involutive

(ii) Δ contains the distribution $sp\{g_1,\ldots,g_m\}$

(iii) Δ is invariant under the vector fields $f,g_1,\ldots,g_m$

then at each point $\bar{x} \in N$ it is possible to find a coordinate transformation defined on a neighborhood U of $\bar{x}$ and a partition of U into slices of dimension d, such that the points reachable at some time T, starting from some initial state $x^o \in U$, along trajectories that stay in U for all $t \in [0,T]$, lie inside a slice of U. Now we want to investigate the actual "thickness" of the subset of points of a slice reached at time T.

The obvious suggestion that comes from the decomposition (5.5) is to look at the "minimal" distribution, if any, that satisfies (ii), (iii) and, then, to examine what can be said about the properties of points which belong to the same slice in the corresponding local decomposition of N. It turns out that this program can be carried out in a rather satisfactory way.

We need first some additional results on invariant distributions. If $\mathcal{D}$ is a family of distributions on N, we define the *smallest* or *minimal* element as the member of $\mathcal{D}$ (when it exists) which is contained in every other element of $\mathcal{D}$.

(6.1) *Lemma*. Let Δ be a given smooth distribution and $\tau_1,\ldots,\tau_q$ a given set of vector fields. The family of all distributions which are invariant under $\tau_1,\ldots,\tau_q$ and contain Δ has a minimal element, which is a smooth distribution.

Proof. The family in question is nonempty because the distribution $sp\{V(N)\}$ clearly belongs to it. Let Δ_1 and Δ_2 be two elements of this family, then it is easily seen that their intersection $\Delta_1 \cap \Delta_2$ contains Δ and, being invariant under $\tau_1,\ldots,\tau_q$, is an element of the same family. This argument shows that the intersection $\hat{\Delta}$ of all elements in the family contains Δ, is invariant under $\tau_1,\ldots,\tau_q$ and is contained in any other element of the family. Thus is its minimal element. $\hat{\Delta}$ must be smooth because otherwise $\mathrm{smt}(\hat{\Delta})$ would be a smooth distribution containing Δ (because Δ is smooth by assumption), invariant under $\tau_1,\ldots,\tau_q$ (see Remark (4.6)) and possibly contained in $\hat{\Delta}$. □

In what follows, the smallest distribution which contains Δ and is invariant under the vector fields $\tau_1,\ldots,\tau_q$ will be denoted by the symbol

$$\langle \tau_1,\ldots,\tau_q | \Delta \rangle$$

While the existence of a minimal element in the family of distributions which satisfy (ii) and (iii) is always guaranteed, the nonsingularity and the involutivity require some additional assumptions. We deal with the problem in the following way. Given a distribution Δ and a set $\tau_1,\ldots,\tau_q$ of vector fields we define the nondecreasing sequence of distributions

(6.2a) $$\Delta_0 = \Delta$$

(6.2b) $$\Delta_k = \Delta_{k-1} + \sum_{i=1}^{q} [\tau_i, \Delta_{k-1}]$$

There is a simple consequence of this definition

(6.3) *Lemma*. The distributions $\Delta_0, \Delta_1, \ldots$ generated with the algorithm (6.2) are such that

$$\Delta_k \subset \langle \tau_1,\ldots,\tau_q | \Delta \rangle$$

for all k. If there exists an integer k^* such that $\Delta_{k^*} = \Delta_{k^*+1}$, then

$$\Delta_{k^*} = \langle \tau_1,\ldots,\tau_q | \Delta \rangle$$

Proof. If Δ' is any distribution which contains Δ and is invariant under τ_i, then it is easy to see that $\Delta' \supset \Delta_k$ implies $\Delta' \supset \Delta_{k+1}$. For, we have

$$\Delta_{k+1} = \Delta_k + \sum_{i=1}^{q} [\tau_i, \Delta_k] = \Delta_k + \sum_{i=1}^{q} \mathrm{sp}\{[\tau_i, \tau] : \tau \in \Delta_k\}$$

$$\subset \Delta_k + \sum_{i=1}^{q} \mathrm{sp}\{[\tau_i, \tau] : \tau \in \Delta'\} \subset \Delta'$$

Since $\Delta' \supset \Delta_0$, by induction we see that $\Delta' \supset \Delta_k$ for all k.

If $\Delta_{k^*} = \Delta_{k^*+1}$ for some k^* we easily see that $\Delta_{k^*} \supset \Delta$ (by definition) and Δ_{k^*} is invariant under $\tau_1,\ldots,\tau_q$ (because $[\tau_i,\Delta_{k^*}] \subset \Delta_{k^*+1} = \Delta_{k^*}$ for all $1 \le i \le q$). Thus Δ_{k^*} must coincide with $\langle \tau_1,\ldots,\tau_q|\Delta\rangle$. □

The property $\Delta_{k^*} = \Delta_{k^*+1}$ expresses a sort of finiteness quality of the sequence $\Delta_0,\Delta_1,\ldots$, and such a property is clearly useful from a computational point of view. The simplest practical situation in which the chain of distributions (6.2) satisfies the assumption of Lemma (6.3) arises when all the distributions of the chain are nonsingular. In this case, in fact, since by construction

$$\dim \Delta_k \le \dim \Delta_{k+1} \le n$$

it is easily seen that there exists an integer $k^* < n$ such that $\Delta_{k^*} = \Delta_{k^*+1}$.

If the distributions $\Delta_0,\Delta_1,\ldots$ are singular, one has the following weaker result.

(6.4) *Lemma*. There exist an open and dense subset N^* of N with the property that at each point $p \in N^*$

$$\langle \tau_1,\ldots,\tau_q|\Delta\rangle(p) = \Delta_{n-1}(p)$$

Proof. Suppose U is an open set with the property that, for some k^*, $\Delta_{k^*}(p) = \Delta_{k^*+1}(p)$ for all $p \in U$. Then, it is possible to show that $\langle \tau_1,\ldots,\tau_q|\Delta\rangle(p) = \Delta_{k^*}(p)$ for all $p \in U$. For, we already know from Lemma (6.3) that $\langle \tau_1,\ldots,\tau_q|\Delta\rangle \supset \Delta_{k^*}$. Suppose the inclusion is proper at some $\bar{p} \in U$ and define a new distribution $\bar{\Delta}$ by setting

$$\bar{\Delta}(p) = \Delta_{k^*}(p) \qquad \text{if } p \in U$$

$$\bar{\Delta}(p) = \langle \tau_1,\ldots,\tau_q|\Delta\rangle(p) \qquad \text{if } p \notin U$$

This distribution contains Δ and is invariant under $\tau_1,\ldots,\tau_q$.

For, if τ is a vector field in $\bar{\Delta}$, then $[\tau_i,\tau] \in \langle \tau_1,\ldots,\tau_q|\Delta\rangle$ (because $\bar{\Delta} \subset \langle \tau_1,\ldots,\tau_q|\Delta\rangle$) and, moreover, $[\tau_i,\tau](p) \in \Delta_{k^*}(p)$ for all $p \in U$ (because, in a neighborhood of p, $\tau \in \Delta_{k^*}$ and $[\tau_i,\Delta_k^*] \subset \Delta_k^*$). Since $\bar{\Delta}$ is properly contained in $\langle \tau_1,\ldots,\tau_q|\Delta\rangle$, this would contradict the minimality of $\langle \tau_1,\ldots,\tau_q|\Delta\rangle$.

Now, let N_k be the set of regular points of Δ_k. This set is an open and dense submanifold of N (see Lemma (2.9)) and so is the set $N^* = N_0 \cap N_1 \cap \ldots \cap N_{n-1}$. In a neighborhood of every point $p \in N^*$ the distributions $\Delta_0,\ldots,\Delta_{n-1}$ are nonsingular. This, together with the previous discussion and a dimensionality argument, shows that $\Delta_{n-1} = \langle \tau_1,\ldots,\tau_q|\Delta\rangle$ on N^* and completes the proof. □

(6.5) *Remark*. If the distribution Δ is spanned by some of the vector fields of the set $\{\tau_1,\ldots,\tau_q\}$, then, it is possible to show that there exists an open and dense submanifold N^* of N with the following property. For each $p \in N^*$ there exist a neighborhood U of p and d vector fields (with $d = \dim\langle \tau_1,\ldots,\tau_q|\Delta\rangle(p)$) $\theta_1,\ldots,\theta_d$ of the form

$$\theta_i = [v_r,[v_{r-1},\ldots,[v_1,v_0]]]$$

where $r \leq n-1$ is an integer which may depend on i and $v_0,\ldots,v_r$ are vector fields in the set $\{\tau_1,\ldots,\tau_q\}$, such that

$$\langle \tau_1,\ldots,\tau_q|\Delta\rangle(q) = \text{span}\{\theta_1(q),\ldots,\theta_d(q)\}$$

for all $q \in U$.

This fact may be proved by induction using as N^* the subset of N defined in the proof of Lemma (6.4). Let d_0 denote the dimension of Δ_0 (which may depend on p but is constant locally around p). Since, by assumption, Δ_0 is the span of some vector fields in the set $\{\tau_1,\ldots,\tau_q\}$, there exist exactly d_0 vector fields in this set that span Δ_0 locally around p. Let d_k denote the dimension of Δ_k (constant around p) and suppose Δ_k is spanned locally around p by d_k vector fields $\theta_1,\ldots,\theta_{d_k}$ of the form

$$\theta_i = [v_r,[v_{r-1},\ldots,[v_1,v_0]]]$$

where $v_0,\ldots,v_r$ (with $r \leq k$ and possibly depending on i) are vector fields in the set $\{\tau_1,\ldots,\tau_q\}$. Then, a similar result holds for Δ_{k+1}. For, let τ be any vector field in Δ_k. From Lemma (2.7) it is known that there exists real-valued smooth functions $c_1,\ldots,c_{d_k}$ defined locally around p such that τ may be expressed, locally around p, as $\tau = c_1\theta_1 + \ldots + c_{d_k}\theta_{d_k}$. If τ_j is any vector in the set $\{\tau_1,\ldots,\tau_q\}$ we have

$$[\tau_j, c_1\theta_1+\ldots+c_{d_k}\theta_{d_k}] = c_1[\tau_j,\theta]+\ldots+c_{d_k}[\tau_j,\theta_{d_k}]+(L_{\tau_j}c_1)\theta_1+\ldots+(L_{\tau_j}c_{d_k})\theta_{d_k}$$

As a consequence

$$\Delta_{k+1} = \Delta_k + [\tau_1,\Delta_k] + \ldots + [\tau_q,\Delta_k] =$$

$$sp\{\theta_i,[\tau_1,\theta_i],\ldots,[\tau_q,\theta_i]: i = 1,\ldots,d_k\}$$

Since Δ_{k+1} is nonsingular around p, then it is possible to find exactly d_{k+1} vector fields of the form

$$\theta_i = [v_r,[v_{r-1},\ldots,[v_1,v_o]]]$$

where $v_o,\ldots,v_r$ (with $r \leq k+1$ and possibly depending on i) are vector fields in the set $\{\tau_1,\ldots,\tau_q\}$, which span Δ_{k+1} locally around p. □

The previous remark is useful in getting involutivity for the distribution $\langle\tau_1,\ldots,\tau_q|\Delta\rangle$.

(6.6) *Lemma*. Suppose Δ is spanned by some of the vector fields $\tau_1,\ldots,\tau_q$ and that $\langle\tau_1,\ldots,\tau_q|\Delta\rangle$ is nonsingular. Then $\langle\tau_1,\ldots,\tau_q|\Delta\rangle$ is involutive.

Proof. We use first the conclusion of Remark (6.5) to prove that if τ_1 and τ_2 are two vector fields in Δ_{n-1}, then their Lie bracket $[\tau_1,\tau_2]$ is such that $[\tau_1,\tau_2](p) \in \Delta_{n-1}(p)$ for all $p \in N^*$. Using again Lemma (2.7) and the previous result we deduce, in fact, that in a neighborhood U of p

$$[\tau_1,\tau_2] = [\sum_{i=1}^{d} c_i^1\theta_i , \sum_{j=1}^{d} c_j^1\theta_j] \in sp\{\theta_i,\theta_j,[\theta_i,\theta_j]: i,j = 1,\ldots,d\}$$

where θ_i,θ_j are vector fields of the form described before.

In order to prove the claim, we have only to show that $[\theta_i,\theta_j](p)$ is a tangent vector in $\Delta_{n-1}(p)$. For this purpose, we recall that on N^* the distribution Δ_{n-1} is invariant under the vector fields $\tau_1,\ldots,\tau_q$ (see Lemma (6.4)) and that any distribution invariant under vector fields τ_1 and τ_2 is also invariant under their Lie bracket $[\tau_1,\tau_2]$ (see Lemma (4.5)). Since each θ_i is a repeated Lie bracket of the vector fields $\tau_1,\ldots,\tau_q$, $[\theta_i,\Delta_{n-1}](p) \subset \Delta_{n-1}(p)$ for all $1\leq i\leq d$ and, thus, in particular $[\theta_i,\theta_j](p)$ is a tangent vector which belongs to $\Delta_{n-1}(p)$.

Thus the Lie bracket of two vector fields τ_1,τ_2 in Δ_{n-1} is such that $[\tau_1,\tau_2](p) \in \Delta_{n-1}(p)$. Moreover, it has already been observed that $\langle\tau_1,\ldots,\tau_q|\Delta\rangle = \Delta_{n-1}$ in a neighborhood of p and, therefore, we conclude that at any point p of N^* the Lie bracket of any two vector fields τ_1,τ_2 in $\langle\tau_1,\ldots,\tau_q|\Delta\rangle$ is such that $[\tau_1,\tau_2](p) \in \langle\tau_1,\ldots,\tau_q|\Delta\rangle(p)$.

Consider now the distribution

$$\bar{\Delta} = \langle \tau_1,\dots,\tau_q|\Delta\rangle + \text{sp}\{[\theta_i,\theta_j]:\theta_i,\theta_j \in \langle \tau_1,\dots,\tau_q|\Delta\rangle\}$$

which, by construction, is such that

$$\bar{\Delta} \supset \langle \tau_1,\dots,\tau_q|\Delta\rangle$$

From the previous result it is seen that $\bar{\Delta}(p) = \langle \tau_1,\dots,\tau_q|\Delta\rangle(p)$ at each point p of N^*, which is a dense set in N. By assumption, $\langle \tau_1,\dots,\tau_q|\Delta\rangle$ is nonsingular. So, by Lemma (2.11) we deduce that $\bar{\Delta} = \langle \tau_1,\dots,\tau_q|\Delta\rangle$, and, therefore, that $[\theta_i,\theta_j] \in \langle \tau_1,\dots,\tau_q|\Delta\rangle$ for all pair $\theta_i,\theta_j \in \langle \tau_1,\dots,\tau_q|\Delta\rangle$. This concludes the proof. □

(6.7) *Remark.* From Lemmas (6.4),(6.6) and (2.11) it may also be deduced that if Δ is spanned by some of the vector fields $\tau_1,\dots,\tau_q$ and Δ_{n-1} is nonsingular, then

$$\langle \tau_1,\dots,\tau_q|\Delta\rangle = \Delta_{n-1}$$

and $\langle \tau_1,\dots,\tau_q|\Delta\rangle$ is involutive. □

We now come back to the original problem of the study the smallest distribution which contains $\text{sp}\{g_1,\dots,g_m\}$ and is invariant under the vector fields $f,g_1,\dots,g_m$. From the previous Lemma it is seen that if $\langle f,g_1,\dots,g_m|\ \text{sp}\{g_1,\dots,g_m\}\rangle$ is nonsingular, then it is also involutive and, therefore, the decomposition (5.5) may be performed. We will see later that the minimality of $\langle f,g_1,\dots,g_m|\text{sp}\{g_1,\dots,g_m\}\rangle$ makes it possible to deduce an interesting topological property of the set of points reached at some fixed time T starting from a given point x^o. However, before doing this, it is convenient to analyze some other characteristics of the decomposition (5.5).

Consider the distribution $\langle f,g_1,\dots,g_m|\text{sp}\{f,g_1,\dots,g_m\}\rangle$, i.e. the smallest distribution invariant under $f,g_1,\dots,g_m$ and which contains $\text{sp}\{f,g_1,\dots,g_m\}$ (note that now not only the vector fields $g_1,\dots,g_m$ but also the vector field f is assumed to belong to this distribution).

If this distribution is nonsingular, and therefore involutive by Lemma (6.6), it may indeed be used in defining a local decomposition of the control system (5.1) similar to the decomposition (5.5). We are going to see in which way this new decomposition is related to the decomposition (5.5) and why it may be of interest.

In order to simplify the notation, we set

(6.8a) $$P = \langle f,g_1,\dots,g_m|\text{sp}\{g_1,\dots,g_m\}\rangle$$

(6.8b) $$R = \langle f, g_1, \ldots, g_m | sp\{f, g_1, \ldots, g_m\} \rangle$$

The relation between P and R is described in the following statement

(6.9) *Lemma*. The distributions P and R are such that

(a) $P + sp\{f\} \subset R$

(b) if x is a regular point of $P + sp\{f\}$, then

$$(P + sp\{f\})(x) = R(x)$$

•

Proof. By definition, $P \subset R$ and $f \in R$, so (a) is true.

It is known from the proof of Lemma (6.6) that, around each point x of an open dense submanifold N^* of N, R is spanned by vector fields of the form

$$\theta_i = [v_r, \ldots, [v_1, v_0]]$$

where $r \leq n-1$ is an integer which may depend on i, and $v_r, \ldots, v_1, v_0$ are vector fields in the set $\{f, g_1, \ldots, g_m\}$.

It is easy to see that all such vector fields belong to $P+sp\{f\}$. For, if θ_i is just one of the vector fields in the set $\{f, g_1, \ldots, g_m\}$ it either belongs to P (which contains $g_1, \ldots, g_m$) or to $sp\{f\}$. If θ_i has the general form shown above we may, without loss of generality, assume that v_0 is in the set $\{g_1, \ldots, g_m\}$. For, if $v_0 = f$ and $v_1 = f$, then $\theta_i = 0$. Otherwise, if $v_0 = f$ and $v_1 = g_j$, then $-\theta_i = [v_r, \ldots, [f, g_j]]$ has the desired form. Any vector of the form

$$\theta_i = [v_r, \ldots, [v_1, g_j]]$$

with $v_r, \ldots, v_1$ in the set $\{f, g_1, \ldots, g_m\}$ is in P because P contains g_j and is invariant under $f, g_1, \ldots, g_m$ and so the claim is proved.

From this fact we deduce that on an open and dense submanifold N^* of N,

$$R \subset P + sp\{f\}$$

and therefore, since $R \supset P + sp\{f\}$ on N, that on N^*

$$R = P + sp\{f\}$$

Suppose that P + span f has constant dimension on some neighborhood U. Then, from Lemma (2.11) we conclude that the two distributions R and P + sp{f} coincide on U. □

(6.10) *Corollary.* If P and P + sp{f} are nonsingular, then

$$\dim(R) - \dim(P) \leq 1. \quad \square$$

If P and P + sp{f} are both nonsingular, so is R and, by Lemma (6.6), both P and R are involutive. Suppose that P is properly contained in R. Then, using Theorem (3.10), one can find, locally around each $\bar{x} \in N$, a neighborhood U of $\bar{x}$ and a coordinate transformation $\xi = \xi(x)$ defined on U such that

$$P(x) = sp\{(\frac{\partial}{\partial \xi_1})_x, \ldots, (\frac{\partial}{\partial \xi_{r-1}})_x\} \tag{6.11a}$$

$$R(x) = sp\{(\frac{\partial}{\partial \xi_1})_x, \ldots, (\frac{\partial}{\partial \xi_{r-1}})_x, (\frac{\partial}{\partial \xi_r})_x\} \tag{6.11b}$$

for all $x \in U$, where $r = \dim(R)$.

In the ξ coordinates the control system (5.1a) is represented by equations of the form

$$\begin{aligned}
\dot{\xi}_1 &= f_1(\xi_1,\ldots,\xi_n) + \sum_{i=1}^{m} g_{i1}(\xi_1,\ldots,\xi_n)u_i \\
&\cdots \\
\dot{\xi}_{r-1} &= f_{r-1}(\xi_1,\ldots,\xi_n) + \sum_{i=1}^{m} g_{i,r-1}(\xi_1,\ldots,\xi_n)u_i \\
\dot{\xi}_r &= f_r(\xi_r,\ldots,\xi_n) \\
\dot{\xi}_{r+1} &= 0 \\
&\cdots \\
\dot{\xi}_n &= 0
\end{aligned} \tag{6.12}$$

The last components of the vector field f are vanishing because, by construction, $f \in R$. In the particular case where $R = P$ also the r-th component of f vanishes and the corresponding equation for ξ_r is

$$\dot{\xi}_r = 0$$

From the equation (6.12) we see that any trajectory x(t) evolving on the neighborhood U actually belongs to an r-dimensional slice of U passing through the initial point. This slice is in turn partitioned into $(r-1)$-dimensional slices, each one including the set of points reached at a prescribed time T.

(6.13) *Remark*. A further change of local coordinates makes it possible to better understand the role of the time in the behavior of the control system (6.12). We may assume, without loss of generality, that the initial point x^o is such that $\xi(x^o) = 0$. Therefore we have $\xi_i(t) = 0$ for all $i = r+1,\ldots,n$ and

$$\dot{\xi}_r = f_r(\xi_r, 0, \ldots, 0)$$

Moreover, if we make the assumption that $f \notin P$, then the function f_r is nonzero everywhere on the neighborhood U. Now, let $\xi_r(t)$ denote the solution of this differential equation which passes through 0 at $t = 0$. Clearly, the mapping

$$\mu : t \longmapsto \xi_r(t)$$

is a diffeomorphism from an open interval $(-\varepsilon, \varepsilon)$ of the time axis onto the open interval of the ξ_r axis $(\xi_r(-\varepsilon), \xi_r(\varepsilon))$. If its inverse μ^{-1} is used as a local coordinate transformation on the ξ_r axis one easily sees that the new coordinate

$$\bar{\xi}_r = \mu^{-1}(\xi_r) = t$$

satisfies the differential equation

$$\dot{\bar{\xi}}_r = 1$$

In these new coordinates, points on the r-dimensional slice of U passing through the initial state are parametrized by $(\xi_1, \ldots, \xi_{r-1}, t)$. In particular, the points reached at time T belong to the $(r-1)$-dimensional slice

$$S = \{x \in U: \xi_r(x) = T, \ \xi_{r+1}(x) = 0, \ldots, \xi_n(x) = 0\}. \ \Box$$

(6.14) *Remark*. If f is a vector field of P then the local representation (6.12) is such that f_r vanishes on U. Therefore, starting from a point x^o such that $\xi(x^o) = 0$ we shall have $\xi_i(t) = 0$ for all $i=r,\ldots,n$ and the state $x(t)$ shall evolve on a $(r-1)$-dimensional slice of U passing through x^o. $\Box$

By definition the distribution R is the smallest distribution which contains $f, g_1, \ldots, g_m$ and is invariant under $f, g_1, \ldots, g_m$. Thus, we may say that in the associated decomposition (6.12) the dimension r is "minimal", in the sense that it is not possible to find another set of local coordinates $\tilde{\xi}_1, \ldots, \tilde{\xi}_{\tilde{r}}, \ldots, \tilde{\xi}_n$, with $\tilde{r}$ strictly less than r, with the property that the last $n-\tilde{r}$ coordinates remain constant with the time. We shall now show that, from the point of view of the interaction between input and state, the decomposition (6.12) has even stronger properties. Actually, we are going to prove that the states reachable from the initial state x^o fill up at least an open subset of the r-dimensional slice of in which they are contained.

(6.15) *Theorem*. Suppose the distribution R (i.e. the smallest distribution invariant under $f, g_1, \ldots, g_m$ which contains $f, g_1, \ldots, g_m$) is nonsingular. Let r denote the dimension of R. Then, for each $x^o \in N$ it is possible to find a neighborhood U of x^o and a coordinate transformation $\xi = \xi(x)$ defined on U with the following properties

(a) the set $R(x^o)$ of states reachable starting from x^o along trajectories entirely contained in U and under the action of piecewise constant input functions is a subset of the slice

$$S_{x^o} = \{x \in U : \xi_{r+1}(x) = \xi_{r+1}(x^o), \ldots, \xi_n(x) = \xi_n(x^o)\}$$

(b) the set $R(x^o)$ contains an open subset of S_{x^o}.

Proof. The proof of the statement (a) follows from the previous discussion. We proceed directly to the proof of (b), assuming throughout the proof to operate on the neighborhood U on which the coordinate transformation $\xi(x)$ is defined. For convenience, we break up the proof in several steps.

(i) Let $\theta_1, \ldots, \theta_k$ be a set of vector fields, with $k < r$, and let $\Phi_t^1, \ldots, \Phi_t^k$ denote the corresponding flows. Consider the mapping

$$F : (-\varepsilon, \varepsilon)^k \to N$$

$$(t_1, \ldots, t_k) \longmapsto \Phi_{t_k}^k \circ \ldots \circ \Phi_{t_1}^1 (x^o)$$

where x^o is a point of N and suppose that its differential has rank k at some $s_1, \ldots, s_k$, with $0 \le s_i < \varepsilon$ for $1 \le i \le k$. For ε sufficiently small the mapping

(6.16) $$\bar{F} : (s_1,\varepsilon)\times\ldots\times(s_k,\varepsilon) \to N$$

$$(t_1,\ldots,t_k) \longmapsto F(t_1,\ldots,t_k)$$

is an embedding.

Let M denote the image of the mapping (6.16) (which depends on the point x^o). Consider the slice of U

$$S_{x^o} = \{x \in U : \xi_i(x) = \xi_i(x^o),\ r+1 \leq i \leq n\}$$

If the vector fields $\theta_1,\ldots,\theta_k$ have the form

$$\theta_j = f + \sum_{i=1}^{m} g_i u_i^j$$

with $u_i^j \in \mathbb{R}$ for $1 \leq i \leq m$ and $1 \leq j \leq k$, then for ε small M is an embedded submanifold of S_{x^o}. This implies, in particular, that for each $x \in M$

(6.17) $$T_xM \subset R(x)$$

where R, as before, is the smallest distribution invariant under $f,g_1,\ldots,g_m$ which contains $f,g_1,\ldots,g_m$ (recall that R(x) is the tangent space to S_{x^o} at x).

(ii) Suppose that the vector fields $f,g_1,\ldots,g_m$ are such that

(6.18a) $$f(x) \in T_xM$$

(6.18b) $$g_i(x) \in T_xM \qquad 1 \leq i \leq m$$

for all $x \in M$. We shall show that this contradicts the assumption $k < r$. For, consider the distribution $\bar{\Delta}$ defined by setting

$$\bar{\Delta}(x) = T_xM \qquad \text{for all } x \in M$$

$$\bar{\Delta}(x) = R(x) \qquad \text{for all } x \in (N\setminus M)$$

This distribution is contained in R (because of (6.17)) and contains the vector fields $f,g_1,\ldots,g_m$ (because these vector fields are in R and, moreover, it is assumed that (6.18) are true).

Let τ be any vector field of $\bar{\Delta}$. Then $\tau \in R$ and since R is invariant under $f,g_1,\ldots,g_m$, then for all $x \in (N\setminus M)$

$$(6.19a) \qquad [f,\tau](x) \in \bar{\Delta}(x)$$

$$(6.19b) \qquad [g_i,\tau](x) \in \bar{\Delta}(x) \qquad 1 \le i \le m$$

Moreover since $\tau, f, g_1, \ldots, g_m$ are vector fields which are tangent to M at each $x \in M$, we have also that (6.19) hold for all $x \in M$, and therefore for all $x \in N$.

Having shown $\bar{\Delta}$ is invariant under $f, g_1, \ldots, g_m$ and contains $f, g_1, \ldots, g_m$, we deduce that $\bar{\Delta}$ must coincide with R. But this is a contradiction since for all $x \in M$

$$\dim \bar{\Delta}(x) = k$$

$$\dim R(x) = r > k$$

(iii) If (6.18) are not true, then it is possible to find m real numbers $u_1^{k+1}, \ldots, u_m^{k+1}$ and a point $\bar{x} \in M$ such that the vector field

$$\theta_{k+1} = f + \sum_{i=1}^{m} g_i u_i^{k+1}$$

satisfies the condition $\theta_{k+1}(\bar{x}) \notin T_{\bar{x}} M$.

Let $\bar{x} = \bar{F}(s_1', \ldots, s_k')$ be this point and Φ_t^{k+1} denote the flow of θ_{k+1}. Then the mapping

$$F' : (-\varepsilon, \varepsilon)^{k+1} \to N$$

$$(t_1, \ldots, t_k, t_{k+1}) \longmapsto \Phi_{t_{k+1}}^{k+1} \circ F(t_1, \ldots, t_k)$$

at the point $(s_1', \ldots, s_k', 0)$ has rank k+1.

For, note that

$$(F')_* (\frac{\partial}{\partial t_i})(s_1', \ldots, s_k', 0) = (F)_* (\frac{\partial}{\partial t_i})(s_1', \ldots, s_k')$$

for $i = 1, \ldots, k$ and that

$$(F')_* (\frac{\partial}{\partial t_{k+1}})(s_1', \ldots, s_k', 0) = \theta_{k+1}(\bar{x})$$

The first k tangent vectors at $\bar{x}$ are linearly independent, because F has rank k at all points of $(s_1, \varepsilon) \times \ldots \times (s_k, \varepsilon)$. The (k+1)-th one is independent from the first k by construction and therefore F' has rank k+1 at $(s_1, \ldots, s_k, 0)$.

Since $s_i' > s_i$, we may conclude that the mapping F' has rank k+1 at a point $(s_1',\ldots,s_{k+1}')$, with $0 \leq s_i' < \varepsilon$ for $1 \leq i \leq k+1$.

Note that given any real number $T > 0$ it is always possible to choose the point $\bar{x}$ in such a way that

$$(s_1' - s_1)+\ldots+(s_k' - s_k) < T$$

For, otherwise, we had that any vector field of the form

$$\theta = f + \sum_{i=1}^{m} g_i u_i$$

would be tangent to the image under $\bar{F}$ of the open set

$$\{(t_1,\ldots,t_k) \in (s_1,\varepsilon)\times\ldots\times(s_k,\varepsilon):(t_1-s_1)+\ldots+(t_k-s_k) < T\}$$

and this, as in (ii), would be a contradiction.

(iv) We can now construct a sequence of mappings of the form (6.16).

Let $\theta_1 = f + \sum_{i=1}^{m} g_i u_i^1$ be a vector field which is not zero at x^o (such a vector field can always be found because, otherwise, we would have $R(x^o) = \{0\}$) and let M_1 denote the image of the mapping

$$\bar{F}_1 : (0,\varepsilon) \to N$$

$$t_1 \longmapsto \Phi_{t_1}^1 (x^o)$$

Let $\bar{x} = \bar{F}_1(s_1^1)$ be a point of M_1 in which a vector field of the form $\theta_2 = f + \sum_{i=1}^{m} g_i u_i^2$ is such that $\theta_2(\bar{x}) \notin T_{\bar{x}} M_1$. Then we may define the mapping

$$\bar{F}_2 : (s_1^1,\varepsilon)\times(0,\varepsilon) \to N$$

$$(t_1,t_2) \longmapsto \Phi_{t_2}^2 \circ \Phi_{t_1}^1 (x^o)$$

Iterating this procedure, at stage k we start with a mapping

$$\bar{F}_k : (s_1^{k-1},\varepsilon)\times\ldots\times(s_{k-1}^{k-1},\varepsilon)\times(0,\varepsilon) \to N$$

$$(t_1,\ldots,t_{k-1},t_k) \longmapsto \Phi_{t_k}^k \circ\ldots\circ \Phi_{t_1}^1 (x^o)$$

and we find a point $\bar{x} = \bar{F}_k(s_1^k,\ldots,s_k^k)$ of its image M_k and a vector field $\theta_{k+1} = f + \sum_{i=1}^{m} g_i u_i^{k+1}$ such that $\theta_{k+1}(\bar{x}) \notin T_{\bar{x}}M_k$. This makes it possible to define the next mapping $\bar{F}_{k+1}$. Note that $s_i^k > s_i^{k-1}$ for $i = 1,\ldots,k-1$ and $s_k^k > 0$.

The procedure clearly stops at the stage r, when a mapping $\bar{F}_r$ is defined

$$\bar{F}_r : (s_1^{r-1},\varepsilon)\times\ldots\times(s_{r-1}^{r-1},\varepsilon)\times(0,\varepsilon) \to N$$

$$(t_1,\ldots,t_{r-1},t_r) \longmapsto \Phi_{t_r}^r \circ\ldots\circ\Phi_{t_1}^1(x^o)$$

(v) Observe that a point $x = \bar{F}_r(t_1,\ldots,t_r)$ in the image M_r of the embedding $\bar{F}_r$ can be reached, starting from the state x^o at time $t=0$, under the action of the piecewise constant control defined by

$$u_i(t) = u_i^k \text{ for } t \in [t_1+\ldots+t_{k-1}, t_1+t_2+\ldots+t_k)$$

Thus, we know from our previous discussions that M_r must be contained in the slice of U

$$S_{x^o} = \{x \in U: \xi_i(x) = \xi_i(x^o),\ r+1 \leq i \leq n\}$$

The images under $\bar{F}_r$ of the open sets of

$$U_r = (s_1^{r-1},\varepsilon)\times\ldots\times(s_{r-1}^{r-1},\varepsilon)\times(0,\varepsilon)$$

are open in the topology of M_r as a subset of U (because $\bar{F}_r$ is an embedding) and therefore they are also open in the topology of M_r as a subset of S_{x^o} (because S_{x^o} is an embedded submanifold of U). Therefore we have that M_r is an emebedded submanifold of S_{x^o} and a dimensionality argument tell us that M_r is actually an open submanifold of S_{x^o}. □

(6.20) *Theorem*. Suppose the distributions P (i.e. the smallest distribution invariant under $f,g_1,\ldots,g_m$ which contains $g_1,\ldots,g_m$) and P+sp{f} are nonsingular. Let p denote the dimension of P. Then, for each $x^o \in M$ it is possible to find a neighborhood U of x^o and a coordinate transformation $\xi = \xi(x)$ defined on U with the following properties:

(a) the set $R(x^o,T)$ of states reachable at time $t = T$ starting from x^o at $t = 0$, along trajectories entirely contained in U and under the action of piecewise constant input functions, is a subset of the slice

$$S_{x^o,T} = \{x \in U: \xi_{p+1} = \xi_{p+1}(\Phi_T^f(x^o)); \xi_{p+2}(x) = \xi_{p+2}(x^o), \ldots, \xi_n(x) = \xi_n(x^o)\}$$

(b) the set $R(x^o,T)$ contains an open subset of $S_{x^o,T}$.

Proof. We know from Lemma (6.9) that R is nonsingular. Therefore one can repeat the construction used to prove the part (b) of Theorem (6.15). Moreover, from Corollary (6.10) it follows that r, the dimension of R, is equal either to p+1 or to p.

Suppose the first situation happens. Given any real number $T \in (0,\varepsilon)$, consider the set

$$U_r^T = \{(t_1,\ldots,t_r) \in U_r : t_1+\ldots+t_r = T\}$$

where U_r is as defined at the step (v) in the proof of Theorem (6.15). From the last remark at the step (iii) we know that there exists always a suitable choice of $s_1^{r-1},\ldots,s_{r-1}^{r-1}$ after which this set is not empty.

Clearly the image $\bar{F}_r(U_r^T)$ consists of points reachable at time T and therefore is contained in $R(x^o,T)$. Moreover, using the same arguments as in (v), we deduce that the set $\bar{F}_r(U_r^T)$ is an open subset of $S_{x^o,T}$.

If p = r, i.e. if P = R, the proof can be carried out by simply adding an extra state variable satisfying the equation

$$\dot{\xi}_{n+1} = 1$$

and showing that this reduces the problem to the previous one. The details are left to the reader. □

7. Local Observability

We have seen in section 5 that if there is a nonsingular distribution Δ of dimension d with the properties that

(i) Δ is involutive

(ii) Δ is contained in the distribution $sp\{dh_1,\ldots,dh_\ell\}^\perp$

(iii) Δ is invariant under the vector fields $f,g_1,\ldots,g_m$

then, at each point $\bar{x} \in N$ it is possible to find a coordinate transformation defined in a neighborhood U of $\bar{x}$ and a partition of U into slices of dimension d, such that points on each slice produce the same output under any input u which keeps the state trajectory evolving on U. We want now to find conditions under which points belonging to different slices of U produce different outputs, i.e. are distinguishable.

In this case we see from the decomposition (5.7) that the right object to look for is now the "largest" distribution which satisfies (ii),(iii). Since the existence of a nonsingular distribution Δ which satisfies (i),(ii),(iii) implies and is implied by the existence of a codistribution Ω (namely $\Delta^{\perp}$) with the properties that

(i') Ω is spanned, locally around each point $p \in N$, by n-d exact covector fields

(ii') Ω contains the codistribution $sp\{dh_1,\ldots,dh_\ell\}$

(iii') Ω is invariant under the vector fields $f,g_1,\ldots,g_m$

we may as well look for the "smallest" codistribution which satisfies (ii'),(iii').

Like in the previous section, we need some background material. However, most of the results stated below require proofs which is are similar to those of the corresponding results stated before and, for this reason, will be omitted.

(7.1) *Lemma*. Let Ω be a given smooth codistribution and $\tau_1,\ldots,\tau_q$ a given set of vector fields. The family of all codistributions which are invariant under $\tau_1,\ldots,\tau_q$ and contain Ω has a minimal element, which is a smooth codistribution. □

We shall use the symbol $\langle \tau_1,\ldots,\tau_q|\Omega \rangle$ to denote the smallest codistribution which contains Ω and is invariant under $\tau_1,\ldots,\tau_q$.

Given a codistribution Ω and a set of vector fields $\tau_1,\ldots,\tau_q$ one can consider the following dual version of the algorithm (6.2)

(7.2a) $$\Omega_0 = \Omega$$

(7.2b) $$\Omega_k = \Omega_{k-1} + \sum_{i=1}^{q} L_{\tau_i}\Omega_{k-1}$$

and have the following result.

(7.3) *Lemma*. The codistributions $\Omega_0,\Omega_1,\ldots$ generated with the algorithm (7.2) are such that

$$\Omega_k \subset \langle \tau_1,\ldots,\tau_q|\Omega \rangle$$

for all k. If there exists an integer k^* such that $\Omega_{k^*} = \Omega_{k^*+1}$, then

$$\Omega_{k}^{*} = \langle \tau_1,\dots,\tau_q | \Omega \rangle \qquad \square$$

The dual version of Lemma (6.4) is the following one

(7.4) *Lemma.* There exists an open and dense subset N^* of N with the property that at each point $p \in N^*$

$$\langle \tau_1,\dots,\tau_q | \Omega \rangle = \Omega_{n-1}(p)$$

(7.5) *Remark.* If the codistribution Ω is spanned by a set $d\lambda_1,\dots,d\lambda_s$ of exact covector fields, then there exists an open and dense submanifold N^* of N with the following property. For each $p \in N^*$ there exists a neighborhood U of p and d exact covector fields (with $d = \dim\langle \tau_1,\dots,\tau_q | \Omega \rangle(p)$) $\omega_1,\dots,\omega_d$ which have the form

$$\omega_i = d(L_{v_r} \dots L_{v_1} \lambda_j)$$

where $r \leq n-1$ is an integer which may depend on i, $v_1,\dots,v_r$ are vector fields in the set $\{\tau_1,\dots,\tau_q\}$ and λ_j is a function in the set $\{\lambda_1,\dots,\lambda_s\}$, such that

$$\langle \tau_1,\dots,\tau_q | \Omega \rangle(q) = \mathrm{sp}\{\omega_1(q),\dots,\omega_d(q)\}$$

for all $q \in U$.

This may easily be proved by induction as for the corresponding statement in Remark (6.5). $\square$

(7.6) *Lemma.* Suppose Ω is spanned by a set $d\lambda_1,\dots,d\lambda_s$ of exact covector fields and that $\langle \tau_1,\dots,\tau_q | \Omega \rangle$ is nonsingular. Then $\langle \tau_1,\dots,\tau_q | \Omega \rangle^{\perp}$ is involutive.

Proof. From the previous Remark, it is seen that in a neighborhood of each point p in an open and dense submanifold N^*, the codistribution $\langle \tau_1,\dots,\tau_q | \Delta \rangle$ is spanned by exact covector fields.

Therefore, the Lie bracket of any two vector fields τ_1, τ_2 in $\langle \tau_1,\dots,\tau_q | \Omega \rangle^{\perp}$ is such that $[\tau_1,\tau_2](p) \in \langle \tau_1,\dots,\tau_q | \Omega \rangle^{\perp}(p)$ (see Remark (3.9)).

From this result, using again Lemma (2.11) as in the proof of Lemma (6.6), one deduces that $\langle \tau_1,\dots,\tau_q | \Omega \rangle^{\perp}$ is involutive. $\square$

(7.7) *Remark.* From Lemmas (7.4), (7.6) and (2.11) one may also deduce that if Ω is spanned by a set $d\lambda_1,\dots,d\lambda_s$ of exact covector fields and Ω_{n-1} is nonsingular, then

$$\langle \tau_1,\ldots,\tau_q|\Omega\rangle = \Omega_{n-1}$$

and $\langle \tau_1,\ldots,\tau_q|\Omega\rangle^{\perp}$ is involutive. □

In the study of the state-output interactions in a control system of the form (5.1), we consider the distribution

$$Q = \langle f,g_1,\ldots,g_m|\mathrm{sp}\{dh_1,\ldots,dh_\ell\}\rangle^{\perp}$$

From Lemma (4.8) we deduce that this distribution is invariant under $f,g_1,\ldots,g_m$ and we also see that, by definition, it is contained in $\mathrm{sp}\{dh_1,\ldots,dh_\ell\}^{\perp}$. If nonsingular, then, according to Lemma (7.6) is also involutive.

Invoking Proposition (5.6), this distribution may be used in order to find locally around each $\bar{x} \in N$ an open neighborhood U of $\bar{x}$ and a coordinate transformation yielding a decomposition of the form (5.7). Let s denote the dimension of Q. Since $Q^{\perp}$ is the smallest codistribution invariant under $f,g_1,\ldots,g_m$ which contains $dh_1,\ldots,dh_\ell$, then in this case the decomposition we find is maximal, in the sense that it is not possible to find another set of local coordinates $\tilde{\xi}_1,\ldots,\tilde{\xi}_{\tilde{s}},\tilde{\xi}_{\tilde{s}+1},\ldots,\tilde{\xi}_n$ with $\tilde{s}$ strictly larger than s, with the property that only the last $n-\tilde{s}$ coordinates influence the output. We show now that this corresponds to the fact that points belonging to different slices of the neighborhood U are distinguishable.

(7.8) *Theorem.* Suppose the distribution Q (i.e. the annihilator of the smallest codistribution invariant under $f,g_1,\ldots,g_m$ and which contains $dh_1,\ldots,dh_\ell$) is nonsingular. Let s denote the dimension of Q. Then, for each $\bar{x} \in N$ it is possible to find a neighborhood U of $\bar{x}$ and a coordinate transformation $\xi = \xi(x)$ defined on U with the following properties

(a) Any two initial states x^a and x^b of U such that

$$\xi_i(x^a) = \xi_i(x^b) \ , \ i = s+1,\ldots,n$$

produce identical output functions under any input which keeps the state trajectories evolving on U

(b) Any initial state x of U which cannot be distinguished from $\bar{x}$ under piecewise constant input functions belongs to the slice

$$S_{\bar{x}} = \{x \in U : \xi_i(x) = \xi_i(\bar{x}),\ s+1 \le i \le n\}.$$

Proof. We need only to prove (b). For simplicity, we break up the proof in various steps.

(i) consider a piecewise-constant input function

$$u_i(t) = u_i^k \quad \text{for} \quad t \in [t_1+\ldots+t_{k-1}, t_1+\ldots+t_k)$$

Define the vector field

$$\theta_k = f + \sum_{i=1}^{m} g_i u_i^k$$

and let Φ_t^k denote the corresponding flow. Then, the state reached at time t_k starting from x^o at time $t = 0$ under this input may be expressed as

$$x(t_k) = \Phi_{t_k}^k \circ \ldots \circ \Phi_{t_1}^1 (x^o)$$

and the corresponding output y as

$$y_i(t_k) = h_i(x(t_k))$$

Note that this output may be regarded as the value of a mapping

$$F_i^{x^o} : (-\varepsilon, \varepsilon)^k \longrightarrow \mathbb{R}$$

$$(t_1, \ldots, t_k) \longmapsto h_i \circ \Phi_{t_k}^k \circ \ldots \circ \Phi_{t_1}^1 (x^o)$$

If two initial states x^a and x^b are such that they produce two identical outputs for any possible piecewise constant input, we must have

$$F_i^{x^a}(t_1, \ldots, t_k) = F_i^{x^b}(t_1, \ldots, t_k)$$

for all possible $(t_1, \ldots, t_k)$, with $0 \le t_i < \varepsilon$ for $1 \le i \le k$. From this we deduce that

$$\left(\frac{\partial F_i^{x^a}}{\partial t_1 \ldots \partial t_k}\right)_{t_1=\ldots=t_k=0} = \left(\frac{\partial F_i^{x^b}}{\partial t_1 \ldots \partial t_k}\right)_{t_1=\ldots=t_k=0}$$

An easy calculation shows that

$$\left(\frac{\partial F_i^{x^o}}{\partial t_1 \ldots \partial t_k}\right)_{t_1=\ldots=t_k=0} = (L_{\theta_1} \ldots L_{\theta_k} h_i(x))_{x^o}$$

and, therefore, we must have

$$(L_{\theta_1} \ldots L_{\theta_k} h_i(x))_{x^a} = (L_{\theta_1} \ldots L_{\theta_k} h_i(x))_{x^b}$$

(ii) Now, remember that θ_j, $j = 1,\ldots,k$, depends on $(u_1^j,\ldots,u_m^j)$ and that the above equality must hold for all possible choices of $(u_1^j,\ldots,u_m^j) \in \mathbb{R}^m$. By appropriately selecting these $(u_1^j,\ldots,u_m^j)$ one easily arrives at an equality of the form

$$(7.9) \qquad (L_{v_1} \ldots L_{v_k} h_i)_{x^a} = (L_{v_1} \ldots L_{v_k} h_i)_{x^b}$$

where $v_1,\ldots,v_k$ are vector fields belonging to the set $\{f,g_1,\ldots,g_m\}$.

For, set $\gamma_2 = L_{\theta_2} \ldots L_{\theta_k} h$. From the equality $(L_{\theta_1}\gamma_2)_{x^a} = (L_{\theta_1}\gamma_2)_{x^b}$ we obtain

$$(L_f\gamma_2)_{x^a} + \sum_{i=1}^{m} (L_{g_i}\gamma_2)_{x^a} u_i^1 = (L_f\gamma_2)_{x^b} + \sum_{i=1}^{m} (L_{g_i}\gamma_2)_{x^b} u_i^1$$

This, due to the arbitrariness of the $u_1^1,\ldots,u_m^1$, implies that

$$(L_v\gamma_2)_{x_a} = (L_v\gamma_2)_{x_b}$$

where v is any vector in the set $\{f,g_1,\ldots,g_m\}$. This procedure can be iterated, by setting $\gamma_3 = L_{\theta_3} \ldots L_{\theta_k} h$. From the above equality one gets

$$(L_v L_f \gamma_3)_{x^a} + \sum_{i=1}^{m} (L_v L_{g_i}\gamma_3)_{x^a} u_i^2 = (L_v L_f \gamma_3)_{x^b} + \sum_{i=1}^{m} (L_v L_{g_i}\gamma_3)_{x^b} u_i^2$$

and, therefore,

$$(L_{v_1} L_{v_2} \gamma_3)_{x^a} = (L_{v_1} L_{v_2} \gamma_3)_{x^b}$$

for all v_1,v_2 belonging to the set $\{f,g_1,\ldots,g_m\}$. Finally, one arrives at (7.9).

(iii) Let U be a neighborhood of $\bar{x}$ on which a coordinate transformation $\xi(x)$ is defined which makes the condition

$$(7.10) \qquad Q(x) = \mathrm{span}\{(\frac{\partial}{\partial \xi_1})_x, \ldots, (\frac{\partial}{\partial \xi_s})_x\}$$

satisfied for all $x \in U$. From Remark (7.5), we know that there exists an open subset U^* of U, dense in U, with the property that, around each $x' \in U^*$ it is possible to find a set of $n-s$ real-valued functions $\lambda_1,\ldots,\lambda_{n-s}$ which have the form

$$\lambda_i = L_{v_r}\cdots L_{v_1}h_j \tag{7.11}$$

with $v_1,\ldots,v_r$ vector fields in $\{f,g_1,\ldots,g_m\}$ and $1 \le j \le \ell$, such that

$$Q^{\perp}(x') = \text{span}\{d\lambda_1(x'),\ldots,d\lambda_{n-s}(x')\}$$

Since $Q^{\perp}(x')$ has dimension $n-s$, it follows that the tangent covectors $d\lambda_1(x'),\ldots,d\lambda_{n-s}(x')$ are linearly independent.

In the local coordinates which satisfy (7.10), $\lambda_1,\ldots,\lambda_{n-s}$ are functions only of $\xi_{s+1},\ldots,\xi_n$ (see (5.7)). Therefore, we may deduce that the mapping

$$\Lambda : (\xi_{s+1},\ldots,\xi_n) \longmapsto (\lambda_1(\xi_{s+1},\ldots,\xi_n),\ldots,\lambda_{n-s}(\xi_{s+1},\ldots,\xi_n))$$

has a jacobian matrix which is square and nonsingular at $(\xi_{s+1}(x'),\ldots,\xi_n(x'))$.

The mapping Λ is thus locally injective. We may use this property to deduce that, for some suitable neighborhood U' of x', any other point x'' of U' such that

$$\lambda_i(x') = \lambda_i(x'')$$

for $1 \le i \le n-s$, must be such that

$$\xi_{s+i}(x'') = \xi_{s+i}(x')$$

for $1 \le i \le n-s$, i.e. must belong to the slice of U passing through x'. This, in view of the results proved in (ii) completes the proof in the case where $\bar{x} \in U^*$.

(iv) Suppose $\bar{x} \notin U^*$. Let $x(\bar{x},T,u)$ denote the state reached at time $t=T$ under the action of the piecewise constant input function u. If T is sufficiently small, $x(\bar{x},T,u)$ is still in U. Suppose $x(\bar{x},T,u) \in U^*$. Then, using the conclusions of (iii), we deduce that in some neighborhood U' of $x' = x(\bar{x},T,u)$, the states indistinguishable from x' lie on the slice of U passing through x'.

Now, recall that the mapping

$$\Phi : x^o \longmapsto x(x^o,T,u)$$

is a local diffeomorphism. Thus, there exists a neighborhood $\bar{U}$ of $\bar{x}$ whose (diffeomorphic) image under Φ is a neighborhood $U'' \subset U'$ of x'.

Let $\bar{\bar{x}}$ denote a point of $\bar{U}$ indistinguishable from $\bar{x}$ under piecewise constant inputs. Then, clearly, also $x'' = x(\bar{\bar{x}},T,u)$ is indistinguishable from $x(\bar{x},T,u) = x'$. From the previous discussion we know that x'' and x' belong to the same slice of U. But this implies also that $\bar{x}$ and $\bar{\bar{x}}$ belong to the same slice of U. Thus the proof is completed, provided that

$$x(\bar{x},T,u) \in U^* \tag{7.12}$$

(v) All we have to show now is that (7.12) can be satisfied. For, suppose $R(\bar{x})$, the set of states reachable from $\bar{x}$ under piecewise constant control along trajectories entirely contained in U, is such that

$$R(\bar{x}) \cap U^* = \emptyset \tag{7.13}$$

If this is true, we know from Theorem (6.15) that it is possible to find an r-dimensional embedded submanifold V of U entirely contained in $R(\bar{x})$ and therefore such that $V \cap U^* = \emptyset$. For any choice of functions $\lambda_1,\dots,\lambda_{n-s}$ of the form (7.11), at any point $x \in V$ the covectors $d\lambda_1(x),\dots,d\lambda_{n-s}(x)$ are linearly dependent. Thus, without loss of generality, we may assume that there exist $d < n-s$ functions $\gamma_1,\dots,\gamma_d$ still of the form (7.11) such that, for some open subset V' of V,

- $\text{span}\{dh_1(x),\dots,dh_\ell(x)\} \subset \text{span}\{d\gamma_1(x),\dots,d\gamma_d(x)\}$ for all $x \in V'$
- $d\gamma_1(x),\dots,d\gamma_d(x)$ are linearly independent covectors at all $x \in V'$,
- $dL_v\gamma_j(x) \in \text{span}\{d\gamma_1(x),\dots,d\gamma_d(x)\}$ for all $x \in V'$ and $v \in \{f,g_1,\dots,g_m\}$.

Now, we define a codistribution on N as follows

$$\Omega(x) = Q^{\perp}(x) \qquad \text{for } x \notin V'$$

$$\Omega(x) = \text{span}\{d\gamma_1(x),\dots,d\gamma_d(x)\} \text{ for } x \in V'$$

Using the fact that $f,g_1,\dots,g_m$ are tangent to V', it is not difficult to verify that this codistribution is invariant under $f,g_1,\dots,g_m$, contains $\text{sp}\{dh_1,\dots,dh_\ell\}$ and is smaller than $\langle f,g_1,\dots,g_m|\text{sp}\{dh_1,\dots,dh_\ell\}\rangle$. This is a contradiction and therefore (7.13) must be false. □

CHAPTER II
GLOBAL DECOMPOSITIONS OF CONTROL SYSTEMS

1. Sussmann's Theorem and Global Decompositions

In the previous chapter, we have shown that a nonsingular and involutive distribution induces a local partition of the manifold N into lower dimensional submanifolds and we have used this result to obtain local decompostions of control systems. The decompositions thus obtained are very useful to understand the behavior of control systems from the point of view of input-state and, respectively, state-output interaction. However, it must be stressed that the existence of decompositions of this type is strictly related to the assumption that the dimension of the distribution is constant at least over a neighborhood of the point around which we want to investigate the behaviour of our control system.

In this section we shall see that the assumption that Δ is nonsingular can be removed and that global partitions of N can be obtained. To begin with, we need the following definitions. A submanifold S of N is said to be an *integral submanifold* of the distribution Δ if, for every $p \in S$, the tangent space T_pS to S at p coincides with the subspace $\Delta(p)$ of T_pN. A *maximal* integral submanifold of Δ is a connected integral submanifold S of Δ with the property that every other connected integral submanifold of Δ which contains S coincides with S.

We see immediately from this definition that any two maximal integral submanifolds of Δ passing through a point $p \in N$ must coincide. This motivates the following notion. A distribution Δ on N has the *maximal integral manifolds property* if through every point $p \in N$ passes a maximal integral submanifold of Δ or, in other words, if there exists a *partition* of N into maximal integral submanifolds of Δ.

It is easily seen that this is a global version of the notion of complete integrability for a distribution. As a matter of fact, a nonsingular and completely integrable distribution is such that for each $p \in N$ there exists a neighborhood U of p with the property that Δ restricted to U has the maximal integral manifolds property.

A simple consequence of the previous definitions is the following one.

(1.1) *Lemma*. A distribution Δ which has the maximal integral manifolds property is involutive.

Proof. If τ is a vector field which belongs to a distribution Δ with the maximal integral manifolds property, then τ must be tangent to every maximal integral submanifold S of Δ. As a consequence, the Lie bracket $[\tau_1,\tau_2]$ of two vector fields τ_1 and τ_2 both belonging to Δ must be tangent to every maximal integral submanifold S of Δ. Thus $[\tau_1,\tau_2]$ belongs to Δ. □

Thus, involutivity is a necessary condition for Δ to have the maximal integral manifolds property but, unlike the notion of complete integrability, this condition is no longer sufficient.

(1.2) *Example*. Let $N = \mathbb{R}^2$ and let Δ be a distribution defined by

$$\Delta(x) = \text{span}\{(\frac{\partial}{\partial x_1})_x, \lambda(x_1)(\frac{\partial}{\partial x_2})_x\}$$

where $\lambda(x_1)$ is a C^∞ function such that $\lambda(x_1) = 0$ for $x_1 \leq 0$ and $\lambda(x_1) > 0$ for $x_1 > 0$. This distribution is involutive and

$$\dim \Delta(x) = 1 \quad \text{if } x \text{ is such that } x_1 \leq 0$$

$$\dim \Delta(x) = 2 \quad \text{if } x \text{ is such that } x_1 > 0$$

Clearly, the open subset of N

$$\{(x_1,x_2) \in \mathbb{R}^2 : x_1 > 0\}$$

is an integral submanifold of Δ (actually a maximal integral submanifold) and so is any subset of the form $(a,b)\times\{c\}$ with $a < b < 0$. However, it is not possible to find integral submanifolds of Δ passing through a point $(0,c)$. □

Another important point to be stressed, which emphasizes the difference between the general problem here considered and its local version described in section 1.3, is that the elements of a global partition of N induced by a distribution which has the integral manifolds property are *immersed* submanifolds. On the contrary, local partitions induced by a nonsingular and completely integrable distribution are always made of slices of a coordinate neighborhood, i.e. of *imbedded* submanifolds.

(1.3) *Example*. Consider a torus $T_2 = S_1 \times S_2$. We define a vector field on the torus in the following way.

Let τ a vector field on $\mathbb{R}^2$ defined by setting

$$\tau(x_1,x_2) = -x_2\left(\frac{\partial}{\partial x_1}\right)_x + x_1\left(\frac{\partial}{\partial x_2}\right)_x$$

At each point $(x_1,x_2) \in S_1$ this mapping defines a tangent vector in $T_{(x_1,x_2)}S_1$, and therefore a vector field on S_1 whose flow is given by

$$\Phi_t^\tau(x_1^o,x_2^o) = (x_1^o \cos t - x_2^o \sin t, x_1^o \sin t - x_2^o \cos t)$$

In order to simplify the notation we may represent a point (x_1,x_2) of S_1 with the complex number $z = x_1 + jx_2$, $|z| = 1$, and have $\Phi_t^\tau(z) = e^{jt}z$. Similarly, by setting

$$\theta(x_1,x_2) = -x_2\alpha\left(\frac{\partial}{\partial x_1}\right)_x + x_1\alpha\left(\frac{\partial}{\partial x_2}\right)_x$$

we define another vector field on S_1, whose flow is now given by $\Phi_t^\theta(z) = e^{j\alpha t}z$.

From τ and θ we may define a vector field f on T_2 by setting

$$f(z_1,z_2) = (\tau(z_1),\theta(z_2))$$

and we readily see that the flow of f is given by

$$\Phi_t^f(z_1,z_2) = (e^{jt}z_1, e^{j\alpha t}z_2)$$

If α is a rational number, then there exists a T such that $\Phi_t^f = \Phi_{t+kT}^f$ for all $t \in \mathbb{R}$ and all $k \in \mathbf{Z}$. Otherwise, if α is irrational, for each fixed $p = (z_1,z_2) \in T_2$ the mapping $F_p : t \mapsto \Phi_t^f(z_1,z_2)$ is an injective immersion of $\mathbb{R}$ into T_2, and $F_p(\mathbb{R})$ is an immersed submanifold of T_2.

From the vector field f we can define the one-dimensional distribution $\Delta = \mathrm{sp}\{f\}$ and see that, if α is irrational, the maximal integral submanifold of Δ passing through a point $p \in T_2$ is exactly $F_p(\mathbb{R})$ and Δ has the maximal integral manifold property.

$F_p(\mathbb{R})$ is an immersed but not an imbedded submanifold of T_2. For, it is easily seen that given any point $p \in T_2$ and any open (in the topology of T_2) neighborhood U of p, the intersection $F_p(\mathbb{R}) \cap U$ is dense in U and this excludes the possibility of finding a coordinate cube (U,φ) around p with the property that $F_p(\mathbb{R}) \cap U$ is a slice of U. □

The following theorem establishes the desired necessary and sufficient condition.

(1.4) *Theorem* (Sussmann). A distribution Δ has the maximal integral manifolds property if and only if, for every vector field $\tau \in \Delta$ and for every pair $(t,p) \in \mathbb{R} \times N$ such that the flow $\Phi_t^\tau(p)$ of τ is defined, the differential $(\Phi_t^\tau)_*$ at p maps the subspace $\Delta(p)$ into the subspace $\Delta(\Phi_t^\tau(p))$. □

We are not going to give the proof of this Theorem, that can be found in the literature. Nevertheless, some remarks are in order.

(1.5) *Remark*. An intuitive understanding of the constructions that are behind the statement of Sussmann's theorem may be obtained in this way.

Let $\tau_1, \ldots, \tau_k$ be a collection of vector fields of Δ and let $\Phi_{t_1}^{\tau_1}, \ldots, \Phi_{t_k}^{\tau_k}$ denote the corresponding flows. It is clear that if p is a point of N, and S is an integral manifold of Δ passing through p, then $\Phi_{t_i}^{\tau_i}(p)$ should be a point of S for all values of t_i for which $\Phi_{t_i}^{\tau_i}(p)$ is defined. Thus, S should include all points of N that can be expressed in the form

$$\Phi_{t_k}^{\tau_k} \circ \Phi_{t_{k-1}}^{\tau_{k-1}} \circ \ldots \circ \Phi_{t_1}^{\tau_1}(p) \tag{1.6}$$

In particular, if τ and θ are vector fields of Δ, the smooth curve

$$\sigma : (-\varepsilon, \varepsilon) \longrightarrow N$$

$$t \longmapsto \Phi_{t_1}^{\tau} \circ \Phi_t^{\theta} \circ \Phi_{-t_1}^{\tau}(p)$$

passing through p at $t = 0$, should be contained in S and its tangent vector at p should be contained in $\Delta(p)$. Computing this tangent vector we obtain

$$(\Phi_{t_1}^{\tau})_* \theta(\Phi_{-t_1}^{\tau}(p)) \in \Delta(p)$$

i.e. setting $q = \Phi_{-t_1}^{\tau}(p)$

$$(\Phi_{t_1}^{\tau})_* \theta(q) \in \Delta(\Phi_{t_1}^{\tau}(q))$$

and this motivates the necessity of Sussmann's condition. □

According to the statement of Theorem (1.4), in order to "test" whether or not a given distribution Δ is integrable, one should check that $(\Phi_t^\tau)_*$ maps $\Delta(p)$ into $\Delta(\Phi_t^\tau(p))$ for all vector fields τ in Δ. Actually one could limit oneself to make this test only on some suitable subset of vector fields in Δ, because the statement of Theorem (1.4) can be given the following weaker version, also due to Sussmann.

(1.7) *Theorem.* A distribution Δ has the maximal integral manifolds property if and only if there exists a set of vector fields T, which spans Δ, with the property that for every $\tau \in T$ and every pair $(t,p) \in \mathbb{R} \times N$ such that the flow $\Phi_t^\tau(p)$ of τ is defined, the differential $(\Phi_t^\tau)_*$ at p maps the subspace $\Delta(p)$ into the subspace $\Delta(\Phi_t^\tau(p))$.

(1.8) *Remark.* At this point it is clear the proof of the "if" part of Theorem (1.4) comes directly from the "if" part of Theorem (1.7), because the set of all vector fields in Δ is indeed a set of vector fields which spans Δ. Conversely, the "only if" part Theorem (1.7) comes from the "only if" part of Theorem (1.4). □

We have seen that involutivity is a necessary but not sufficient condition for a distribution Δ to have the maximal integral manifolds property. However, the involutivity is something easier to test - in principle - because it involves only the computation of the Lie bracket of vector fields in Δ whereas the test of the condition stated in the Theorem (1.7) requires the knowledge of the flows Φ_t^τ associated with all the vector fields τ of the subset T which spans Δ. Therefore, one might wish to identify some special *classes* of distributions for which the involutivity becomes a sufficient condition for them to have the maximal integral manifolds property. Actually, this is possible with a relatively little effort.

A set T of vector fields is *locally finitely generated* if, for every $p \in N$ there exist a neighborhood U of p and a finite set $\{\tau_1,\dots,\tau_k\}$ of vector fields of T with the property that every other vector field belonging to T can be represented on U in the form

$$\tau = \sum_{i=1}^{k} c_i \tau_i \tag{1.9}$$

where each c_i is a real-valued smooth function defined on U.

The class of the distributions which are spanned by locally finitely generated sets of vector fields is actually one of the classes we were looking for, as it will be shown hereafter.

We prove first a slightly different result, which will be also used independently.

(1.10) *Lemma.* Let T be a locally finitely generated set of vector fields which spans Δ and θ another vector field such that $[\theta,\tau] \in T$ for all $\tau \in T$. Then, for every pair $(t,p) \in \mathbb{R} \times N$ such that the flow $\Phi_t^\theta(p)$ is defined, the differential $(\Phi_t^\theta)_*$ at p maps the subspace $\Delta(p)$ into the subspace $\Delta(\Phi_t^\theta(p))$.

Proof. The reader will have no difficulty in finding that the same arguments used for the statement (ii) in the proof of Theorem I.(3.1) can be used. □

Note that in the above statement the vector field θ may possibly not belong to T. If the set T is *involutive*, i.e. if the Lie bracket $[\tau_1,\tau_2]$ of any two vector fields $\tau_1 \in T$, $\tau_2 \in T$ is again a vector field in T, from the previous Lemma and from Sussmann's Theorem we derive immediately the following result.

(1.11) *Theorem.* A distribution Δ spanned by an involutive and locally finitely generated set of vector fields T has the maximal integral manifolds property. □

The existence of an involutive and locally finitely generated set of vector fields appears to be something easier to prove, at least in principle. In particular, there are some classes of distributions in which the existence on a locally finitely generated set of vector fields is automatically guaranteed. This yields the following corollaries of Theorem (1.11).

(1.12) *Corollary.* A nonsingular distribution has the maximal integral manifolds property if and only if it is involutive.

Proof. In this case, the set of all vector fields which belong to the distribution is involutive and, as a consequence of Lemma I.(2.7), locally finitely generated. □

(1.13) *Corollary.* An analytic distribution on a real analytic manifold has the maximal integral manifolds property if and only if it is involutive.

Proof. It depends on the fact that any set of analytic vector fields defined on a real analytic manifold is locally finitely generated. □

We conclude this section with another interesting consequence of the previous results, which will be used later on.

(1.14) *Lemma.* Let Δ be a distribution with the maximal integral manifolds property and let S be a maximal integral submanifold of Δ. Then, given any two points p and q in S, there exist vector fields $\tau_1,\ldots,\tau_k$ in Δ and real numbers $t_1,\ldots,t_k$ such that $q = \Phi_{t_1}^{\tau_1} \circ \ldots \circ \Phi_{t_k}^{\tau_k}(p)$.

(1.15) *Theorem.* Let Δ be an involutive distribution invariant under a complete vector field θ. Suppose the set of all vector fields in Δ is locally finitely generated. Let p_1 and p_2 be two points belonging to the same maximal integral submanifold of Δ. Then, for all T, $\Phi^\theta_T(p_1)$ and $\Phi^\theta_T(p_2)$ belong to the same maximal integral submanifold of Δ.

Proof. Observe, first of all, that Δ has the maximal integral submanifold property (see Theorem (1.11)).

Let τ be a vector field in Δ. Then, for ε sufficienty small the mapping

$$\sigma : (-\varepsilon,\varepsilon) \longrightarrow N$$

$$t \longmapsto \Phi^\theta_T \circ \Phi^\tau_t \circ \Phi^\theta_{-T}(p)$$

defines a smooth curve on N which passes through p at t = 0. Computing the tangent vector to this curve at t we get

$$\sigma_*(\tfrac{d}{dt})_t = (\Phi^\theta_T)_* \tau(\Phi^\tau_t \circ \Phi^\theta_{-T}(p)) =$$

$$= (\Phi^\theta_T)_* \tau(\Phi^\theta_{-T}(\sigma(t)))$$

But since $\tau \in \Delta$, we know from Lemma (1.10) that for all q $(\Phi^\theta_T)_*\tau(\Phi^\theta_{-T}(q)) \subseteq \Delta(q)$ and therefore we get

$$\sigma_*(\tfrac{d}{dt})_t \in \Delta(\sigma(t))$$

for all $t \in (-\varepsilon,\varepsilon)$. This shows that the smooth curve σ lies on an integral submanifold of Δ. Now, let $p_1 = \Phi^\theta_{-T}(p)$ and $p_2 = \Phi^\tau_t(p_1)$. Then p_2 and p_1 are two points belonging to a maximal integral submanifold of Δ, and the previous result shows that $\Phi^\theta_T(p_1)$ and $\Phi^\theta_T(p_2)$ again are two points belonging to a maximal integral submanifold of Δ. Thus the Theorem is proved for points p_1,p_2 such that $p_2 = \Phi^\tau_t(p_1)$. If this is not the case, using Lemma (1.14) we can always find vector fields $\tau_1,\ldots,\tau_k$ of Δ such that $p_2 = \Phi^{\tau_1}_{t_1} \circ \ldots \circ \Phi^{\tau_k}_{t_k}(p_1)$ and use the above result in order to prove the Theorem. □

2. The Control Lie Algebra

The notions developed in the previous section are useful in dealing with the study of input-state interaction properties from a global point of view. As in chapter 1, we consider here control systems described by equations of the form

$$\dot{x} = f(x) + \sum_{i=1}^{m} g_i(x)u_i \tag{2.1}$$

Recall that the local analysis of these properties was based upon the consideration of the smallest distribution, denoted R, invariant under the vector fields $f,g_1,\ldots,g_m$ and which contains $f,g_1,\ldots,g_m$. It was also shown that if this distribution is nonsingular, then it is involutive (Lemma I.(6.6)). This property makes it possible to use immediately one of the results discussed in the previous section and find a global decomposition of the state space N.

(2.2) *Lemma*. Suppose R is nonsingular, then R has the maximal integral manifolds property.

Proof. Just use Corollary (1.12). □

The decomposition of N into maximal integral submanifolds of R has the following interpretation from the point of view of the study of interaction between inputs and states. It is known that each of the vector fields $f,g_1,\ldots,g_m$ is in R, and therefore tangent to each maximal integral submanifold of R. Let S_{x^o} be the maximal integral submanifold of R passing through x^o. From what we have said before we know that any vector field of the form $\tau = \sum_{i=1}^{m} g_i u_i$, where $u_1,\ldots,u_m$ are real numbers, will be tangent to S_{x^o} and, therefore, that the integral curve of τ passing through x^o at time $t = 0$ will belong to S_{x^o}. We conclude that any state trajectory emanating from the point x^o, under the action of a piecewise constant control, will stay in S_{x^o}.

Putting together this observation with the part (b) of the statement of Theorem I.(6.15), one obtains the following result.

(2.3) *Theorem*. Suppose R in nonsingular. Then there exists a partition of N into maximal integral submanifolds of R, all with the same dimension. Let S_{x^o} denote the maximal integral submanifold of R passing through x^o. The set $R(x^o)$ of states reachable from x^o under piecewise constant input functions

(a) is a subset of S_{x^o}

(b) contains an open set of S_{x^o}. □

The result might be interpreted as a global version of Theorem I.(6.15). However, there are more general versions, which do not require the assumption that R is nonsingular. Of course, since one is interested in having global decompositions, it is necessary to work with distributions having the maximal integral manifolds property. From the discussions of the previous section, we see that a reasonable situation is the one in which the distributions are spanned by a set of vector fields which is involutive and locally finitely generated. This motivates the interest in the following considerations.

Let $\{\tau_i : 1 \le i \le q\}$ be a finite set of vector fields and L_1, L_2 two subalgebras of V(N) which both contain the vector fields $\tau_1,\dots,\tau_q$. Clearly, the intersection $L_1 \cap L_2$ is again a subalgebra of V(N) and contains $\tau_1,\dots,\tau_q$. Thus we conclude that there exists a unique subalgebra L of V(N) which contains $\tau_1,\dots,\tau_q$ and has the property of being contained in all the subalgebras of V(N) which contain the vector fields $\tau_1,\dots,\tau_q$. We refer to this as to the *smallest* subalgebra of V(N) which contains the vector fields $\tau_1,\dots,\tau_q$.

(2.4) *Remark*. One may give a description of the subalgebra L also in the following terms. Consider the set

$$L_o = \{\tau \in V(N) : \tau = [\tau_{i_k},[\tau_{i_{k-1}},\dots,[\tau_{i_2},\tau_{i_1}]]]; \quad 1 \le i_k \le q,\ 1 < k < \infty\}$$

and let $LC(L_o)$ denote the set of all finite $\mathbb{R}$-linear combinations of elements of L_o. Then, it is possible to see that $L = LC(L_o)$. For, by construction, every element of L_o is an element of L because L, being a subalgebra of V(N) which contains $\tau_1,\dots,\tau_q$, must contain every vector field of the form $[\tau_{i_k},[\tau_{i_{k-1}},\dots,[\tau_{i_2},\tau_{i_1}]]]$. Therefore $LC(L_o) \subset L$ and also $\tau_i \in LC(L_o)$ for $1 \le i \le q$. To prove that $L = LC(L_o)$ we only need to show that $LC(L_o)$ is a subalgebra of V(N). This follows from the fact that the Lie bracket of any two vector fields in L_o is an $\mathbb{R}$-linear combination of elements of L_o. □

With the subalgebra L we may associate a distribution Δ_L in a natural way, by setting

$$\Delta_L = sp\{\tau : \tau \in L\}$$

Clearly, Δ_L need not to be nonsingular. Thus, in order to be able to operate with Δ_L, we have to set explicitly some suitable assumptions. In view of the results discussed at the end of the previous section we shall assume that the subalgebra L is spanned by a locally finitely generated set of vector fields.

An immediate consequence of this assumption is the following one.

(2.5) *Lemma*. If the subalgebra L is locally finitely generated, the distribution Δ_L has the maximal integral manifolds property.

Proof. The set L is involutive by construction (because is a subalgebra of V(N)). Then, using Theorem (1.11) we see that Δ_L has the maximal integral manifolds property. □

When dealing with control systems of the form (2.1), we take into consideration the smallest subalgebra of V(N) which contains the vector fields $f,g_1,\ldots,g_m$. This subalgebra will be denoted by C and called the *Control Lie Algebra*. With C we associate the distribution

$$\Delta_C = \mathrm{sp}\{\tau : \tau \in C\}$$

(2.6) *Remark*. It is not difficult to prove that the codistribution $\Delta_C^{\perp}$ is invariant under the vector fields $f,g_1,\ldots,g_m$. For, let τ be any vector field in C and ω a covector field in $\Delta_C^{\perp}$. Then $\langle \omega,\tau \rangle = 0$ and $\langle \omega,[f,\tau] \rangle = 0$ because $[f,\tau]$ is again a vector field in C. Therefore, from the equality

$$\langle L_f\omega,\tau \rangle = L_f\langle \omega,\tau \rangle - \langle \omega,[f,\tau] \rangle = 0$$

we deduce that $L_f\omega$ annihilates all vector fields in C. Since Δ_C is spanned by vector fields in C, it follows that $L_f\omega$ is a covector field in $\Delta_C^{\perp}$, i.e. that $\Delta_C^{\perp}$ is invariant under f. In the same way it is proved that $\Delta_C^{\perp}$ is invariant under $g_1,\ldots,g_m$.

If the codistribution $\Delta_C^{\perp}$ is smooth (e.g. when the distribution Δ_C is nonsingular), then using Lemma I.(4.8) one concludes that Δ_C itself is invariant under $f,g_1,\ldots,g_m$.

(2.7) *Remark*. The distribution Δ_C, and the distributions P and R introduced in the previous chapter are related in the following way

(a) $\Delta_C \subset P + \mathrm{sp}\{f\} \subset R$

(b) if x is a regular point of Δ_C, then $\Delta_C(x) = (P+\mathrm{sp}\{f\})(x) = R(x)$.

We leave to the reader the proof of this statement. □

The role of the Control Lie Algebra C in the study of interactions between input and state depends on the following consideration. Suppose Δ_C has the maximal integral manifolds property and let S_{x^o} be the maximal integral submanifold of Δ_C passing through x^o. Since the vector fields $f,g_1,\ldots,g_m$, as well as any vector field τ of the form

$\tau = f + \sum_{i=1}^{m} g_i u_i$ with $u_1,\ldots,u_m$ real numbers, are in Δ_C (and therefore tangent to S_{x^o}), then any state trajectory of the control system (2.1) passing through x^o at $t = 0$, due to the action of a piecewise constant control, will stay in S_{x^o}.

As a consequence of this we see that, when studying the behavior of a control system intialized at $x^o \in N$, we may regard as a natural state space the submanifold S_{x^o} of N instead of the whole N. Since for all $\hat{x} \in S_{x^o}$ the tangent vectors $f(\hat{x}), g_1(\hat{x}),\ldots,g_m(\hat{x})$ are elements of the tangent space to S_{x^o} at $\hat{x}$, by taking the *restrictions* to S_{x^o} of the original vector fields $f, g_1,\ldots,g_m$ one may define a set of vector fields $\hat{f}, \hat{g}_1,\ldots,\hat{g}_m$ on S_{x^o} and a control system evolving on S_{x^o}

$$\dot{\hat{x}} = \hat{f}(\hat{x}) + \sum_{i=1}^{m} \hat{g}_i(\hat{x}) u_1 \tag{2.8}$$

which behaves exactly as the original one.

By construction, the smallest subalgebra $\hat{C}$ of $V(S_{x^o})$ which contains $\hat{f}, \hat{g}_1,\ldots,\hat{g}_m$ spans, at each $\hat{x} \in S_{x^o}$, the whole tangent space $T_{\hat{x}} S_{x^o}$. This may easily be seen using for $\hat{C}$ and C the description illustrated in the Remark (2.4).

Therefore, one may conclude that for the control system (2.8) (which evolves on S_{x^o}), the dimension of $\Delta_{\hat{C}}$ is equal to that of S_{x^o} at each point or, also, that the smallest distribution $\hat{R}$ invariant under $\hat{f}, \hat{g}_1,\ldots,\hat{g}_m$ which contains $\hat{f}, \hat{g}_1,\ldots,\hat{g}_m$ is nonsingular (see Remark (2.7)), with a dimension equal to that of S_{x^o}.

The control system (2.8) is such that the assumptions of Theorem (2.3) are satisfied, and this makes it possible to state the following result.

(2.9) *Theorem*. Suppose the distribution Δ_C has the maximal integral manifolds property. Let S_{x^o} denote the maximal integral submanifold of Δ_C passing through x^o. The set $R(x^o)$ of states reachable from x^o under piecewise constant input functions

(a) is a subset of S_{x^o}

(b) contains an open set of S_{x^o}.

(2.10) *Remark*. Note that, if Δ_C has the maximal integral manifolds property but is *singular*, then the dimensions of different maximal integral submanifolds of Δ_C may be different. Thus, it may happen that at two different initial states x^1 and x^2 one obtains two control systems of the form (2.8) which evolves on two manifolds S_{x^1} and S_{x^o} of different dimensions. We will see examples of this in section 4. □

(2.11) *Remark*. Note that the assumption "the distribution Δ_C has the maximal integral manifolds property"is implied by the assumption "the distribution Δ_C is nonsingular". In this case, in fact, $\Delta_C = R$ (see Remark (2.7)) and R has the maximal integral manifolds property (Lemma (2.2)). □

We conclude this section by the illustration of some terminology which is frequently used. The control system (2.1) is said to satisfy the *controllability rank condition* at x^o if

$$\dim \Delta_C(x^o) = n \tag{2.12}$$

Clearly, if this is the case, and if Δ_C has the maximal integral manifolds property, then the maximal integral submanifold of Δ_C passing through x^o has dimension n and, according to Theorem (2.9), the set of states reachable from x^o fill up at least an open set of the state space N.

The following Corollary of Theorem (2.9) describes the situation which holds when one is free to choose arbitrarily the initial state x^o. A control system of the form (2.1) is said to be *weakly controllable* on N if for every initial state $x^o \in N$ the set of states reachable under piecewise constant input functions contains at least an open set of N.

(2.13) *Corollary*. A sufficient condition for a control system of the form (2.1) to be weakly controllable on N is that

$$\dim \Delta_C(x) = n$$

for all $x \in N$. If the distribution Δ_C has the maximal integral manifolds property then this condition is also necessary.

Proof. If this condition is satisfied, Δ_C is nonsingular, involutive and therefore, from the previous discussions, we conclude that the system is weakly controllable. Conversely, if the distribution Δ_C has the maximal integral manifolds property and $\dim \Delta_C(x^o) < n$ at some $x^o \in N$ then the set of states reachable from x^o belongs to a submanifold of N whose dimension is strictly less than n (Theorem (2.9)).So this set cannot contain an open subset of N. □

3. The Observation Space

In this section we study output-state interaction properties from a global point of view, for a system described by equations of the form (2.1), together with an output map

$$y = h(x) \tag{3.1}$$

The presentation will be closely analogue to the one given in the previous section. First of all, recall that the local analysis carried out in section I.7 was based upon the consideration of the smallest codistribution invariant under the vector fields $f, g_1, \ldots, g_m$ and containing the covector fields $dh_1, \ldots, dh_\ell$. If the annihilator Q of this codistribution is nonsingular, then it is also involutive (Lemma I.(7.6)) and may be used to perform a global decomposition of the state space. Parallel to Lemma (2.2) we have the following result.

(3.2) *Lemma*. Suppose Q is nonsingular. Then Q has the maximal integral manifolds property. □

The role of this decomposition in explaining the output-state interaction may be explained as follows. Observe that Q, being nonsingular and involutive, satisfies the assumptions of Theorem (1.15) (because the set of all vector fields in a nonsingular distribution is locally finitely generated). Let S be any maximal integral submanifold of Q. Since Q is invariant under $f, g_1, \ldots, g_m$ and also under any vector field of the form $\tau = f + \sum_{i=1}^{m} g_i u_i$, where $u_1, \ldots, u_m$ are real numbers, using Theorem (1.15) we deduce that given any two points x^a and x^b in S and any vector field of the form $\tau = f + \sum_{i=1}^{m} g_i u_i$, the points $\Phi_t^\tau(x^a)$ and $\Phi_t^\tau(x^b)$ for all t belong to the same maximal integral submanifold of Q. In other words, we see that from any two initial states on some maximal integral submanifold of Q, under the action the same piecewise constant control one obtains two trajectories which, at any time, pass through the same maximal integral submanifold of Q.

Moreover, it is easily seen that the functions $h_1, \ldots, h_\ell$ are constant on each maximal integral submanifold of Q. For, let S be any of these submanifolds and let $\hat{h}_i$ denote the restriction of h_i to S. At each point p of S the derivative of $\hat{h}_i$ along any vector v of T_pS is zero, because $Q \subset sp(dh_i)^{\perp}$, and therefore the function $\hat{h}_i$ is a constant.

As a conclusion, we immediately see that if x^a and x^b are two initial states belonging to the same integral manifold of Q then under the action of the same piecewise constant control one obtains two trajectories which, at any time, produce identical values on each component of the output, e.g. are indistinguishable.

These considerations enables us to state the following global version of Theorem I.(7.8).

(3.3) *Theorem*. Suppose Q is nonsingular. Then there exists a partition of N into maximal integral submanifolds of Q, all with the same dimension. Let S denote the maximal integral submanifold of Q passing through x^o. Then

(a) no other point of S can be distinguished from x^o under piecewise constant input functions

(b) there exists an open neighborhood U of x^o in N with the property that any point $x \in U$ which cannot be distinguished from x^o under piecewise constant input functions necessarily belongs to $U \cap S$. □

Proof. The statement (a) has already been proved. The statement (b) requires some remark. Since Q is nonsingular, we know that around any point x^o we can find a neighborhood U and a partition of U into slices each of which is clearly an integral submanifold of Q. But also the intersection of S with U, which is a nonempty open subset of S is an integral submanifold of Q. Therefore, since S is maximal, we deduce that the slice of U passing through x^o is contained into $U \cap S$. From the statement (b) of Theorem I.(7.7) we deduce that any other state x of U which cannot be distinguished from x^o under piecewise constant inputs belongs to the slice of U passing through x^o, and therefore to $U \cap S$. □

If the distribution Q is singular, one may approach the problem on the basis of the following considerations. Let $\{\lambda_i : 1 \leq i \leq \ell\}$ be a finite set of real-valued functions and $\{\tau_i : 1 \leq i \leq q\}$ be a finite set of vector fields. Let S_1 and S_2 be two subspaces of $C^\infty(N)$ which both contain the functions $\lambda_1,\dots,\lambda_\ell$ and have the property that, for all $\lambda \in S_i$ and for all $1 \leq j \leq q$, $L_{\tau_j}\lambda \in S_i$, $i = 1,2$. Clearly the intersection $S_1 \cap S_2$ is again a subspace of $C^\infty(N)$ which contains $\lambda_1,\dots,\lambda_\ell$ and is such that, for all $\lambda \in S_1 \cap S_2$ and for all $1 \leq j \leq q$, $L_{\tau_j}\lambda \in S_1 \cap S_2$. Thus we conclude that there exists a unique subspace S of $C^\infty(N)$ which contains $\lambda_1,\dots,\lambda_\ell$ and is such that, for all $\lambda \in S$

and for all $1 \leq j \leq q$, $L_{\tau_j}\lambda \in S$. This is the smallest subspace of $C^{\infty}(N)$ which contains $\lambda_1,\ldots,\lambda_\ell$ and is closed under differentiation along $\tau_1,\ldots,\tau_q$.

(3.4) *Remark*. The subspace S may be described as follows. Consider the set

$$S_o = \{\lambda \in C^{\infty}(N): \lambda = \lambda_j \text{ or } \lambda = L_{\tau_{i_k}}\ldots L_{\tau_{i_1}}\lambda_j;\ 1 \leq j \leq \ell,\ 1 \leq i_k \leq q,\ 1 \leq k < \infty\}$$

and let $LC(S_o)$ denote the set of all $\mathbb{R}$-linear combinations of elements of S_o. Then, $LC(S_o) = S$. As a matter of fact, it is easily checked that every element of $LC(S_o)$ is an element of S, so $LC(S_o) \subset S$, that $\lambda_j \in LC(S_o)$ for $1 \leq j \leq \ell$ and that $LC(S_o)$ is closed under differentiation along $\tau_1,\ldots,\tau_q$. □

With the subspace S we may associate a codistribution Ω_S, in a natural way, by setting

$$\Omega_S = \text{sp}\{d\lambda : \lambda \in S\}$$

The codistribution Ω_S is smooth by construction, but - as we know - the distribution $\Omega_S^{\perp}$ may fail to be so. Since we are interested in smooth distributions because we use them to partition the state space into maximal integral submanifolds, we should rather be looking at the distribution $\text{smt}(\Omega_S^{\perp})$ (see section I.2).

The following result is important when looking at $\text{smt}(\Omega_S^{\perp})$ for the purpose of finding global decompositions of N.

(3.5) *Lemma*. Suppose the set of all vector fields in $\text{smt}(\Omega_S^{\perp})$ is locally finitely generated. Then $\text{smt}(\Omega_S^{\perp})$ has the maximal integral manifolds property.

Proof. In view of Theorem (1.11), we have only to show that $\text{smt}(\Omega_S^{\perp})$ is involutive. Let τ_1 and τ_2 be vector fields in $\text{smt}(\Omega_S^{\perp})$ and λ any function in S. Since $\langle d\lambda,\tau_1\rangle = 0$ and $\langle d\lambda,\tau_2\rangle = 0$ we have

$$\langle d\lambda,[\tau_1,\tau_2]\rangle = L_{\tau_1}\langle d\lambda,\tau_2\rangle - L_{\tau_2}\langle d\lambda,\tau_1\rangle = 0$$

The vector field $[\tau_1,\tau_2]$ is thus in $\Omega_S^{\perp}$. But $[\tau_1,\tau_2]$, being smooth, is also in $\text{smt}(\Omega_S^{\perp})$. □

In order to study observability we consider the smallest subspace of $C^{\infty}(N)$ which contains the functions $h_1,\ldots,h_\ell$ and is closed under differentiation along the vector fields $f,g_1,\ldots,g_m$. This subspace will be denoted by $\mathcal{O}$ and called the *observation space*. Moreover, with $\mathcal{O}$ we associate the codistribution

$$\Omega_{\mathcal{O}} = \mathrm{sp}\{d\lambda : \lambda \in \mathcal{O}\}$$

(3.6) *Remark*. It is possible to prove that the distribution $\Omega_{\mathcal{O}}^{\perp}$ is invariant under the vector fields $f, g_1, \ldots, g_m$. For, let λ be any function in $\mathcal{O}$ and τ a vector field in $\Omega_{\mathcal{O}}^{\perp}$. Then $\langle d\lambda, \tau \rangle = 0$ and $\langle dL_f\lambda, \tau \rangle = 0$ because $L_f\lambda$ is again a function in $\mathcal{O}$. Therefore, from the equality

$$\langle d\lambda, [f,\tau] \rangle = L_f \langle d\lambda, \tau \rangle - \langle dL_f\lambda, \tau \rangle = 0$$

we deduce that $[f,\tau]$ annihilates all functions in $\mathcal{O}$. Since $\Omega_{\mathcal{O}}$ is spanned by differentials of functions in $\mathcal{O}$, it follows that $[f,\tau]$ is a vector field in $\Omega_{\mathcal{O}}^{\perp}$. In the same way one proves invariance under $g_1, \ldots, g_m$.

If the distribution $\Omega_{\mathcal{O}}^{\perp}$ is smooth (e.g. when the codistribution $\Omega_{\mathcal{O}}$ is nonsingular) then using Lemma I.(4.8) one concludes that $\Omega_{\mathcal{O}}$ itself is invariant under $f, g_1, \ldots, g_m$.

(3.7) *Remark*. The distribution $\Omega_{\mathcal{O}}^{\perp}$ and the distribution Q introduced in the previous chapter are related in the following way

(a) $\Omega_{\mathcal{O}}^{\perp} \supset Q$

(b) if x is a regular point of $\Omega_{\mathcal{O}}$, then $\Omega_{\mathcal{O}}^{\perp}(x) = Q(x)$.

We leave to the reader the proof of this statement. □

From the previous Remark and from Remark I.(4.6) it is deduced that the distribution $\mathrm{smt}(\Omega_{\mathcal{O}}^{\perp})$ is invariant under the vector fields $f, g_1, \ldots, g_m$ and so under any vector field τ of the form

$\tau = f + \sum_{i=1}^{m} g_i u_i$, where $u_1, \ldots, u_m$ are real numbers. Now suppose that the set of all vector fields in $\mathrm{smt}(\Omega_{\mathcal{O}}^{\perp})$ is locally finitely generated, so that $\mathrm{smt}(\Omega_{\mathcal{O}}^{\perp})$ has the maximal integral manifolds property. Using Theorem (1.15), as we did before in the case of nonsingular Q, we may conclude that from any two states on the same integral submanifold of $\mathrm{smt}(\Omega_{\mathcal{O}}^{\perp})$, under the action of the same piecewise constant control one obtains two trajectories that at any time lie on the same maximal integral submanifold of $\mathrm{smt}(\Omega_{\mathcal{O}}^{\perp})$. Observe now that $\mathrm{smt}(\Omega_{\mathcal{O}}^{\perp})$ is also contained in $\mathrm{sp}\{dh_i\}^{\perp}$, $1 \le i \le \ell$, because every tangent vector in $\mathrm{smt}(\Omega_{\mathcal{O}}^{\perp})(x)$ is also in $\Omega_{\mathcal{O}}^{\perp}(x)$ and every tangent vector v in $\Omega_{\mathcal{O}}^{\perp}(x)$ is such that $\langle dh_i(x), v \rangle = 0$. Therefore one may deduce that the functions h_i are constant on each maximal integral submanifold of $\mathrm{smt}(\Omega_{\mathcal{O}}^{\perp})$.

This, together with the previous observations, shows that any two initial states x^a and x^b on the same maximal integral submanifold of $smt(\Omega_0^\perp)$ are indistinguishable under piecewise constant inputs. This extends the statement (a) of Theorem (3.3). As for the statement (b), some regularity is required, as it seen hereafter.

(3.8) *Theorem*. Suppose the set of all vector fields contained in $smt(\Omega_0^\perp)$ is locally finitely generated. Let S denote the maximal integral submanifold of $smt(\Omega_0^\perp)$ passing through x^o. Then

(a) any other point of S cannot be distinguished from x^o under piecewise constant inputs

(b) If x^o is a regular point of Ω_0 , then there exists an open neighborhood U of x^o in N with the property that any point $x \in U$ which cannot be distinguished from x^o under piecewise constant inputs necessarily belongs to $U \cap S$.

Proof. The statement (a) has already been proved. The statement (b) is proved essentially in the same way as in the statement (b) of Theorem (3.3). □

The following example illustrates the need for the "regularity" assumption in the statement (b) of the previous theorem.

(3.9) *Example*. Consider the following system with $N = \mathbb{R}$ and

$$\dot{x} = 0$$

$$y = h(x)$$

where h(x) is defined as

$$h(x) = \exp(-\frac{1}{x^2})\sin(\frac{1}{x}) \quad \text{for } x \neq 0$$

$$h(0) = 0$$

For this system, two states x^a and x^b are indistinguishable if and only if $h(x^a) = h(x^b)$. In particular, the set of states which are indistinguishable from the state $x = 0$ coincides with the set of the roots of the equation $h(x) = 0$. Each point in this set is isolated but the point $x = 0$. Thus, no matter how small we choose an open neighborhood U of $x = 0$, U contains points indistinguishable from $x = 0$.

It is also seen that the codistribution $\Omega_0 = sp\{dh\}$ has dimension 1 everywhere but at the points x in which $\frac{dh}{dx} = 0$, where its dimension is 0. Thus, any smooth vector field belonging to $\Omega_0^\perp$ must vanish identically on $\mathbb{R}$ and $\mathrm{smt}(\Omega_0^{\perp}) = \{0\}$. The maximal integral submanifold of $\mathrm{smt}(\Omega_0^{\perp})$ passing through x is the point x itself.

At the point $x = 0$, which is not a regular point of Ω_0, we have that $U \cap S = \{0\}$ for all U, whereas we know there are other points of U indistinguishable from $x = 0$. □

We conclude this section with some global considerations. The control system (2.1)-(3.1) is said to satisfy the *observability rank condition* at x^o if

$$\dim \Omega_0(x^o) = n \tag{3.10}$$

Clearly, if this is the case then x^o is a regular point of Ω_0 and from the previous discussion it is seen that any point x in a suitable neighborhood U of x^o can be distinguished under piecewise constant inputs. A control system of the form (2.1)-(3.1) is said to be *locally observable* on N if for every state x^o there is neighborhood U of x^o in which every point can be distinguished from x^o under piecewise constant inputs.

(3.11) *Corollary*. A sufficient condition for a control system of the form (2.1)-(3.1) to be locally observable on N is that

$$\dim \Omega_0(x) = n$$

for all $x \in N$.

4. Linear Systems, Bilinear Systems and Some Examples

In this section we describe some elementary examples, in order to make the reader more familiar with the ideas introduced so far.

As a first application, we shall compute the Lie algebra C and the distribution Δ_C for a *linear system*

$$\dot{x} = Ax + Bu$$
$$y = Cx$$

We may easily interpret this system as a system of the form (2.1)-(3.1). The manifold N on which the system evolves is the whole of $\mathbb{R}^n$ and, in the standard (single) coordinate chart of $\mathbb{R}^n$, the vector fields $f(x)$ and $g_1(x),\dots,g_m(x)$ have the expressions

(4.1) $$f(x) = Ax$$

(4.2) $$g_i(x) = b_i \qquad 1 \leq i \leq m$$

where b_i is the i-th column of the matrix B. The functions $h_1(x),\ldots,h_\ell(x)$ are expressed as

$$h_i(x) = c_i x \qquad 1 \leq i \leq \ell$$

where c_i is the i-th row of the matrix C.

We want to prove first that the Lie algebra $\mathcal{C}$ is the subspace of V(N) consisting of all vector fields which are $\mathbb{R}$-linear combinations of the vector fields in the set

(4.3) $$\{Ax\} \cup \{A^k b_i : 1 \leq i \leq m,\ 0 \leq k \leq n-1\}$$

For, observe that this set contains the vector fields Ax and $b_1,\ldots,b_m$ (i.e. the vector fields f and $g_1,\ldots,g_m$) and also that this set is contained in $\mathcal{C}$, because any of its elements is a repeated Lie bracket of f and $g_1,\ldots,g_m$. As a matter of fact,

$$A^k b_i = \underbrace{[[[g_i,f],\ldots],f]}_{\text{k-times}}$$

Moreover, it is easy to see that the set

(4.4) $$LC(\{Ax\} \cup \{A^k b_i : 1 \leq i \leq m,\ 0 \leq k \leq n-1\})$$

of all $\mathbb{R}$-linear combinations of vector fields in the set (4.3) is already a Lie subalgebra, i.e. is closed under Lie bracketing.

For, one easily sees that if $\tau_1(x)$ and $\tau_2(x)$ are vector fields of the form

$$\tau_1(x) = A^k b_i$$

$$\tau_2(x) = A^h b_j$$

then $[\tau_1,\tau_2](x) = 0$. On the other hand, if

$$\tau_1(x) = A^k b_i$$
$$\tau_2(x) = Ax$$

then

$$[\tau_1, \tau_2] = A^{k+1} b_i$$

If $k < n-1$, this vector field is in the set (4.3) and, if $k = n-1$,this vector field is an $\mathbb{R}$-linear combination of vector fields in the set (4.3) (by Cayley-Hamilton theorem).

If τ_1 and τ_2 are $\mathbb{R}$-linear combinations of vector fields of (4.3), then their Lie bracket is still an $\mathbb{R}$-linear combination of vector fields of (4.3), and this proves that the set (4.4) is a Lie subalgebra.

The set (4.4) is a Lie algebra which contains $f, g_1, \ldots, g_m$ and is contained in C, the smallest Lie subalgebra which contains $f, g_1, \ldots, g_m$. Then, the set (4.4) coincides with C.

Evaluating the distribution Δ_C we get, at a point $x \in \mathbb{R}^n$,

$$(4.5) \quad \Delta_C(x) = \text{span}\{Ax\} + \text{span}\{A^k b_i : 1 \leq i \leq m,\ 0 \leq k \leq n-1\}$$

$$= \text{span}\{Ax\} + \sum_{k=0}^{n-1} \text{Im}(A^k B)$$

We are also interested in the distribution P, the smallest distribution which contains $g_1, \ldots, g_m$ and is invariant under $f, g_1, \ldots, g_m$. By means of arguments similar to the ones used before or, else, by means of the recursive algorithm presented at the beginning of section I.6, it is not difficult to discover that, at any point $x \in \mathbb{R}^n$,

$$(4.6) \quad P(x) = \text{span}\{A^k b_i : 1 \leq i \leq m,\ 0 \leq k \leq n-1\}$$

Thus, we see that

$$\Delta_C = \text{sp}\{f\} + P$$

The distribution Δ_C is spanned by a set of vector fields which is locally finitely generated (because any vector field in C is *analytic* on $\mathbb{R}^n$), and therefore - by Lemma (2.5) - the distribution Δ_C has the maximal integral manifolds property. The distribution P is nonsingular and involutive and thus - by Corollary (1.12) - it also has the maximal integral manifolds property.

The maximal integral submanifolds of P, all of the same dimension, have the form $x+V$, where

$$V = \mathrm{Im}(B)+\mathrm{Im}(AB)+\ldots+\mathrm{Im}(A^{n-1}B)$$

(see Example I.(3.4) and Remark I.(4.2)). The maximal integral submanifolds of Δ_C may have different dimensions, because Δ_C may have singularities.

If, at some point $x \in \mathbb{R}^n$, $f(x) \in P(x)$, then the maximal integral submanifold of Δ_C passing through x coincides with the one of the distribution P, i.e. is a subset of the form $x+V$. Otherwise, if such a condition is not verified, the maximal integral submanifold of Δ_C is a submanifold whose dimension exceeds by 1 that of P and this submanifold, in turn, is partitioned into subsets of the form $x'+V$

(4.7) *Example*. The following simple example illustrates the case of a singular Δ_C. Let the system described by

$$\dot{x} = \begin{pmatrix} 1 & 0 & 0 \\ 0 & -1 & 0 \\ 0 & 0 & 1 \end{pmatrix} x + \begin{pmatrix} 1 \\ 0 \\ 0 \end{pmatrix} u$$

Then we easily see that

$$V = \{x \in \mathbb{R}^3 : x_2 = x_3 = 0\}$$

and that

$$P = \mathrm{sp}\{\frac{\partial}{\partial x_1}\}$$

The tangent vector $f(x)$ belongs to P only at those x in which $x_2 = x_3 = 0$, i.e. only on V. Thus, the maximal integral submanifolds of Δ_C will have dimension 2 everywhere but on V. A direct computation shows that these submanifolds may be described in the following way:

(i) if x^o is such that $x_2^o = 0$ (resp. $x_3^o = 0$) then the maximal submanifold passing through x^o is the half open plane

$$\{x \in \mathbb{R}^n : x_2 = 0 \text{ and } \mathrm{sgn}(x_3) = \mathrm{sgn}(x_3^o)\}$$

$$(\text{resp. } \{x \in \mathbb{R}^n : x_3 = 0 \text{ and } \mathrm{sgn}(x_2) = \mathrm{sgn}(x_2^o)\}$$

(ii) if x^o is such that both $x_2^o \neq 0$ and $x_3^o \neq 0$, then the maximal submanifold passing through x^o is the surface

$$\{x \in \mathbb{R}^n : x_2x_3 = x_2^ox_3^o\}. \quad \square$$

We turn now on the computation of the subspace $\mathcal{O}$ and the codistribution $\Omega_{\mathcal{O}}$. It is easy to prove that $\mathcal{O}$ is the subspace of $C^{\infty}(N)$ consisting of all $\mathbb{R}$-linear combinations of functions of the form c_iA^kx or $c_iA^kb_j$, namely that

$$(4.8) \qquad \mathcal{O} = LC\{\lambda \in C^{\infty}(N):\lambda(x) = c_iA^kx \text{ or } \lambda(x) = c_iA^kb_j ; \\ 1 \le i \le \ell,\ 1 \le j \le m,\ 0 \le k \le n-1\}$$

For, note that functions of the form c_iA^kx or $c_iA^kb_j$ are such that

$$c_iA^kx = \underbrace{L_f \cdots L_f}_{k\text{-times}} h_i(x)$$

$$c_iA^kb_j = L_{g_j}\underbrace{L_f \cdots L_f}_{k\text{-times}} h_i(x)$$

and this implies that the right-hand-side of (4.8) is contained in $\mathcal{O}$. Moreover, the functions $h_1,\ldots,h_\ell$ are elements of the right-hand-side of (4.8). Then, the proof of (4.8) is completed as soon as we show that its right-hand-side is closed under differentiation along $f,g_1,\ldots,g_m$.

If $\lambda(x) = c_iA^kx$, then $L_f\lambda = c_iA^{k+1}x$ and $L_{g_j}\lambda(x) = c_iA^kb_j$. If $\lambda(x) = c_iA^kb_j$, then $L_f\lambda(x) = L_{g_j}\lambda(x) = 0$. Thus, using again Cayley-Hamilton Theorem, it easily seen that the right-hand-side of (4.8) is closed under differentiation along $f,g_1,\ldots,g_m$.

At each point x, the codistribution $\Omega_{\mathcal{O}}$ is given by $\Omega_{\mathcal{O}}(x) = \text{span}\{c_iA^k :\ 1 \le i \le \ell,\ 0 \le k \le n-1\}$ and therefore

$$\Omega_{\mathcal{O}}^{\perp}(x) = \bigcap_{k=0}^{n-1} \ker(CA^k)$$

The codistribution $\Omega_{\mathcal{O}}$ is nonsingular, and so is the distribution $\Omega_{\mathcal{O}}^{\perp}$. Moreover, $\Omega_{\mathcal{O}}^{\perp} = \text{smt}(\Omega_{\mathcal{O}}^{\perp})$. From Remark (3.7) we see that $\Omega_{\mathcal{O}}^{\perp} = Q$ and so this distribution has the maximal integral manifolds property (Lemma (3.2)). The maximal integral submanifolds of Q have now the form x+W where

$$W = \ker(C) \cap \ker(CA) \ldots \cap \ker(CA^{n-1})$$

As a second application we consider a *bilinear system*, i.e. a system described by equations of the form

$$\dot{x} = Ax + \sum_{i=1}^{m} (N_i x) u_i$$

$$y = Cx$$

Here also the manifold on which the system evolves is the whole of $\mathbb{R}^n$, we set f and $h_1,\ldots,h_\ell$ as before, and

$$g_i(x) = N_i x \qquad 1 \leq i \leq m$$

In order to compute the subalgebra C we note first that any vector field τ in the set $\{f,g_1,\ldots,g_m\}$ has the form $\tau(x) = Tx$, where T is an $n\times n$ matrix. If we want to take the Lie bracket of two vector fields τ_1,τ_2 of the form

$$\tau_1(x) = T_1 x \quad , \quad \tau_2(x) = T_2 x$$

we have

$$[\tau_1,\tau_2](x) = (T_2T_1-T_1T_2)x = [T_1,T_2]x$$

where $[T_1,T_2] = (T_2T_1-T_1T_2)$ is the *commutator* of T_1 and T_2.

On the basis of this observation, it is easy to set up a recursive procedure yielding the smallest Lie subalgebra which contains a set of vector fields of the form $\tau_1(x) = T_1x,\ldots,\tau_r(x) = T_rx$.

(4.9) *Lemma*. Consider the nondecreasing sequence of subspaces of $\mathbb{R}^{n\times n}$, the $\mathbb{R}$-vector space of all $n\times n$ matrices of real numbers, defined by setting

$$M_o = \mathrm{span}\{T_1,\ldots,T_r\}$$

$$M_k = M_{k-1} + \mathrm{span}\{[T_1,T],\ldots,[T_r,T] : T \in M_{k-1}\}$$

Then, there exists an integer k^* such that

$$M_k = M_{k^*}$$

for all $k > k^*$. The set of vector fields

$$L = \{\tau \in V(\mathbb{R}^n) : \tau(x) = Tx,\ T \in M_{k^*}\}$$

is the smallest Lie subalgebra of vector fields which contains

$\tau_1(x) = T_1x, \ldots, \tau_r(x) = T_rx$.

Proof. The proof is rather simple and consists in the following steps. A dimensionality argument proves the existence of the integer k^* such that $M_k = M_{k*}$ for all $k > k^*$. Then, one checks that the subspace M_{k*} contains $T_1, \ldots, T_r$ and any repeated commutator of the form $[T_{i_1}, \ldots, [T_{i_{h-1}}, T_{i_h}]]$ and is such that $[P,Q] \in M_{k*}$ for all $P \in M_{k*}$ and $Q \in M_{k*}$. From these properties, it is straightforward to deduce that L is the desired Lie algebra. □

Based on this result, it is easy to construct the Lie algebra C by simply initializing the algorithm described in the above Lemma with the matrices $A, N_1, \ldots, N_m$.

In this case, unlike the previous one, we cannot anymore give a simple expression of $\Delta_C(x)$ and/or its maximal integral submanifolds. In some special situations, however, like the one illustrated in the following example, a rather satisfactory analysis is possible.

(4.10) *Example*. Consider the system

$$\dot{x} = Ax + Nxu$$

where $x \in \mathbb{R}^3$ and

$$A = \begin{pmatrix} 0 & 1 & 0 \\ -1 & 0 & 0 \\ 0 & 0 & 0 \end{pmatrix} \qquad N = \begin{pmatrix} 0 & 0 & 1 \\ 0 & 0 & 0 \\ -1 & 0 & 0 \end{pmatrix}$$

An easy computation shows that

$$[A,N] = \begin{pmatrix} 0 & 0 & 0 \\ 0 & 0 & 1 \\ 0 & -1 & 0 \end{pmatrix}$$

$$[N,[A,N]] = A$$

$$[A,[A,N]] = -N$$

Therefore, we have

$$C = \{\tau \in V(\mathbb{R}^3) : \tau(x) = Tx,\ T \in \operatorname{span}\{A, N, [A,N]\}\}$$

To compute the dimension of Δ_C we evaluate the rank of the matrix

$$(Ax,Nx,[A,N]x) = \begin{pmatrix} x_2 & x_3 & 0 \\ -x_1 & 0 & x_3 \\ 0 & -x_1 & -x_2 \end{pmatrix}$$

and we find the following result

$$\dim \Delta_C(x) = 0 \qquad \text{if} \quad x = 0$$

$$\dim \Delta_C(x) = 2 \qquad \text{if} \quad x \neq 0$$

A direct computation shows that the maximal integral submanifold of Δ_C passing through x^o is the set

$$\{x \in \mathbb{R}^3 : x_1^2+x_2^2+x_3^2 = (x_1^o)^2+(x_2^o)^2+(x_2^o)^2\}$$

i.e. the sphere centered at the origin passing through x^o.

Therefore, we can say that the state of the system is not free to evolve on the whole of $\mathbb{R}^n$, but rather on the sphere centered at the origin which passes through the initial state.

Around any point $x \neq 0$ the distribution Δ_C is nonsingular, so we can obtain locally a decomposition of the form I.(6.12), by means of a suitable coordinates transformation.

To this end, we may make use of the construction introduced in the proof of Theorem I.(3.3) and find a set of three vector fields τ_1, τ_2, τ_3 with the property that τ_1 and τ_2 belong to Δ_C and $\tau_1(x^o)$, $\tau_2(x^o)$, $\tau_3(x^o)$ are linearly independent. If we consider an initial point on the line

$$\{x \in \mathbb{R}^3 : x_1 = x_2 = 0\}$$

we may take the vector fields

$$\tau_1(x) = (Nx)$$

$$\tau_2(x) = ([A,N]x)$$

$$\tau_3(x) = (0 \quad 0 \quad 1)'$$

Accordingly, we get

$$\Phi_t^1(x) = \begin{pmatrix} (\cos t)x_1 + (\sin t)x_3 \\ x_2 \\ -(\sin t)x_1 + (\cos t)x_3 \end{pmatrix}$$

$$\Phi_t^2(x) = \begin{pmatrix} x_1 \\ (\cos t)x_2 + (\sin t)x_3 \\ -(\sin t)x_2 + (\cos t)x_3 \end{pmatrix}$$

$$\Phi_t^3(x) = \begin{pmatrix} x_1 \\ x_2 \\ t + x_3 \end{pmatrix}$$

The local coordinate chart around the point x^o is given by the inverse of the function

$$F : (z_1, z_2, z_3) \longmapsto \Phi_{z_1}^1 \circ \Phi_{z_2}^2 \circ \Phi_{z_3}^3 (x^o)$$

For $x_1^o = x_2^o = 0$ and $x_3^o = a$ we have

$$F(z_1, z_2, z_3) = \begin{pmatrix} (\sin z_1)(\cos z_2)(z_3+a) \\ (\sin z_2)(z_3+a) \\ (\cos z_1)(\cos z_2)(z_3+a) \end{pmatrix}$$

The local representations of the vector fields f and g in the new coordinate chart are given by

$$\tilde{f}(z) = (F_*)^{-1} f(F(z)) = (F_*)^{-1} AF(z)$$

$$\tilde{g}(z) = (F_*)^{-1} g(F(z)) = (F_*)^{-1} NF(z)$$

A simple but tedious computation yields

$$\tilde{f}(z) = \begin{pmatrix} \cos z_1 \operatorname{tg} z_2 \\ -\sin z_1 \\ 0 \end{pmatrix} ; \qquad \tilde{g}(z) = \begin{pmatrix} 1 \\ 0 \\ 0 \end{pmatrix}$$

We conclude that around x^o the system, in the z coordinates, is described by the equations

$$\dot{z}_1 = \cos z_1 \operatorname{tg} z_2 + u$$

$$\dot{z}_2 = -\sin z_1$$

$$\dot{z}_3 = 0 \qquad \square$$

The study of the observability of a bilinear system is much simpler. By means of arguments similar to those used in the case of linear systems it is easy to prove that $\mathcal{O}$ is given by

$$\mathcal{O} = LC\{\lambda \in C^\infty(N) : \lambda(x) = c_i N_{j_1} \dots N_{j_k} x \; ;$$

$$1 \leq i \leq \ell,\ 1 \leq k \leq n-1;\ 0 \leq j_1, \dots, j_k \leq m\}$$

(with $N_o = A$). Therefore

$$\Omega_{\mathcal{O}}^{\perp}(x) = \bigcap_{k=0}^{n-1} \bigcap_{j_1,\dots,j_k=0}^{m} \ker(CN_{j_1} \dots N_{j_k})$$

The distribution $\Omega_{\mathcal{O}}^{\perp} = Q$ is nonsingular and its maximal integral submanifolds have the form x+W, where now

$$W = \bigcap_{k=0}^{n-1} \bigcap_{j_1,\dots,j_k=0}^{m} \ker(CN_{j_1} \dots N_{j_k})$$

It may be worth observing that the subspace W thus defined is invariant under $A, N_1, \dots, N_m$, is contained in ker(C) and is the largest subspace of $\mathbb{R}^n$ having these properties. From linear algebra we know that by making a suitable change of coordinates in $\mathbb{R}^n$ (see e.g. section I.1) the matrices $A, N_1, \dots, N_m$ become block triangular and, therefore, the dynamics of the system becomes described by equations of the form

$$\dot{x}_1 = A_{11}x_1 + A_{12}x_2 + \sum_{i=1}^{m} (N_{i,11}x_1 + N_{i,12}x_2)u_i$$

$$\dot{x}_2 = A_{22}x_2 + \sum_{i=1}^{m} N_{i,22}x_2u_i$$

Moreover, the output y depends only on the x_2 coordinates,

$$y = C_2x_2$$

The above equations are exactly of the form I.(5.7), this time obtained by means of standard linear algebra arguments.

CHAPTER III

INPUT-OUTPUT MAPS AND REALIZATION THEORY

1. Fliess Functional Expansions

The purpose of this section and of the following section is to describe representations of the input-output behavior of a nonlinear system. We consider, as usual, systems described by differential equations of the form

(1.1a) $$\dot{x} = f(x) + \sum_{i=1}^{m} g_i(x)u_i$$

(1.1b) $$y_j = h_j(x) \qquad j = 1,\ldots,\ell$$

Throughout the chapter, we systematically assume that the manifold N on which the state evolves is an open set of $\mathbb{R}^n$ and that the vector fields $f, g_1, \ldots, g_m$ are *analytic* vector fields defined on N. Likewise, the output functions $h_1, \ldots, h_\ell$ are analytic functions defined on N.

For the sake of notational convenience most of the times we represent the output of the system as a vector-valued function

$$y = h(x) = (h_1(x) \ldots h_\ell(x))'$$

We require first some combinational notations. Consider the set of m+1 indexes $I = \{0,1,\ldots,m\}$ (we represent here, as usual, indexes with integer numbers, but we could as well represent the m+1 indexes with elements of any set Z with card(Z) = m+1). Let I_k be the set of all sequences $(i_k \ldots i_1)$ of k elements $i_k, \ldots, i_1$ of I. An element of this set I_k will be called a multiindex of lenght k. For consistency we define also a set I_0 whose unique element is the empty sequence (i.e. a multiindex of lenght 0), denoted $\emptyset$. Finally, let

$$I^* = \bigcup_{k \geq 0} I_k$$

It is easily seen that the set I^* can be given a structure of free monoid, with composition rule

$$(i_k \ldots i_1)(j_h \ldots j_1) \longmapsto (i_k \ldots i_1 j_h \ldots j_1)$$

with neutral element $\emptyset$.

A *formal power series* in m+1 noncommutative indeterminates and coefficients in $\mathbb{R}$ is a mapping

$$c : I^* \longmapsto \mathbb{R}$$

In what follows we represent the value of c at some element $i_k \dots i_0$ of I^* with the symbol $c(i_k \dots i_0)$.

The second relevant object we have to introduce is called an *iterated integral* of a given set of functions and is defined in the following way. Let T be a fixed value of the time and suppose $u_1, \dots, u_m$ are real-valued piecewise continuous functions defined on $[0,T]$. For each multindex $(i_k \dots i_0)$ the corresponding iterated integral is a real-valued function of t

$$E_{i_k \dots i_1 i_0}(t) = \int_0^t d\xi_{i_k} \dots d\xi_{i_1} d\xi_{i_0}$$

defined for $0 \leq t \leq T$ by recurrence on the lenght, setting:

$$\xi_0(t) = t$$

$$\xi_i(t) = \int_0^t u_i(\tau) d\tau \quad \text{for} \quad 1 \leq i \leq m$$

and

$$\int_0^t d\xi_{i_k} \dots d\xi_{i_0} = \int_0^t d\xi_{i_k}(\tau) \int_0^\tau d\xi_{i_{k-1}} \dots d\xi_{i_0}$$

The iterated integral corresponding to the multindex $\emptyset$ is the real number 1.

(1.2) *Example*. Just for convenience, let us compute the first few iterated integrals, in a case where m = 1.

$$\int_0^t d\xi_0 = t \ ; \quad \int_0^t d\xi_1 = \int_0^t u_1(\tau) d\tau$$

$$\int_0^t d\xi_0 d\xi_0 = \frac{t^2}{2!} \ ; \quad \int_0^t d\xi_0 d\xi_1 = \int_0^t \int_0^\tau u_1(\theta) d\theta d\tau$$

$$\int_0^t d\xi_1 d\xi_0 = \int_0^t u_1(\tau)\tau d\tau; \quad \int_0^t d\xi_1 d\xi_1 = \int_0^t u_1(\tau)\int_0^\tau u_1(\theta)d\theta d\tau, \text{ etc. } \square$$

Given a formal power series in m+1 non-commutative indeterminates, it is possible to associate with this series a functional of $u_1,\ldots,u_m$ by taking the sum over I^* of all the products of the form

$$c(i_k\ldots i_0)\int_0^t d\xi_{i_k}\ldots d\xi_{i_0}$$

The convergence of a sum of this kind is guaranteed by some growth condition on the "coefficients" $c(i_k\ldots i_0)$, as stated below.

(1.3) *Lemma*. Suppose there exist real numbers $K > 0$, $M > 0$ such that

$$|c(i_k\ldots i_0)| < K(k+1)!M^{k+1} \tag{1.4}$$

for all $k \geq 0$ and all multiindexes $i_k\ldots i_0$.

Then, there exists a real number $T > 0$ such that, for each $0 \leq t \leq T$ and each set of piecewise continuous functions $u_1,\ldots,u_m$ defined on $[0,T]$ and subject to the constraint

$$\max_{0\leq\tau\leq T} |u_i(\tau)| < 1, \tag{1.5}$$

the series

$$y(t) = c(\emptyset) + \sum_{k=0}^{\infty} \sum_{i_0,\ldots,i_k=0}^{m} c(i_k\ldots i_0)\int_0^t d\xi_{i_k}\ldots d\xi_{i_0} \tag{1.6}$$

is absolutely and uniformly convergent.

Proof. It is easy to see, from the definition of iterated integral, that, if the functions $u_1,\ldots,u_m$ satisfy the constraint (1.5) then

$$\int_0^t d\xi_{i_k}\ldots d\xi_{i_0} \leq \frac{t^{k+1}}{(k+1)!}$$

If the growth condition is satisfied, then

$$\left|\sum_{i_0,\ldots,i_k=0}^{m} c(i_k\ldots i_0)\int_0^t d\xi_{i_k}\ldots d\xi_{i_0}\right| \leq K[M(m+1)t]^{k+1}$$

As a consequence, if T is sufficiently small, the series (1.6) converges absolutely and uniformly on $[0,T]$. □

The expression (1.6) clearly defines a functional of $u_1,\ldots,u_m$. This functional is *causal*, in the sense that $y(t)$ depends only on the restrictions of $u_1,\ldots,u_m$ to the time interval $[0,t]$.

A representation of the form (1.6) is unique.

(1.7) *Lemma*. Let c^a and c^b be two formal power series in m+1 noncommutative indeterminates and let the associated functionals of the form (1.6) be defined on the same interval $[0,T]$. Then the two functionals coincides if and only if $c^a = c^b$. □

Proof. Let c^a,c^b be two formal power series and $y^a(t),y^b(t)$ the associated functionals of the form (1.6). Note that

$$y(t) = y^a(t)-y^b(t)$$

is still a functional of the form (1.6) associated with a formal power series c whose coefficients are defined as differences between the corresponding coefficients of c^a and c^b. To prove the lemma, all we need is to show that if $y(t) = 0$ for all $t \in [0,T]$ and for all input functions, all the coefficients of the series c vanish.

If, in particular, $u_1=\ldots=u_m=0$ on $[0,T]$, then $y(t) = 0$ for all $t \in [0,T]$ implies

$$c(\emptyset) + c(0)t + c(00)\frac{t^2}{2!} + \ldots = 0$$

for all $t \in [0,T]$, i.e.

$$c(\emptyset) = 0$$

$$c(\underbrace{0\ldots0}_{k\text{-times}}) = 0 \qquad 1 \leq k \leq \infty$$

Taking the derivative of (1.6) with respect to time and evaluating it at $t = 0$, one obtains

$$\left(\frac{dy}{dt}\right)_{t=0} = \sum_{i=1}^{m} c(i)u_i(0)$$

Therefore, $\left(\frac{dy}{dt}\right)_{t=0} = 0$ for all $u_1(0),\ldots,u_m(0)$ implies

$$c(i) = 0 \qquad 1 \leq i \leq m$$

Continuing this way, one may compute the second derivative of $y(t)$ at

$t = 0$ and get

$$\left(\frac{d^2y}{dt^2}\right)_{t=0} = \sum_{i_0,i_1=1}^{m} c(i_1 i_0) u_{i_1}(0) u_{i_0}(0) + \sum_{i=1}^{m} (c(0i)+c(i0)) u_i(0)$$

If this is zero for all $u_1(0),\ldots,u_m(0)$, then

$$c(i_1 i_0) = 0 \qquad 1 \leq i_1, i_0 \leq m$$

$$c(0i) = -c(i0) \qquad 1 \leq i \leq m$$

In the third derivative, the contribution of terms

$$\sum_{i=1}^{m} (c(0i) \int_0^t d\xi_0 d\xi_i + c(i0) \int_0^t d\xi_i d\xi_0)$$

is

$$\sum_{i=1}^{m} \left[\frac{1}{6} c(0i) + \frac{1}{3} c(i0)\right] \left(\frac{du_i}{dt}\right)_{t=0}$$

If this is zero for all $\left(\frac{du_i}{dt}\right)_{t=0}$, then $c(0i) = -2c(i0)$ which, together with the previous equality $c(0i) = -c(i0)$ implies

$$c(0i) = 0 \qquad 1 \leq i \leq m$$

Continuing in the same way, one may complete the proof. □

We are now going to show that the output $y(t)$ of the nonlinear system (1.1) can be represented as a functional of the inputs $u_1,\ldots,u_m$ in the form (1.6). To this end we need some preliminary results.

(1.8) *Lemma*. Let $g_0, g_1, \ldots, g_m$ be a set of analytic vector fields and λ a real-valued analytic function defined on N. Given a point $x^o \in N$, consider the formal power series defined by

$$c(\emptyset) = \lambda(x^o)$$

(1.9)

$$c(i_k \ldots i_1 i_0) = L_{g_{i_0}} L_{g_{i_1}} \ldots L_{g_{i_k}} \lambda(x^o)$$

Then, there exist real numbers $K > 0$ and $M > 0$ such that the growth condition (1.4) is satisfied.

Proof. The reader is referred to the literature. □

In view of this result and of Lemma (1.3), one may associate with $g_0,g_1,\ldots,g_m$ and λ the functional

$$(1.10)\qquad v(t) = \lambda(x^o) + \sum_{k=0}^{\infty} \sum_{i_0,\ldots,i_k=0}^{m} L_{g_{i_0}} L_{g_{i_1}} \ldots L_{g_{i_k}} \lambda(x^o) \int_0^t d\xi_{i_k} \ldots d\xi_{i_1} d\xi_{i_0}$$

(1.11) *Lemma*. Let $g_0,g_1,\ldots,g_m$ be as in the previous Lemma and let $\lambda_1,\ldots,\lambda_\ell$ be real-valued analytic functions defined on N. Moreover, let γ be a real-valued analytic function defined on $\mathbb{R}^\ell$.
Let $v_1(t),\ldots,v_\ell(t)$ denote the functionals defined by setting, in (1.10), $\lambda = \lambda_1,\ldots,\lambda = \lambda_\ell$. The composition $\gamma(v_1(t),\ldots,v_\ell(t))$ is again a functional of the form (1.10), corresponding to the setting $\lambda = \gamma(\lambda_1,\ldots,\lambda_\ell)$.

Proof. We will only give a trace to the reader for the proof. Let c_1,c_2 denote the formal power series defined by setting, in (1.9), $\lambda = \lambda_1$ and respectively $\lambda = \lambda_2$, and let $v_1(t),v_2(t)$ denote the associated functionals (1.10). Then, it is immediately seen that with the formal power series defined by setting $\lambda = \alpha_1\lambda_1 + \alpha_2\lambda_2$, where α_1 and α_2 are real numbers, there is associated the functional $\alpha_1 v_1(t) + \alpha_2 v(t)$.

With a little work, it is also seen that with the formal power series defined by setting $\lambda = \lambda_1\lambda_2$, there is associated the functional $v_1(t)v_2(t)$. We show only the very first computations needed for that. For, consider the product

$$v_1(t)v_2(t) = (\lambda_1+L_{g_0}\lambda_1\int_0^t d\xi_0+L_{g_1}\lambda_1\int_0^t d\xi_1+L_{g_0}L_{g_0}\lambda_1\int_0^t d\xi_0 d\xi_0+\ldots)$$
$$(\lambda_2+L_{g_0}\lambda_2\int_0^t d\xi_0+L_{g_1}\lambda_2\int_0^t d\xi_1+L_{g_0}L_{g_0}\lambda_2\int_0^t d\xi_0 d\xi_0+\ldots)$$

where, for simplicity, we have omitted specifying that the values of all the functions of x are to be taken at $x = x^o$. Multiplying term-by-term we have

$$v_1(t)v_2(t)=\lambda_1\lambda_2+(\lambda_1L_{g_0}\lambda_2+\lambda_2L_{g_0}\lambda_1)\int_0^t d\xi_0+(\lambda_1L_{g_1}\lambda_2+\lambda_2L_{g_1}\lambda_1)\int_0^t d\xi_1 +$$
$$(\lambda_1+L_{g_0}L_{g_0}\lambda_2+\lambda_2L_{g_0}L_{g_0}\lambda_1)\int_0^t d\xi_0 d\xi_0 +$$
$$(L_{g_0}\lambda_1)(L_{g_0}\lambda_2)(\int_0^t d\xi_0)(\int_0^t d\xi_0)+\ldots$$

The factors that multiply $\int_0^t d\xi_0$ and $\int_0^t d\xi_1$ are clearly $L_{g_0}\lambda_1\lambda_2$ and respectively $L_{g_1}\lambda_1\lambda_2$. For the other three, we have

$$L_{g_0}L_{g_0}\lambda_1\lambda_2 = \lambda_1 L_{g_0}L_{g_0}\lambda_2+\lambda_2 L_{g_0}L_{g_0}\lambda_1+2(L_{g_0}\lambda_1)(L_{g_0}\lambda_2)$$

but also

$$(\int_0^t d\xi_0)(\int_0^t d\xi_0) = 2\int_0^t d\xi_0 d\xi_0$$

so that the three terms in question give exactly

$$L_{g_0}L_{g_0}\lambda_1\lambda_2 \int_0^t d\xi_0 d\xi_0$$

It is not difficult to set up a recursive formalism which makes it possible to completely verify the claim.

If now γ is any real-valued analytic function defined on $\mathbb{R}^\ell$, we may take its Taylor series expansion at the origin and use recursively the previous results in order to show that the composition $\gamma(v_1(t),\ldots,v_\ell(t))$ may be represented as a series like the (1.10) with λ replaced by the Taylor series expansion of $\gamma(\lambda_1,\ldots,\lambda_\ell)$. □

At this point, it is easy to obtain the desired representation of $y(t)$ as a functional of the form (1.10).

(1.12) *Theorem*. Suppose the inputs $u_1,\ldots,u_m$ of the control system (1.1) satisfy the constraint (1.5). If T is sufficiently small, then for all $0 \leq t \leq T$ the j-th output $y_j(t)$ of the system (1.1) may be expanded in following way

$$(1.13) \qquad y_j(t)=h_j(x^o) + \sum_{k=0}^{\infty} \sum_{i_0,\ldots,i_k=0}^{m} L_{g_{i_0}}\ldots L_{g_{i_k}} h_j(x^o)\int_0^t d\xi_{i_k}\ldots d\xi_{i_0}$$

where $g_0 = f$.

Proof. We first show that the j-th component of the solution of the differential equation (1.1a) may be expressed as

$$(1.14) \qquad x_j(t)=x_j(x^o)+ \sum_{k=0}^{\infty} \sum_{i_0,\ldots,i_k=0}^{m} L_{g_{i_0}}\ldots L_{g_{i_k}} x_j(x^o)\int_0^t d\xi_{i_k}\ldots d\xi_{i_0}$$

where the function $x_j(x)$ stands for

$$x_j : (x_1,\dots,x_n) \longmapsto x_j$$

Note that, by definition of iterated integral

$$\frac{d}{dt}\int_0^t d\xi_0 d\xi_{i_{k-1}}\dots d\xi_{i_0} = \int_0^t d\xi_{i_{k-1}}\dots d\xi_{i_0}$$

and

$$\frac{d}{dt}\int_0^t d\xi_i d\xi_{i_{k-1}}\dots d\xi_{i_0} = u_i(t)\int_0^t d\xi_{i_{k-1}}\dots d\xi_{i_0}$$

for $1 \leq i \leq m$. Then, taking the derivative of the right-hand-side of (1.14) with respect to the time and rearranging the terms we have

$$\dot{x}_j(t) = L_f x_j(x^o) + \sum_{k=0}^{\infty} \sum_{i_0,\dots,i_k=0}^{m} L_{g_{i_0}}\dots L_{g_{i_k}} L_f x_j(x^o) \int_0^t d\xi_{i_k}\dots d\xi_{i_0} +$$

$$+ \sum_{i=1}^{m} [L_{g_i} x_j(x^o) + \sum_{k=0}^{\infty} \sum_{i_0,\dots,i_k=0}^{m} L_{g_{i_0}}\dots L_{g_{i_k}} L_{g_i} x_j(x^o) \int_0^t d\xi_{i_k}\dots d\xi_{i_0}] u_i(t)$$

Now, let f_j and g_{ij} denote the j-th components of f and g_i, $1 \leq j \leq n$, $1 \leq i \leq m$ and observe that

$$L_f x_j = f_j(x_1,\dots,x_n)$$

Therefore, on the basis of the Lemma (1.11), we may write

$$L_f x_j(x^o) + \sum_{k=0}^{\infty} \sum_{i_0,\dots,i_k=0}^{m} L_{g_{i_0}}\dots L_{g_{i_k}} L_f x_j(x^o) \int_0^t d\xi_{i_k}\dots d\xi_{i_0} =$$

$$f_j(x^o) + \sum_{k=0}^{\infty} \sum_{i_0,\dots,i_k=0}^{m} L_{g_{i_0}}\dots L_{g_{i_k}} f_j(x^o) \int_0^t d\xi_{i_k}\dots d\xi_{i_0} =$$

$$f_j(x_1(t),\dots,x_n(t))$$

A similar substitution can be performed on the other terms thus yielding

$$\dot{x}_j(t) = f_j(x_1(t),\dots,x_n(t)) + \sum_{i=1}^{m} g_{ij}(x_1(t),\dots,x_n(t))u_i(t)$$

Moreover, the $x_j(t)$ satisfy the condition

$$x_j(0) = x_j^o$$

and therefore are the components of the solution $x(t)$ of the differential equation (1.1a).

A further application of Lemma (1.11) shows that the output (1.1b) can be expressed in the form (1.13). □

The development (1.13) will be from now on referred to as the *fundamental formula* or *Fliess functional expansion* of $y_j(t)$. Obviously, one may deal directly with the case of a vector-valued output with the same formalism, by just replacing the scalar-valued function $h_j(x)$ with the vector-valued function $h(x)$. We stress that, from Lemma (1.3), it is known that the series (1.13) converges absolutely and uniformly on $[0,T]$.

(1.15) *Remark*. The reader will immediately observe that the functions $h_j(x)$ and $L_{g_{i_0}} \ldots L_{g_{i_k}} h_j(x)$, with $1 \leq j \leq \ell$ and $(i_k \ldots i_0) \in (I^* \backslash I_0)$, whose values at x^o characterize the functional (1.13), span the observation space $\mathcal{O}$ defined in section II.3. □

(1.16) *Examples*. In the case of a linear system, the formal power series which characterizes the functional (1.13) takes the form

$$c(\emptyset) = c_j x^o$$

$$c(i_k \ldots i_0) = \begin{cases} c_j A^{k+1} x^o & \text{if} \quad i_0 = \ldots = i_k = 0 \\ c_j A^k b_{i_0} & \text{if} \quad i_0 \neq i_1 = \ldots = i_k = 0 \\ 0 & \text{elsewhere} \end{cases}$$

In the case of a bilinear system, the formal power series which characterizes the functional (1.13) takes the form

$$c(\emptyset) = c_j x^o$$

$$c(i_k \ldots i_0) = c_j N_{i_k} \ldots N_{i_0} x^o$$

where $N_0 = A$. □

2. Volterra Series Expansions

The input-output behavior of a nonlinear system of the form (1.1) may also be represented by means of a series of *generalized convolution integrals*. A generalized convolution integral of order k is defined as follows. Let $(i_k \dots i_1)$ be a multiindex of lenght k, with $i_k, \dots, i_1$ elements of the set $\{1,\dots,m\}$. With this multiindex there is associated a real-valued continuous function $w_{i_k \dots i_1}$, defined on the subset of $\mathbb{R}^{k+1}$

$$S_k = \{(t,\tau_k,\dots,\tau_1) \in \mathbb{R}^{k+1} : T \geq t \geq \tau_k \dots \geq \tau_1 \geq 0\}$$

where T is a fixed number. If $u_1,\dots,u_m$ are real-valued piecewise continuous functions defined on $[0,T]$, the generalized convolution integral of order k of $u_1,\dots,u_m$ with kernel $w_{i_k \dots i_1}$ is defined as

$$\int_0^t \int_0^{\tau_k} \dots \int_0^{\tau_2} w_{i_k \dots i_1}(t,\tau_k,\dots,\tau_1) u_{i_k}(\tau_k) \dots u_{i_1}(\tau_1) d\tau_1 \dots d\tau_k$$

for $0 \leq t \leq T$.

For consistency, if $k = 0$, rather than a generalized convolution integral, one considers simply a continuous real-valued function w_0 defined on the set

$$S_0 = \{t \in \mathbb{R} : T \geq t \geq 0\}$$

The sum of a series of generalized convolution integrals may describe a functional of $u_1,\dots,u_m$, under the conditions stated below.

(2.1) *Lemma*. Suppose there exist real numbers $K > 0$, $M > 0$ such that

$$|w_{i_k \dots i_1}(t,\tau_k,\dots,\tau_1)| < K(k)!M^k \tag{2.2}$$

for all $k > 0$, for all multiindexes $(i_k \dots i_1)$, and all $(t,\tau_k,\dots,\tau_1) \in S_k$.

Then, there exists a real number $T > 0$ such that, for each $0 \leq t \leq T$ and each set of piecewise continuous functions $u_1,\dots,u_m$ defined on $[0,T]$ and subject to the constraint

$$\max_{0 \leq \tau \leq T} |u_i(\tau)| < 1, \tag{2.3}$$

the series

$$(2.4)\qquad y(t) = w_0(t) + \sum_{k=1}^{\infty} \sum_{i_1,\ldots,i_k=1}^{m} \int_0^t \int_0^{\tau_k} \cdots \int_0^{\tau_2} w_{i_k\ldots i_1}(t,\tau_k,\ldots,\tau_1) u_{i_k}(\tau_k)\ldots u_{i_1}(\tau_1) d\tau_1 \ldots d\tau_k$$

is absolutely and uniformly convergent.

Proof. It is similar to that of Lemma (1.3). □

The expression (2.4) clearly defines a functional of $u_1,\ldots,u_m$, which is causal, and is called a *Volterra series expansion.*

As in the previous section, we are interested in the possibility of using an expansion of the form (2.4) for the output of the nonlinear system (1.1). The existence of such an expansion and the expressions of the kernels may be described in the following way.

(2.5) *Lemma.* Let $f, g_1,\ldots,g_m$ be a set of analytic vector fields and λ a real-valued analytic function defined on N. Let Φ_t^f denote the flow of f. For each pair $(t,x) \in \mathbb{R} \times N$ for which the flow $\Phi_t^f(x)$ is defined, let $Q_t(x)$ denote the function

$$(2.6)\qquad Q_t(x) = \lambda \circ \Phi_t^f(x)$$

and $P_t^1(x),\ldots,P_t^m(x)$ the vector fields

$$(2.7)\qquad P_t^i(x) = (\Phi_{-t}^f)_* \, g_i \circ \Phi_t^f(x)$$

$1 \le i \le m$. Moreover, let

$$(2.8')\qquad w_0(t) = Q_t(x^o)$$

$$(2.8'')\qquad w_{i_k\ldots i_1}(t,\tau_k,\ldots,\tau_1) = (L_{P_{\tau_1}^{i_1}(x)} \ldots L_{P_{\tau_k}^{i_k}(x)} Q_t(x))_{x=x^o}$$

Then, there exist real numbers $K > 0$ and $M > 0$ such that the condition (2.2) is satisfied. □

From this result it is easy to obtain the desired representation of y(t) in the form of a Volterra series expansion.

(2.9) *Theorem.* Suppose the inputs $u_1,\ldots,u_m$ of the control system (1.1) satisfy the constraint (2.3). If T is sufficiently small, then for all $0 \le t \le T$ the output $y_j(t)$ of the system (1.1) may be expanded in the form of a Volterra series, with kernels (2.8), where $Q_t(x)$ and $P_t^i(x)$ are as in (2.6)-(2.7) and $\lambda = h_j$. □

This result may be proved either directly, by showing that the Volterra series in question satisfies the equations (1.1), or indirectly, after establishing a correspondence between the functional expansion described at the beginning of the previous section and the Volterra series expansion. We take the second way.

For, observe that for all $(i_k \dots i_1)$ the kernel $w_{i_k \dots i_1}(t,\tau_k,\dots,\tau_1)$ is analytic in a neighborhood of the origin, and consider the Taylor series expansion of this kernel as a function of the variables $t-\tau_k, \tau_k-\tau_{k-1}, \dots, \tau_2-\tau_1, \tau_1$. This expansion has clearly the form

$$w_{i_k \dots i_1}(t,\tau_k,\dots,\tau_1) = \sum_{n_0 \dots n_k=0}^{\infty} c_{i_k \dots i_1}^{n_0 \dots n_k} \frac{(t-\tau_k)^{n_k} \dots (\tau_2-\tau_1)^{n_1} \tau_1^{n_0}}{n_k! \dots n_1! n_0!}$$

where

$$c_{i_k \dots i_1}^{n_0 \dots n_k} = \left[\frac{\partial^{n_0+\dots+n_k} w_{i_k \dots i_1}}{\partial(t-\tau_k)^{n_k} \dots \partial(\tau_2-\tau_1)^{n_1} \partial\tau_1^{n_0}} \right]_{t-\tau_k=\dots=\tau_2-\tau_1=\tau_1=0}$$

If we substitute this expression in the convolution integral associated with $w_{i_k \dots i_1}$, we obtain an integral of the form

$$\sum_{n_0 \dots n_k=0}^{\infty} c_{i_k \dots i_1}^{n_0 \dots n_k} \int_0^t \int_0^{\tau_k} \dots \int_0^{\tau_2} \frac{(t-\tau_k)^{n_k}}{n_k!} u_{i_k}(\tau_k) \dots \frac{(\tau_2-\tau_1)^{n_1}}{n_1!} u_{i_1}(\tau_1) \frac{\tau_1^{n_0}}{n_0!} d\tau_k \dots d\tau_1$$

The integral which appears in this expression is actually an iterated integral of $u_1,\dots,u_m$, and precisely the integral

$$\int_0^t (d\xi_0)^{n_k} d\xi_{i_k} \dots (d\xi_0)^{n_1} d\xi_{i_1} (d\xi_0)^{n_0} \tag{2.10}$$

(where $(d\xi_0)^n$ stands for n-times $d\xi_0$).

Thus, the expansion (2.4) may be replaced with the expansion

$$\begin{aligned} y(t) &= \sum_{n=0}^{\infty} c_0^n \int_0^t (d\xi_0)^n \\ &+ \sum_{k=1}^{\infty} \sum_{i_1 \dots i_k=1}^{m} \sum_{n_0 \dots n_k=0}^{\infty} c_{i_k \dots i_1}^{n_0 \dots n_k} \int_0^t (d\xi_0)^{n_k} d\xi_{i_k} \dots (d\xi_0)^{n_1} d\xi_{i_1} (d\xi_0)^{n_0} \end{aligned} \tag{2.11}$$

which is clearly an expansion of the form (1.6). Of course, one could rearrange the terms and establish a correspondance between the coefficients $c_0^n, c_{i_k \ldots i_1}^{n_0 \ldots n_k}$ (i.e. the values of the derivatives of w_0 and $w_{i_k \ldots i_1}$ at $t-\tau_k = \ldots = \tau_2 - \tau_1 = \tau_1 = 0$) and the coefficients $c(\emptyset), c(i_k \ldots i_0)$ of the expansion (1.6), but this is not needed at this point.

On the basis of these considerations it is very easy to find Taylor series expansions of the kernels which characterize the Volterra series expansion of $y_j(t)$. We see from (2.11) that the coefficient $c_{i_k \ldots i_1}^{n_0 \ldots n_k}$ of the Taylor series expansion of $w_{i_k \ldots i_1}$ coincides with the coefficient of the iterated integral (2.10) in an expansion (1.6), but we know also from (1.13), that the coefficient of the iterated integral (2.10) has the form

$$L_f^{n_0} L_{g_{i_1}} L_f^{n_1} \ldots L_f^{n_{k-1}} L_{g_{i_k}} L_f^{n_k} h_j(x^o)$$

This makes it possible to write down immediately the expressions of the Taylor series expansions of all the kernels which characterize the Volterra series expansion of $y_j(t)$.

$$(2.12a) \quad w_0(t) = \sum_{n=0}^{\infty} L_f^n h_j(x^o) \frac{t^n}{n!}$$

$$(2.12b) \quad w_i(t,\tau_1) = \sum_{n_1=0}^{\infty} \sum_{n_0=0}^{\infty} L_f^{n_0} L_{g_i} L_f^{n_1} h_j(x^o) \frac{(t-\tau_1)^{n_1}}{n_1!} \frac{\tau_1^{n_0}}{n_0!}$$

$$(2.12c) \quad w_{i_2 i_1}(t,\tau_2,\tau_1) = \sum_{n_2=0}^{\infty} \sum_{n_1=0}^{\infty} \sum_{n_0=0}^{\infty} L_f^{n_0} L_{g_{i_1}} L_f^{n_1} L_{g_{i_2}} L_f^{n_2} h_j(x^o) \frac{(t-\tau_2)^{n_2} (\tau_2-\tau_1)^{n_1} \tau_1^{n_0}}{n_2! n_1! n_0!}$$

and so on.

The last step needed in order to prove Theorem (2.9) is to show that the Taylor series expansions of the kernels (2.8), with $Q_t(x)$ and $P_t^i(x)$ defined as in (2.6), (2.7) for $\lambda = h_j(t)$ coincide with the expansions (2.12).

This is only a routine computation, which may be carried out with a little effort by keeping in mind the well-known Campbell-Baker-Hausdorff formula, which provides a Taylor series expansion of $P_t^i(x)$. According to this formula it is possible to expand $P_t^i(x)$ in the following way

$$P_t^i(x) = (\Phi_{-t}^f)_* \, g_i \circ (\Phi_t(x)) = \sum_{n=0}^{\infty} ad_f^n g_i(x) \frac{t^n}{n!}$$

where, as usual, $ad_f^n g \triangleq [f, ad_f^{n-1} g]$ and $ad_f^0 g = g$.

(2.13) *Example*. In the case of bilinear systems, the flow Φ_t^f may be clearly given the following closed form expression

$$\Phi_t^f(x) = (\exp At)x$$

From this it is easy to find the expressions of the kernels of the Volterra series expansion of $y_i(t)$. In this case

$$Q_t(x) = c_j(\exp At)x$$

$$P_t^i(x) = (\exp(-At))N_i(\exp At)x$$

and, therefore,

$$w_0(t) = c_j(\exp At)x^o$$

$$w_i(t,\tau_1) = c_j(\exp A(t-\tau_1))N_i(\exp A\tau_1)x^o$$

$$w_{i_2 i_1}(t,\tau_2,\tau_1) = c_j(\exp A(t-\tau_2))N_{i_2}(\exp A(\tau_2-\tau_1))N_{i_1}\exp(A\tau_1)x^o$$

and so on.

3. Output Invariance

In this section we want to find the conditions under which the output is not affected by the input. These conditions will be used later on in the next chapter when dealing with the disturbance decoupling or with the noninteracting control.

Consider again a system of the form

$$\dot{x} = f(x) + \sum_{i=1}^{m} g_i(x)u_i$$

$$y_j = h_j(x) \qquad (j = 1,\ldots,\ell)$$

and let

$$y_j(t;x^o;u_1,\ldots,u_m)$$

denote the value at time t of the j-th output, corresponding to an initial state x^o and to a set of input functions $u_1,\ldots,u_m$. We say that the output y_j is *unaffected* by (or *invariant* under) the input u_i, if for every initial state $x^o \in N$, for every set of input functions $u_1,\ldots,u_{i-1},u_{i+1},\ldots,u_m$, and for all t

$$(3.1)\quad y_j(t;x^o;u_1,\ldots,u_{i-1},v^a,u_{i+1},\ldots,u_m) = y_j(t;x^o;u_1,\ldots,u_{i-1},v^b,u_{i+1},\ldots,u_m)$$

for every pair of functions v^a and v^b.

There is a simple test that identifies the systems having the output y_j unaffected by the input u_i.

(3.2) *Theorem*. The output y_j is unaffected by the input u_i if and only if, for all $r \geq 1$ and for any choice of vector fields $\tau_1,\ldots,\tau_r$ in the set $\{f,g_1,\ldots,g_m\}$

$$L_{g_i}h_j(x) = 0$$

$$(3.3)\quad L_{g_i}L_{\tau_1}\ldots L_{\tau_r}h_j(x) = 0$$

for all $x \in N$.

Proof. Suppose the above condition is satisfied. Then, one easily sees that the function

$$(3.4)\quad L_{\tau_1}\ldots L_{\tau_r}h_j(x) = 0$$

is identically zero whenever at least one of the vector fields $\tau_1,\ldots,\tau_r$ coincides with g_i. If we now look, for instance, at the Fliess expansion of $y_j(t)$, we observe that under these circumstances

$$c(i_k\ldots i_0) = 0$$

whenever one of the indexes $i_0,\ldots,i_k$ is equal to i, and this, in turn, implies that any iterated integral which involves the input function u_i is multiplied by a zero factor. Thus, the condition (3.1) is satisfied and the output y_j is decoupled from the input u_i.

Conversely, suppose the condition (3.1) is satisfied, for every $x^o \in N$, for every set of inputs $u_1,\ldots,u_{i-1},u_{i+1},\ldots,u_m$ and every pair of functions v^a and v^b. Take in particular $v^a(t) = 0$ for all t. Then in the Fliess expansion of $y_j(t;x^o;u_1,\ldots,u_{i-1},v^a,u_{i+1},\ldots,u_m)$

an iterated integral of the form

$$\int_0^t d\xi_{i_k} \dots d\xi_{i_0}$$

will be zero whenever one of the indexes $i_0,\dots,i_k$ is equal to i. All other iterated integrals of this expansion (i.e. the ones in which none of the indexes $i_0,\dots,i_k$ is equal to i) will be equal to the corresponding iterated integrals in the expansion of $y_j(t;x^o;u_1,\dots,u_{i-1},v^b,u_{i+1},\dots,u_m)$ because the inputs $u_1,\dots,u_{i-1},u_{i+1},\dots,u_m$ are the same. Therefore, we deduce that the difference between the right-hand-side and left-hand-side of (3.1) is a series of the form

$$\sum_{k=0}^{\infty} \sum_{i_0,\dots,i_k=0}^{m} c(i_k\dots i_0) \int_0^t d\xi_{i_k} \dots d\xi_{i_0}$$

in which the only nonzero coefficients are those with at least one of the indexes $i_0,\dots,i_k$ equal to i. The sum of this series is zero for every input $u_1,\dots,u_{i-1},v^b,u_{i+1},\dots,u_m$. Therefore, according to Lemma (1.7), all its coefficients must vanish, for all $x^o \in N$. We conclude that (3.4) and, accordingly, (3.3) are satisfied for all $x \in N$. □

The condition (3.3) can be given other formulations, in geometric terms. Remember that, in section 1, we have already observed that the coefficients of the Fliess expansion of y(t) coincide with the values at x^o of functions that span the observation space $\mathcal{O}$. The differentials of these functions span, by definition, the codistribution

$$\Omega_{\mathcal{O}} = \text{sp}\{d\lambda : \lambda \in \mathcal{O}\}$$

If we fix our attention only on the j-th output, we may in particular define an observation space $\mathcal{O}_j$ as the smallest subspace of $C^\infty(N)$ which contains the function h_j and is closed under differentiation along $f,g_1,\dots,g_m$. Therefore, the set of differentials $dh_j, dL_{g_{i_0}}\dots L_{g_{i_k}}h_j(x)$ with $i_k,\dots,i_0 \in I$ and j fixed spans the codistribution

$$\Omega_{\mathcal{O}_j} = \text{sp}\{d\lambda : \lambda \in \mathcal{O}_j\}$$

Now, observe that the condition (3.3) can be written as

$$\langle dh_j, g_i \rangle (x) = 0$$

$$\langle dL_{g_{i_k}} \cdots L_{g_{i_0}} h_j, g_i \rangle (x) = 0$$

for all $k \geq 0$ and for all $i_k, \ldots, i_0 \in I$. From the above discussion we conclude that the condition stated in Theorem (3.2) is equivalent to the condition

(3.5)
$$g_i \in \Omega^{\perp}_{O_j}$$

Other formulations are possible. For, remember that we have shown in section II.3 that the distribution $\Omega^{\perp}_{O}$ is invariant under the vector fields $f, g_1, \ldots, g_m$. For the same reasons, also the distribution $\Omega^{\perp}_{O_j}$ is invariant under $f, g_1, \ldots, g_m$.

Now, let $\langle f, g_1, \ldots, g_m | sp\{g_i\} \rangle$ denote, as usual, the smallest distribution invariant under $f, g_1, \ldots, g_m$ which contains $sp\{g_i\}$. If (3.5) is true, then, since $\Omega^{\perp}_{O_j}$ is invariant under $f, g_1, \ldots, g_m$, we must have

(3.6)
$$\langle f, g_1, \ldots, g_m | sp\{g_i\} \rangle \subset \Omega^{\perp}_{O_j}$$

Moreover, since

$$\Omega^{\perp}_{O_j} \subset sp\{dh_j\}^{\perp}$$

we see also that if (3.6) is true, we must have

(3.7)
$$\langle f, g_1, \ldots, g_m | sp\{g_i\} \rangle \subset (sp\{dh_j\})^{\perp}$$

Thus, we have seen that (3.5) implies (3.6) and this, in turn, implies (3.7). We will show now that (3.7) implies (3.5) thus proving that the three conditions are in fact equivalent.

For, observe that any vector field of the form $[\tau, g_i]$ with $\tau \in \{f, g_1, \ldots, g_m\}$ is by definition in the left-hand-side of (3.7). Therefore, if (3.7) is true,

$$0 = \langle dh_j, [\tau, g_i] \rangle = L_\tau L_{g_i} h_j - L_{g_i} L_\tau h_j$$

But, again from (3.7), $g_i \in (sp\{dh_j\})^{\perp}$ so we can conclude

$$L_{g_i} L_\tau h_j = 0$$

i.e.

$$g_i \in (\mathrm{sp}\{dL_\tau h_j\})^\perp$$

By iterating this argument it is easily seen that if $\tau_k,\ldots,\tau_1$ is any set of k vector fields belonging to the set $\{f,g_1,\ldots,g_m\}$, then

$$g_i \in (\mathrm{sp}\{dL_{\tau_k}\ldots L_{\tau_1}h_j\})^\perp \tag{3.8}$$

From the Remark II.(3.4), we know that $\mathcal{O}_j$ consists of ℝ-linear combinations of functions of the form h_j or $L_{\tau_k}\ldots L_{\tau_1}h_j$, with $\tau_i \in \{f,g_1,\ldots,g_m\}$, $1 \le i \le k$, $1 \le k < \infty$. Thus, from (3.8) we deduce that g_i annihilates the differential of any function in $\mathcal{O}_j$, i.e. that (3.5) is satisfied.

Summing up we may state following result

(3.9) *Theorem.* The output y_j is unaffected by the input u_i if and only if any one of the following (equivalent) conditions is satisfied

i) $$g_i \in \Omega^\perp_{\mathcal{O}_j}$$

ii) $$\langle f,g_1,\ldots,g_m|\mathrm{sp}\{g_i\}\rangle \subset (\mathrm{sp}\{dh_j\})^\perp$$

iii) $$\langle f,g_1,\ldots,g_m|\mathrm{sp}\{g_i\}\rangle \subset \Omega^\perp_{\mathcal{O}_j}$$

(3.10) *Remark.* It is clear that the statement of Theorem (3.2) can be slightly modified (and weakened) by asking that

$$L_{g_i}h_j(x) = 0$$

$$L_{g_i}L_{\tau_1}\ldots L_{\tau_r}h_j(x) = 0$$

for all $r \ge 1$ and any choice of vector fields $\tau_1,\ldots,\tau_r$ in the set $\{f,g_1,\ldots,g_{i-1},g_{i+1},\ldots,g_m\}$.

Accordingly, the statement of Theorem (3.9) could be modified by taking into consideration, instead of $\langle f,g_1,\ldots,g_m|\mathrm{sp}\{g_i\}\rangle$, the smallest distribution containing g_i and invariant under the vector fields $f,g_1,\ldots,g_{i-1},g_{i+1},\ldots,g_m$. Consistently, instead of $\mathcal{O}_j$, one should consider the smallest subspace of $C^\infty(N)$ containing h_j and closed under differentiation along the vector fields $f,g_1,\ldots,g_{i-1},g_{i+1},\ldots,g_m$.

(3.11) *Remark.* Suppose $\langle f,g_1,\ldots,g_m|\mathrm{sp}\{g_i\}\rangle$ and $\Omega^\perp_{\mathcal{O}_j}$ are nonsingular.

Then both distributions are also involutive (see Lemmas I.(6.6),I.(7.6) and Remark II.(3.7)).If the condition (iii) of Theorem (3.9) is satisfied, then around each point $x \in N$ it is possible to find a coordinate neighborhood U on which the nonlinear system is locally represented by equations of the form

$$\dot{x}_1 = f_1(x_1,x_2) + \sum_{\substack{k=1\\k\neq i}}^{m} g_{k1}(x_1,x_2)u_k + g_i(x_1,x_2)u_i$$

$$\dot{x}_2 = f_2(x_2) + \sum_{\substack{k=1\\k\neq i}}^{m} g_{k2}(x_2)u_k$$

$$y_j = h_j(x_2)$$

from which one sees that the input u_i has no influence on the output y_j. □

Suppose there is a distribution Δ which is invariant under the vector fields $f,g_1,\ldots,g_m$, contains the vector field g_i and is contained in the distribution $(sp\{dh_j\})^{\perp}$. Then

$\langle f,g_1,\ldots,g_m|sp\{g_i\}\rangle \subset \Delta \subset (sp\{dh_j\})^{\perp}$.

We conclude from the above inequality that the condition (ii) of Theorem (3.9) is satisfied. Conversely, if condition (i) of Theorem (3.9) is satisfied, we have a distribution, $\Omega_{0_j}^{\perp}$, which is invariant under the vector fields $f,g_1,\ldots,g_m$, contains g_i and is contained in $(sp\{dh_j\})^{\perp}$. Therefore we may give another different and useful formulation to the invariance condition.

(3.12) *Theorem*. The output y_j is unaffected by the input u_i if and only if there exists a distribution Δ with the following properties

(i) Δ is invariant under $f,g_1,\ldots,g_m$

(ii) $g_i \in \Delta \subset (sp\{dh_j\})^{\perp}$ □

(3.13) *Remark*. Again the condition (i) may be weakened by simply asking that

(i') Δ is invariant under $f,g_1,\ldots,g_{i-1},g_{i+1},\ldots,g_m$

Note that this implies that if there exists a distribution Δ with the properties (i') and (ii) there exists another distribution Δ with the properties (i) and (ii). □

We leave to the reader the task of extending the previous result to the situation in which it is required that a specified set of outputs $y_{j_1},\dots,y_{j_r}$ has to be unaffected by a given set of inputs $u_{i_1},\dots,u_{i_s}$. The conditions stated in Theorem (3.2) remain formally the same, while the ones stated in Theorems (3.9) and (3.12) require appropriate modifications.

In concluding this section it may be worth observing that in case the system in question reduces to a linear system of the form

$$\dot{x} = Ax + \sum_{i=1}^{m} b_i u_i$$

$$y_j = c_j x \qquad\qquad j = 1,\dots,\ell$$

then the condition (3.3) becomes

$$c_j A^k b_i = 0 \qquad\qquad \text{for all } k \geq 0$$

The conditions (i),(ii),(iii) of Theorem (3.9) become respectively

$$b_i \in \bigcap_{k=0}^{n-1} \ker(c_j A^k)$$

$$\sum_{k=0}^{n-1} \operatorname{Im}(A^k b_i) \subset \bigcap_{k=0}^{n-1} \ker(c_j A^k)$$

$$\sum_{k=0}^{n-1} \operatorname{Im}(A^k b_i) \subset \ker(c_j)$$

These clearly imply and are implied by the existence of a subspace V invariant under A and such that

$$b_i \subset V \subset \ker(c_j).$$

4. Left-Invertibility

In this section we consider the problem of finding conditions which ensures that, in a given system, different input functions produce different output functions. If this is the case then the input-output map is invertible from the left and it is possible to reconstruct uniquely the input acting on the system from the knowledge of the corresponding output. Since, as we know, the input-output map of a nonlinear system depends on the initial state x^o, one has to incorporate the dependence on the initial state into a precise definition of invertibility.

A system is *left-invertible* at x^o if whenever u^a and u^b are two different input functions

$$y(t;x^o;u^a) \neq y(t;x^o;u^b)$$

for at least a value of $t \geq 0$.

We restrict our attention to systems with a scalar-valued input (but possibly vector-valued output) because this case can be dealt with relative ease. Thus our system will be described by the equations

$$\dot{x} = f(x) + g(x)u$$

(4.1)

$$y_j = h_j(x) \qquad 1 \leq j \leq \ell$$

A simple sufficient condition for invertibility at x^o is the following one.

(4.2) *Lemma*. The system (4.1) is left-invertible at x^o if for some integer $k_o \geq 0$ and some $1 \leq j \leq \ell$

$$L_g L_f^{k_o} h_j(x^o) \neq 0 \tag{4.3}$$

$$L_g L_f^k h_j(x) = 0 \tag{4.4}$$

for all $x \in N$ and for all $0 \leq k < k_o$

Proof. Suppose that u^a and u^b are two different analytic input functions. Then, there exists an integer r such that

$$\left(\frac{d^r u^a}{dt^r}\right)_{t=0} \neq \left(\frac{d^r u^b}{dt^r}\right)_{t=0} \tag{4.5}$$

Now, let r_o denote the smallest integer such that (4.4) is satisfied. We will show that the $(k_o + r_o + 1)$-th derivatives of $y_i(t;x^o;u^a)$ and of $y_i(t;x^o;u^b)$ with respect to the time t are different at $t = 0$, so that we may conclude that the two output functions, which are analytic, are different.

For, remember that the coefficients of the Fliess expansion of $y_j(t;x^o;u^a)$ have the expression

$$h_j(x^o)$$

$$L_{g_{j_0}} L_{g_{j_1}} \dots L_{g_{j_k}} h_j(x^o)$$

where, in this case, $0 \leq j_0, \dots, j_k \leq 1$ and $g_0 = f$, $g_1 = g$. From (4.4) we have that the only possibly nonzero coefficients in the series are those in which:

- either $j_0 = \dots = j_k = 0$
- or $k \geq k_o$ and $j_{k-k_o+1} = \dots = j_k = 0$

These coefficients multiply iterated integrals which either do not contain the input function, or have the form

$$\int_0^t \underbrace{d\xi_0 \dots d\xi_0}_{k_o\text{-times}} d\xi_{j_{k-k_o}} \dots d\xi_{j_0}$$

Let's now take the k-th derivatives of the function $y_j(t;x^o;u)$ with respect to t and evaluate them at $t = 0$. It is clear from the structure of the iterated integrals that only those terms of Fliess series whose index has a lenght smaller than or equal k will contribute, because all terms whose index has a lenght greater than k vanish at $t = 0$. Thus we have

$$\left(\frac{d^{k_o+r_o+1} y_j}{dt^{k_o+r_o+1}}\right)_{t=0} =$$

$$= \sum_{k=0}^{k_o} \sum_{j_0,\dots,j_k=0}^{1} L_{g_{j_0}} \dots L_{g_{j_k}} L_f^{k_o} h_j(x^o) \left(\frac{d^{r_o+1}}{dt^{r_o+1}} \int_0^t d\xi_{j_k} \dots d\xi_{j_0}\right)_{t=0}$$

At this point, we observe that

$$\left(\frac{d^{r_o+1}}{dt^{r_o+1}} \int_0^t d\xi_1\right)_{t=0} = \left(\frac{d^{r_o} u(t)}{dt^{r_o}}\right)_{t=0}$$

and that all other (r_o+1)-th derivatives of the iterated integrals depend only on $u(0), \dot{u}(0)$, up to the (r_o-1)-th derivative of $u(t)$ at $t = 0$.

Therefore, since

$$\left(\frac{d^k u^a}{dt^k}\right)_{t=0} = \left(\frac{d^k u^b}{dt^k}\right)_{t=0}$$

for all $0 \leq k \leq r_o - 1$, we conclude

$$\left(\frac{d^{k_o+r_o+1} y_j(t;x^o;u^a)}{dt^{k_o+r_o+1}}\right)_{t=0} - \left(\frac{d^{k_o+r_o+1} y_j(t;x^o;u^b)}{dt^{k_o+r_o+1}}\right)_{t=0} =$$

$$= L_g L_f^{k_o} h_j(x^o) \left(\left(\frac{d^{r_o} u^a}{dt^{r_o}}\right)_{t=0} - \left(\frac{d^{r_o} u^b}{dt^{r_o}}\right)_{t=0} \right) \neq 0$$

This completes the proof. □

The condition of Theorem (4.3) may fail to be necessary for left invertibility at a given x^o, but it happens to be necessary and sufficient for a stronger notion of invertibility. For, suppose there exists an integer k_o such that the conditions (4.3) and (4.4) are satisfied for some x^o. Then, there exists a neighborhood U of x^o such that

$$L_g L_f^{k_o} h_j(\bar{x}) \neq 0$$

for all $\bar{x} \in U$ and this together with (4.4) implies - according to our previous theorem - that the system is left invertible at all points $\bar{x}$ of U. Conversely, suppose we cannot find an integer k_o such that (4.3) is satisfied for some x^o. This implies that

$$L_g L_f^k h_j(x) = 0$$

for all $k \geq 0$ and for all $1 \leq j \leq \ell$. This in turn implies that all the coefficients of Fliess expansion of $y(t)$ vanish but the ones in which only differentiations along the vector field f occurr. Under these circumstances we have

$$y(t) = \sum_{k=0}^{\infty} L_f^k h(x) \frac{t^k}{k!}$$

and there is no x for which the system is left-invertible.

Thus, we may state the following result

(4.6) *Theorem.* There exists an open subset U of N with the property that the system is left invertible at all points x of U if and only if there exists an integer $k \geq 0$ such that

$$L_g L_f^k h_j(x^o) \neq 0$$

for some $x^o \in N$ and some $1 \leq j \leq \ell$. □

Of course, the system being analytic, if the condition (4.3) is satisfied at some x^o, then it is satisfied on an open subset U of N which contains x^o and is dense in N. Therefore the existence of an integer k such that the condition (4.3) is satisfied for some $x^o \in N$ and some $1 \leq j \leq \ell$ is actually necessary and sufficient for the existence of an open subset U dense in N with the property that the system is invertible at all $x \in U$.

5. Realization Theory

The problem of "realizing" a given input-output behavior is generally known as the problem of finding a dynamical system with inputs and outputs able to reproduce, when initialized in a suitable state, the given input-output behavior. The dynamical system is thus said to "realize", from the chosen initial state, the prescribed input-output map.

Usually, the search for dynamical systems which realize the input-output map is restricted to special classes in the universe of all dynamical systems,depending on the structure and/or properties of the given input-output map. For example, when this map may be represented as a convolution integral of the form

$$y(t) = \int_0^t w(t-\tau)u(\tau)d\tau$$

where w is a prescribed function of t defined for $t \geq 0$, then one usually looks for a linear dynamical system

$$\dot{x} = Ax + Bu$$

$$y = Cx$$

able to reproduce, when initialized in $x^o = 0$, the given behavior. For this to be true, the matrices A,B,C must be such that

$$C \exp(At)B = w(t)$$

We will now describe the fundamentals of the realization theory for the (rather general) class of input-output maps which can be represented like functionals of the form (1.6). In view of the results of the previous sections, the search for "realizations" of this kind of maps will be restricted to the class of dynamical system of the form (1.1).

From a formal point of view, the problem is stated in the following way. Given a formal power series in m+1 noncommutative indeterminates with coefficients in $\mathbb{R}^{\ell}$, find an integer n, an element x^o of $\mathbb{R}^n$, m+1 analytic vector fields $g_0,g_1,\ldots,g_m$ and an analytic ℓ-vector valued function h defined on a neighborhood U of x^o such that

$$h(x^o) = c(\emptyset)$$

$$L_{g_{i_0}} L_{g_{i_1}} \ldots L_{g_{i_k}} h(x^o) = c(i_k \ldots i_1 i_0)$$

If these conditions are satisfied, then it is clear that the dynamical system

$$\dot{x} = g_0(x) + \sum_{i=1}^{m} g_i(x) u_i$$

$$y = h(x)$$

initialized in $x^o \in \mathbb{R}^n$ produces an input-output behavior of the form

$$y(t) = c(\emptyset) + \sum_{k=0}^{\infty} \sum_{j_0 \ldots j_k = 0}^{m} c(j_k \ldots j_0) \int_0^t d\xi_k \ldots d\xi_0$$

In view of this, the set $\{g_0,g_1,\ldots,g_m,h,x^o\}$ will be called a *realization* of the formal power series c.

In order to present the basic results of the realization theory, we need first to develop some notations and describe some simple algebraic concepts related to the formal power series. In view of the need of dealing with sets of series and defining certain operations on these sets it is useful to represent each series as a *formal* infinite *sum* of "monomials". Let $z_0,z_1,\ldots,z_m$ denote a set of m+1 abstract non commutative indeterminates and let $Z = \{z_0,z_1,\ldots,z_m\}$. With each multi-index $(i_k \ldots i_0)$ we associate the monomial $(z_{i_k} \ldots z_{i_0})$ and we represent the series in the form

$$c = c(\emptyset) + \sum_{k=0}^{\infty} \sum_{i_0 \ldots i_k = 0}^{m} c(i_k \ldots i_0) z_{i_k} \ldots z_{i_0} \tag{5.1}$$

The set of all power series in m+1 noncommutative indeterminates (or, in other words, in the noncommutative indeterminates $z_0,\dots,z_m$) and coefficients in $\mathbb{R}^\ell$ is denoted with the symbol $\mathbb{R}^\ell\langle\langle Z\rangle\rangle$. A special subset of $\mathbb{R}^\ell\langle\langle Z\rangle\rangle$ is the set of all those series in which the number of nonzero coefficients (i.e. the number of nonzero terms in the sum (5.1)) is finite. A series of this type is a polynomial in m+1 noncommutative indeterminates and the set of all such polynomials is denoted with the symbol $\mathbb{R}^\ell\langle Z\rangle$. In particular $\mathbb{R}\langle Z\rangle$ is the set of all polynomials in the m+1 noncommutative indeterminates $z_0,\dots,z_m$ and coefficients in $\mathbb{R}$.

An element of $\mathbb{R}\langle Z\rangle$ may be represented in the form

$$p = p(\emptyset) + \sum_{k=0}^{d} \sum_{i_0\dots i_k=0}^{m} p(i_k\dots i_0) z_{i_k}\dots z_{i_0} \tag{5.2}$$

where d is an integer which depends on p and $p(\emptyset), p(i_k\dots i_0)$ are real numbers.

The sets $\mathbb{R}\langle Z\rangle$ and $\mathbb{R}^\ell\langle\langle Z\rangle\rangle$ may be given different algebraic structures. They can clearly be regarded as $\mathbb{R}$-vector spaces, by letting $\mathbb{R}$-linear combinations of polynomials and/or series be defined coefficient-wise. The set $\mathbb{R}\langle Z\rangle$ may also be given a ring structure, by letting the operation of sum of polynomials be defined coefficient-wise (with the neutral element given by the polynomial whose coefficients are all zero) and the operation of product of polynomials defined through the customary product of the corresponding representations (5.2) (in which case the neutral element is the polynomial whose coefficients are all zeros but $p(\emptyset)$ which is equal to 1). Later on, in the proof of Theorem (5.8), we shall also endow $\mathbb{R}\langle Z\rangle$ and $\mathbb{R}^\ell\langle\langle Z\rangle\rangle$ with structures of modules over the ring $\mathbb{R}\langle Z\rangle$ but, for the moment, those additional structures are not required.

What is important at this point is to know that the set $\mathbb{R}\langle Z\rangle$ can also be given a structure of a Lie algebra, by taking the above-mentioned $\mathbb{R}$-vector space structure and defining a Lie bracket of two polynomials p_1, p_2 by setting $[p_1,p_2] = p_2p_1 - p_1p_2$. The smallest subalgebra of $\mathbb{R}\langle Z\rangle$ which contains the monomials $z_0,\dots,z_m$ will be denoted by $L(Z)$. Clearly, $L(Z)$ may be viewed as a subspace of the $\mathbb{R}$-vector space $\mathbb{R}\langle Z\rangle$, which contains $z_0,\dots,z_m$ and is closed under Lie bracketing with $z_0,\dots,z_m$. Actually, it is not difficult to see that $L(Z)$ is the smallest subspace of $\mathbb{R}\langle Z\rangle$ which has these properties.

Now we return to the problem of realizing an input-output map represented by a functional of the form (1.6). As expected, the existence of realizations will be characterized as a property of the formal power series which specifies the functional. We associate with the formal power series c two integers, which will be called, following Fliess, the *Hankel rank* and the *Lie rank* of c. This is done in the following manner. We use the given formal power series c to define a mapping

$$F_c : \mathbb{R}\langle Z \rangle \to \mathbb{R}^{\ell}\langle\langle Z \rangle\rangle$$

in the following way:

a) the image under F_c of any polynomial in the set $Z^* = \{z_{j_k} \ldots z_{j_0} \in \mathbb{R}\langle Z \rangle : (j_k \ldots j_0) \in I^*\}$ (by definition, the polynomial associated with the multiindex $\emptyset \in I^*$ will be the polynomial in which all coefficients are zero but $p(\emptyset)$ which is equal to 1, i.e. the unit of $\mathbb{R}\langle Z \rangle$) is a formal power series defined by setting

$$[F_c(z_{j_k} \ldots z_{j_0})](i_r \ldots i_0) = c(i_r \ldots i_0\, j_k \ldots j_0)$$

for all $j_k \ldots j_0 \in I^*$.

b) the map F_c is an $\mathbb{R}$-vector space morphism of $\mathbb{R}\langle Z \rangle$ into $\mathbb{R}^{\ell}\langle\langle Z \rangle\rangle$.

Note that any polynomial in $\mathbb{R}\langle Z \rangle$ may be expressed as an $\mathbb{R}$-linear combination of elements of Z^* and, therefore, the prescriptions (a) and (b) completely specify the mapping F_c.

Looking at F_c as a morphism of $\mathbb{R}$-vector spaces, we define the *Hankel rank* $\rho_H(c)$ of c as the rank of F_c , i.e. the dimension of the subspace

$$F_c(\mathbb{R}\langle Z \rangle) \subset \mathbb{R}^{\ell}\langle\langle Z \rangle\rangle$$

Moreover, we define the *Lie rank* $\rho_L(c)$ of c as the dimension of the subspace

$$F_c(L(Z)) \subset \mathbb{R}^{\ell}\langle\langle Z \rangle\rangle$$

i.e. the rank of the mapping $F_c|_{L(Z)}$.

(5.3) *Remark*. It is easy to get a matrix representation of the mapping F_c. For, suppose we represent an element p of $\mathbb{R}\langle Z \rangle$ with an infinite column vector of real numbers whose entries are indexed by the elements of I^* and the entry indexed by $j_k \ldots j_0$ is exactly $p(j_k \ldots j_0)$. Of course, p being a polynomial, only finitely many elements of this vector are nonzero. In the same way, we may represent an element c of $\mathbb{R}^{\ell}\langle\langle Z\rangle\rangle$ with an infinite column vector whose entries are ℓ-vectors of real numbers, indexed by the elements of I^* and such that the entry indexed by $i_r\dots i_0$ is $c(i_r\dots i_0)$. Then, any $\mathbb{R}$-vector space morphism defined on $\mathbb{R}\langle Z\rangle$ with values in $\mathbb{R}^{\ell}\langle\langle Z\rangle\rangle$ will be represented by an infinite matrix, whose columns are indexed by elements of I^* and in which each block of ℓ rows is again indexed by elements of I^*. In particular, the mapping F_c will be represented by a matrix, denoted H_c, in which the block of ℓ rows of index $(i_r\dots i_0)$ on the column of index $(j_k\dots j_0)$ is exactly the coefficient

$$c(i_r\dots i_0 j_k\dots j_0)$$

of c. We leave to the reader the elementary check of this statement.

The matrix H_c is called the *Hankel matrix* of the series c. It is clear from the above definitions that the rank of the matrix H_c coincides with the Hankel rank of F_c. □

(5.4) *Example*. If the set I consists of only one element, then it is easily seen that I^* can be identified with the set $\mathbb{Z}^+$ of the nonnegative integers numbers. A formal power series in one indeterminate with coefficients in $\mathbb{R}$, i.e. a mapping

$$c : \mathbb{Z}^+ \to \mathbb{R}$$

may be represented, like in (5.1), as an infinite sum

$$c = \sum_{k=0}^{\infty} c_k z^k$$

and the Hankel matrix associated with the mapping F_c coincides with the classical Hankel matrix associated with the sequence $c_0, c_1 \dots$

$$H_c = \begin{pmatrix} c_0 & c_1 & c_2 & \cdots \\ c_1 & c_2 & c_3 & \cdots \\ c_2 & c_3 & c_4 & \cdots \\ \cdot & \cdot & \cdot & \cdots \end{pmatrix}$$

□

The importance of the Hankel and Lie ranks of the mapping F_c depends on the following basic results.

(5.5) *Lemma*. Let $f, g_1, \dots, g_m, h$ and a point $x^o \in \mathbb{R}^n$ be given. Let Δ_C

be the distribution associated with the control Lie algebra $\mathcal{C}$ and $\Omega_{\mathcal{O}}$ the codistribution associated with the observation space $\mathcal{O}$. Let $K(x^o)$ denote the subset of vectors of $\Delta_{\mathcal{C}}(x^o)$ which annihilate $\Omega_{\mathcal{O}}(x^o)$ i.e. the subspace of $T_{x^o}\mathbb{R}^n$ defined by

$$K(x^o) = \Delta_{\mathcal{C}}(x^o) \cap \Omega_{\mathcal{O}}^{\perp}(x^o) = \{v \in \Delta_{\mathcal{C}}(x^o) : \langle d\lambda(x^o), v \rangle = 0 \quad \forall \lambda \in \mathcal{O}\}$$

Finally, let c be the formal power series defined by

$$c(\emptyset) = h(x^o) \tag{5.6a}$$

$$c(i_k \dots i_0) = L_{g_{i_0}} \dots L_{g_{i_k}} h(x^o) \tag{5.6b}$$

with $g_0 = f$. Then the Lie rank of c has the value

$$\rho_L(c) = \dim \Delta_{\mathcal{C}}(x^o) - \dim K(x^o) = \dim \frac{\Delta_{\mathcal{C}}(x^o)}{\Delta_{\mathcal{C}}(x^o) \cap \Omega_{\mathcal{O}}^{\perp}(x^o)}$$

Proof. Define a morphism of Lie algebras

$$\mu : L(Z) \to V(\mathbb{R}^n)$$

by setting

$$\mu(z_i) = g_i \qquad 0 \leq i \leq m$$

Then, it is easy to check that if p is a polynomial in $L(Z)$ the $(i_k \dots i_0)$-th coefficient of $F_c(p)$ is $L_{\mu(p)} L_{g_{i_0}} \dots L_{g_{i_k}} h(x^o)$. Thus, the series $F_c(p)$ has the expression

$$F_c(p) = L_{\mu(p)} h(x^o) + \sum_{k=0}^{\infty} \sum_{i_0,\dots,i_k=0}^{m} L_{\mu(p)} L_{g_{i_0}} \dots L_{g_{i_k}} h(x^o) z_{i_k} \dots z_{i_0}$$

If we let v denote the value of the vector field $\mu(p)$ at x^o, the above can be rewritten as

$$F_c(p) = \langle dh(x^o), v \rangle + \sum_{k=0}^{\infty} \sum_{i_0,\dots,i_k=0}^{m} \langle dL_{g_{i_0}} \dots L_{g_{i_k}} h(x^o), v \rangle z_{i_k} \dots z_{i_0}$$

When p ranges over $L(Z)$, the tangent vector v takes any value in $\Delta_{\mathcal{C}}(x^o)$. Moreover, the covectors $dh(x^o), \dots, dL_{g_{i_0}} \dots L_{g_{i_k}} h(x^o), \dots$

span $\Omega_0(x^o)$. This implies that the number of $\mathbb{R}$-linearly independent power series in $F_c(L(Z))$ is exactly equal to

$$\dim \Delta_C(x^o) - \dim \Delta_C(x^o) \cap \Omega_0^{\perp}(x^o)$$

and this, in view of the definition of Lie rank of F_c , proves the claim. □

We immediately see from this that if an input-output functional of the form (1.6) is realized by a dynamical system of dimension n, then necessarily the Lie rank of the formal power series which specify the functional is bounded by n. In other words, the *finiteness* of the Lie rank $\rho_L(c)$ is a necessary condition for the existence of finite-dimensional realizations. We shall see later on that this condition is also sufficient. For the moment, we wish to investigate the role of the finiteness of the other rank associated with F_c i.e. the Hankel rank. It comes from the definition that

$$\rho_L(c) \leq \rho_H(c)$$

so the Hankel rank may be infinite when the Lie rank is finite. However, there are special cases in which $\rho_H(c)$ is finite.

(5.7) *Lemma.* Suppose $f, g_1, \ldots, g_m, h$ are linear in x, i.e. that

$$f(x) = Ax, \quad g_1(x) = N_1 x, \ldots, g_m(x) = N_m x \ , \ h(x) = Cx$$

for suitable matrices $A, N_1, \ldots, N_m, C$. Let x^o be a point of $\mathbb{R}^n$. Let V denote the smallest subspace of $\mathbb{R}^n$ which contains x^o and is invariant under $A, N_1, \ldots, N_m$. Let W denote the largest subspace of $\mathbb{R}^n$ which is contained in ker(C) and is invariant under $A, N_1, \ldots, N_m$. The Hankel rank of the formal power series (5.7) has the value

$$\rho_H(c) = \dim V - \dim W \cap V = \dim \frac{V}{W \cap V}$$

Proof. We have already seen, in section II.4, that the subspace W may be expressed in the following way

$$W = (\ker C) \cap [\bigcap_{r=0}^{\infty} \bigcap_{i_0 \ldots i_r = 0}^{m} \ker(CN_{i_r} \ldots N_{i_0}]$$

with $N_0 = A$. With the same kind of arguments one proves that the subspace V may be expressed as

$$V = \text{span}\{x^o\} + \sum_{k=0}^{\infty} \sum_{j_0 \ldots j_k=0}^{m} \text{span}\{N_{j_k} \ldots N_{j_0} x^o\}$$

In the present case the Hankel matrix of F_c is such that the block of ℓ rows of index $(i_r \ldots i_0)$ on the column of index $(j_k \ldots j_0)$, i.e. the coefficient $c(i_r \ldots i_0 j_k \ldots j_0)$ of c has the expression

$$CN_{i_r} \ldots N_{i_0} N_{j_k} \ldots N_{j_0} x^o$$

By factoring out this expression in the form

$$(CN_{i_r} \ldots N_{i_0})(N_{j_k} \ldots N_{j_0} x^o)$$

it is seen that the Hankel matrix can be factored out as the product of two matrices, of which the one on the left-hand-side has a kernel equal to the subspace W, while the one on the right-hand-side has an image equal to the subspace V. From this the claimed result follows immediately. □

Thus, it is seen from this Lemma that if an input output functional of the form (1.3) is realized by a dynamical system of dimension n described by equations of the form

$$\dot{x} = Ax + \sum_{i=1}^{m} N_i x u_i$$

$$y = Cx$$

i.e. by a bilinear dynamical system of dimension n, then the Hankel rank of the formal power series which specifies the functional is bounded by n. The finiteness of the Hankel rank $\rho_H(c)$ is a necessary condition for the existence of bilinear realizations.

We turn now to the problem of showing the sufficiency of the above two conditions. We treat first the case of bilinear realizations, which is simpler. In analogy with the definition given at the beginning of the section, we say that the set $\{N_0, N_1, \ldots, N_m, C, x^o\}$, where $x^o \in \mathbb{R}^n$, $N_i \in \mathbb{R}^{n \times n}$ for $0 \leq i \leq m$ and $C \in \mathbb{R}^{\ell \times n}$ is a *bilinear realization* of the formal power series c if the set $\{g_0, g_1, \ldots, g_m, h, x^o\}$ defined by

$$g_0(x) = N_0 x,\ g_1(x) = N_1(x), \ldots, g_m(x) = N_m x$$

$$h(x) = Cx$$

is a realization of c.

(5.8) *Theorem*. Let c be a formal power series in m+1 noncommutative indeterminates and coefficients in $\mathbb{R}^{\ell}$. There exists a bilinear realization of c if and only if the Hankel rank of c is finite.

Proof. We need only to prove the "if" part. For, consider again the mapping F_c. The sets $\mathbb{R}\langle Z\rangle$ and $\mathbb{R}^{\ell}\langle\langle Z\rangle\rangle$ will now be endowed with structures of modules. The ring $\mathbb{R}\langle Z\rangle$ is regarded as a module over itself. $\mathbb{R}^{\ell}\langle\langle Z\rangle\rangle$ is given an $\mathbb{R}\langle Z\rangle$-module structure by letting the operation of sum of power series be defined coefficient-wise and the product $p\cdot s$ of a polynomial $p \in \mathbb{R}\langle Z\rangle$ by a series $s \in \mathbb{R}^{\ell}\langle\langle Z\rangle\rangle$ be defined in the following way

a) $1\cdot s = s$

b) for all $0 \leq i \leq m$ the series $z_i\cdot s$ is given by

$$(z_i\cdot s)(i_r\ldots i_0) = s(i_r\ldots i_0 i)$$

c) for all $p_1,p_2 \in \mathbb{R}\langle Z\rangle$ and $\alpha_1,\alpha_2 \in \mathbb{R}$

$$(\alpha_1 p_1+\alpha_2 p_2)\cdot s = \alpha_1(p_1\cdot s)+\alpha_2(p_2\cdot s)$$

Note that from (a) and (b) we have that for all $j_k\ldots j_0 \in I^*$

$$(z_{j_k}\ldots z_{j_0}\cdot s)(i_r\ldots i_0) = s(i_r\ldots i_0 j_k\ldots j_0)$$

Note also that since the ring $\mathbb{R}\langle Z\rangle$ is not commutative, the order in which the products are performed is essential.

We leave to the reader the simple proof that the map F_c previously defined becomes an $\mathbb{R}\langle Z\rangle$-module morphism when $\mathbb{R}^{\ell}\langle\langle Z\rangle\rangle$ is endowed with this kind of $\mathbb{R}\langle Z\rangle$-module structure. As a matter of fact, it is trivial to check that $F_c(p) = p\cdot c$.

Now, consider the canonical factorization of F_c

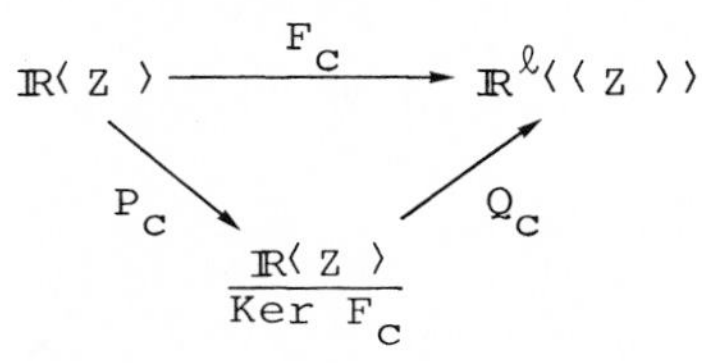

in which, as usual, P_c denotes the canonical projection $p \longmapsto (p+\ker F_c)$ and Q_c the injection $(p+\ker F_c) \mapsto F_c(p)$. P_c and Q_c are $\mathbb{R}$-vector space morphisms, but there is also a canonical $\mathbb{R}\langle Z\rangle$-module structure on

$\mathbb{R}\langle Z \rangle / \ker F_c$ which makes P_c and Q_c $\mathbb{R}\langle Z \rangle$-module morphisms.

Since, by definition, $\mathbb{R}\langle Z \rangle / \ker F_c$ is isomorphic to the image of F_c , we have that the dimension of $\mathbb{R}\langle Z \rangle / \ker F_c$ as an $\mathbb{R}$-vector space is equal to the Hankel rank $\rho_H(c)$ of the formal power series c. Let, for simplicity, denote

$$X = \frac{\mathbb{R}\langle Z \rangle}{\ker F_c}$$

But X is also an $\mathbb{R}\langle Z \rangle$-module, so to each of the indeterminates $z_0,\ldots,z_m$ we may associate mappings

$$M_i : X \to X$$

$$x \mapsto z_i \cdot x$$

The mappings M_i are clearly $\mathbb{R}$-vector space morphisms. We also define an $\mathbb{R}$-vector space morphism

$$H : X \to \mathbb{R}^{\ell}$$

by taking

$$Hx = [Q_c(x)](\emptyset)$$

With the notation on the right-hand-side we mean the coefficient with empty index in the series $Q_c(x)$.

Finally, let x^o be the element of X

$$x^o = P_c(1)$$

where 1 is the unit polynomial in $\mathbb{R}\langle Z \rangle$.

We claim that

$$c(\emptyset) = Hx^o \tag{5.9a}$$

$$c(i_k \ldots i_0) = HM_{i_k} \ldots M_{i_0} x^o \tag{5.9b}$$

For, it is seen immediately that

$$c = F_c(1) = Q_c P_c(1) = Q_c x^o \tag{5.10a}$$

Moreover, suppose that

(5.10b) $$F_c(z_{i_k}\dots z_{i_0}) = Q_c M_{i_k}\dots M_{i_0} x^o$$

then we have

$$F_c(z_i z_{i_k}\dots z_{i_0}) = z_i \cdot F_c(z_{i_k}\dots z_{i_0}) = z_i (Q_c M_{i_k}\dots M_{i_0} x^o)$$

$$Q_c(z_i \cdot M_{i_k}\dots M_{i_0} x^o) = Q_c M_i M_{i_k}\dots M_{i_0} x^o$$

for $0 \leq i \leq m$. Thus (5.10b) is true for all $(i_k\dots i_0) \in I^*$.

Now, keeping in mind the definition of F_c , one has

$$[F_c(z_{i_k}\dots z_{i_0})](\emptyset) = c(z_{i_k}\dots z_{i_0})$$

and therefore, in view of the definition of the mapping H, (5.9) are proved.

Take now a basis in the $\rho_H(c)$-dimensional vector space X. The mappings $M_0,\dots,M_m$ and H will be represented by matrices $N_0,N_1,\dots,N_m$ and C; x^o will be represented by a vector $\hat{x}^o$. These quantities are such that

$$c(i_k\dots i_0) = CN_{i_k}\dots N_{i_0}\hat{x}^o$$

for all $(i_k\dots i_0) \in I^*$. This shows that the set $\{C,N_0,\dots,N_m,\hat{x}^o\}$ is a bilinear realization for our series. □

The result which follows presents a necessary and sufficient condition for the existence of realizations of an input-output functional of the form (1.6), provided that the coefficients of the power series which characterize the functional are suitably bounded.

(5.11) *Theorem*. Let c be a formal power series whose coefficients satisfy the condition

(5.12) $$\|c(i_k\dots i_0)\|_1 \leq C(k+1)!\,r^{(k+1)}$$

for all $(i_k\dots i_0) \in I^*$, for some pair of real numbers $C > 0$ and $r > 0$. Then there exists a realization of c if and only if the Lie rank of c is finite.

Proof. Some more machinery is required. For each polynomial $p \in \mathbb{R}\langle Z\rangle$ we define a mapping $S_p : \mathbb{R}^\ell\langle\langle Z\rangle\rangle \to \mathbb{R}^\ell\langle\langle Z\rangle\rangle$ in the following way

a) if $p \in Z^* = \{z_{j_k}\dots z_{j_0} \in \mathbb{R}\langle Z\rangle : (j_k\dots j_0) \in I^*\}$ then $S_p(c)$ is a formal power series defined by setting

$$[S_{z_{j_k}\dots z_{j_0}}(c)](i_r\dots i_0) = c(j_k\dots j_0 i_r\dots i_0)$$

b) if $\alpha_1,\alpha_2 \in \mathbb{R}$ and $p_1,p_2 \in \mathbb{R}\langle Z\rangle$ then

$$S_p(c) = \alpha_1 S_{p_1}(c) + \alpha_2 S_{p_2}(c)$$

Moreover, suppose that, given a formal power series $s_1 \in \mathbb{R}\langle\langle Z\rangle\rangle$ and a formal power series $s_2 \in \mathbb{R}\langle\langle Z\rangle\rangle$, the sum of the numerical series

$$(5.13) \qquad s_1(\emptyset)s_2(\emptyset) + \sum_{k=0}^{\infty} \sum_{i_0,\dots,i_k=0}^{m} s_1(i_k\dots i_0)s_2(i_k\dots i_0)$$

exists. If this is the case, the sum of this series will be denoted by $\langle s_1,s_2\rangle$.

We now turn our attention to the problem of finding a realization of c. In order to simplify the notation, we assume $\ell = 1$ (i.e. we consider the case of a single-output system). By definition, there exist n polynomials in $L(Z)$, denoted $p_1,\dots,p_n$, with the property that the formal power series $F_c(p_1),\dots,F_c(p_n)$ are $\mathbb{R}$-linearly independent.

With the polynomials $p_1,\dots,p_n$ we associate a formal power series

$$(5.14) \qquad w = \exp\left(\sum_{i=1}^{n} x_i p_i\right) = 1 + \sum_{k=1}^{\infty} \frac{1}{k!}\left(\sum_{i=1}^{n} x_i p_i\right)^k$$

where $x_1,\dots,x_n$ are real variables.

The series c which is to be realized and the series w thus defined are used in order to construct a set of analytic functions of $x_1,\dots,x_n$, defined in a neighborhood of 0 and indexed by the elements of I^*, in the following way

$$h(x) = \langle c,w\rangle$$

$$h_{i_k\dots i_0}(x) = \langle S_{z_{i_k}\dots z_{i_0}}(c),w\rangle$$

The growth condition (5.12) guarantees the convergence of the series on the right-hand-side for all x in a neighborhood of $x = 0$.

It will be shown now that there exist $m + 1$ vector fields, $g_0(x),\dots,g_m(x)$, defined in a neighborhood of 0, with the property that

$$L_{g_i} h_{i_k \ldots i_0}(x) = h_{i_k \ldots i_0 i}(x) \tag{5.15}$$

for all $(i_k \ldots i_0) \in I^*$. This will be actually enough to prove the Theorem because, at $x = 0$, the functions $h_{i_k \ldots i_0}(x)$ by construction are such that

$$h(0) = c(\emptyset)$$

$$h_{i_k \ldots i_0}(0) = c(i_k \ldots i_0)$$

and this shows that the set $\{h, g_0, \ldots, g_m\}$ together with the initial state $x = 0$ is a realization of c.

To find the vector fields $g_0, \ldots, g_m$ one proceeds as follows. Since the n series $F_c(p_1), \ldots, F_c(p_n)$ are $\mathbb{R}$-linear independent, it is easily seen that there exist n monomials $m_1, \ldots, m_n$ in the set Z^* with the property that the $(n \times n)$ matrix of real numbers

$$\begin{pmatrix} [F_c(p_1)](m_1) \ldots [F_c(p_n)](m_1) \\ \cdot \quad \ldots \quad \cdot \\ [F_c(p_1)](m_n) \ldots [F_c(p_n)](m_n) \end{pmatrix} \tag{5.16}$$

has rank n. It is easy to see that

$$[F_c(p_i)](m_j) = \left(\frac{\partial}{\partial x_i} \langle S_{m_j}(c), w \rangle\right)_{x=0}$$

For, if $p_i \in Z^*$, then by definition

$$[F_c(p_i)](m_j) = c(m_j p_i) = [S_{m_j}(c)](p_i) = \left(\frac{\partial}{\partial x_i} \langle S_{m_j}(c), w \rangle\right)_{x=0}$$

From this, using linearity, one concludes that the above expression is true also in the (general) case where p_i is an $\mathbb{R}$-linear combination of elements of Z^*.

Using this property, we conclude that the j-th row of the matrix (5.16) coincides with the value at 0 of the differential of one of the functions $h_{i_k \ldots i_0}$, the one whose multiindex corresponds to the monomial m_j.

Consider now the system of linear equations

$$\begin{pmatrix} \frac{\partial}{\partial x}\langle S_{m_1}(c),w\rangle \\ \vdots \\ \frac{\partial}{\partial x}\langle S_{m_n}(c),w\rangle \end{pmatrix} g_k(x) = \begin{pmatrix} \langle S_{m_1 z_k}(c),w\rangle \\ \vdots \\ \langle S_{m_n z_k}(c),w\rangle \end{pmatrix}$$

in the unknown n-vector $g_k(x)$. The coefficient matrix is nonsingular for all x in a neighborhood of 0 (because at $x = 0$ it coincides - as we have seen - with the matrix (5.16)). Thus, in a neighborhood of 0 it is possible to find a vector field $g_k(x)$ such that

$$L_{g_k}\langle S_{m_i}(c),w\rangle = \langle S_{m_i z_k}(c),w\rangle$$

and this proves that (5.15) can be satisfied, at least for these $h_{i_k \dots i_0}$ whose multiindexes correspond to the monomials $m_1,\dots,m_n$.

The proof that (5.15) holds for all the other functions $h_{i_k \dots i_0}(x)$ depends on the fact that every formal power series in $F_c(L(Z))$ is an $\mathbb{R}$-linear combination of $F_c(p_1),\dots,F_c(p_n)$, and is left for the reader.□

It is seen from the above Theorem that if a formal power series c has a finite Lie rank, and its coefficients satisfy the growth condition (5.12), then it is possible to find a dynamical system of dimension $\delta_L(c)$ which realizes the series.

This fact, together with the result stated before in Lemma (5.5) induces to some further remarks. A realization $\{f,g_1,\dots,g_m,x^o\}$ of a formal power series c is *minimal* if its dimension, i.e. the dimension of the underlying manifold on which $f,g_1,\dots,g_m$ are defined, is less then or equal to the dimension of any other realization of c. Thus, from Lemma (5.5) we immediately deduce the following corollaries.

(5.17) *Corollary*. A realization $\{f,g_1,\dots,g_m,x^o\}$ of a formal power series c is minimal if and only if its dimension is equal to the Lie rank $\delta_L(c)$.

(5.18) *Corollary*. A realization $\{f,g_1,\dots,g_m,x^o\}$ of a formal power series c is minimal if and only if

$$\dim \Delta_C(x^o) = \dim \Omega_O(x^o) = n$$

or, which is the same, the realization satisfies the controllability rank condition and the observability rank condition at x^o.

6. Uniqueness of Minimal Realizations

In this section we prove an interesting uniqueness result, by showing that any two minimal realizations of a formal power series are locally "diffeomorphic".

(6.1) *Theorem*. Let c be a formal power series and let n denotes its Lie rank. Let $\{g_0^a, g_1^a, \ldots, g_m^a, h^a, x^a\}$ and $\{g_0^b, g_1^b, \ldots, g_m^b, h^b, x^b\}$ be two minimal, i.e. n-dimensional realizations of c. Let g_i^a, $0 \leq i \leq m$, and h^a be defined on a neighborhood U^a of x^a in $\mathbb{R}^n$ and g_i^b, $0 \leq i \leq m$, and h^b be defined on a neighborhood U^b of x^b in $\mathbb{R}^n$. Then, there exist open subsets $V^a \subset U^a$ and $V^b \subset U^b$ and a diffeomorphism $F: V^a \rightarrow V^b$ such that

$$(6.2) \qquad g_i^b(x) = F_* g_i^a \circ F^{-1}(x) \qquad 0 \leq i \leq m$$

$$(6.3) \qquad h^b(x) = h^a \circ F^{-1}(x)$$

for all $x \in V^b$.

Proof. We break up the proof in several steps.

(i) Recall that a minimal realization $\{f, g_1, \ldots, g_m, x^o\}$ of c satisfies the observability rank condition at x^o (Corollary (5.18)). From the definitions of $\mathcal{O}$ and $\Omega_{\mathcal{O}}$, one deduces that there exist n real-valued functions $\lambda_1, \ldots, \lambda_n$, defined in a neighborhood U of x^o, having the form

$$\lambda_i(x) = L_{v_r} \ldots L_{v_1} h_j(x)$$

with $v_1, \ldots, v_r$ vector fields in the set $\{f, g_1, \ldots, g_m\}$, r (possibly) depending on i and $1 \leq j \leq \ell$ such that the covectors $d\lambda_1(x^o), \ldots, d\lambda_n(x^o)$ are linearly independent (i.e. span the cotangent space $T^*_{x^o}U$). From this property, using the inverse function theorem, it is deduced that there exists a neighborhood $U_H \subset U$ of x^o such that the mapping

$$H : x \mapsto (\lambda_1(x), \ldots, \lambda_n(x))$$

is a diffeomorphism of U_H onto its image $H(U_H)$.

From any two minimal realizations, labeled "a" and "b", we will construct two of such mappings, denoted H^a and respectively H^b.

(ii) Let $\theta_1,\ldots,\theta_n$ be a set of vector fields, defined in a neighborhood U of x^o, having the form

$$\theta_i = f + \sum_{j=1}^{m} g_j \bar{u}_j^i$$

with $\bar{u}_j^i \in \mathbb{R}$ for $1 \leq j \leq m$. Let Φ_t^i denote the flow of θ_i and G denote the mapping

$$G : (t_1,\ldots,t_n) \longmapsto \Phi_{t_n}^n \circ \ldots \circ \Phi_{t_1}^1 (x^o)$$

defined on a neighborhood $(-\varepsilon,\varepsilon)^n$ of 0.

From any two minimal realizations, labeled "a" and "b" we will construct two of such mappings, denoted G^a and G^b (the same set of $\bar{u}_j^i$'s being used in both G^a and G^b).

Recall that a minimal realization $\{f^a,g_1^a,\ldots,g_m^a,x^a\}$ satisfies the controllability rank condition at x^a (Corollary (5.18)). From the properties of Δ_C and R (see Remark II.(2.7), one deduces that the distribution R is nonsingular and n-dimensional around x^a. Then, using the same arguments as the ones used in the proof of Theorem I.(6.15), it is possible to see that there exist a choice of $\bar{u}_j^i$'s and an open subset W of $(0,\varepsilon)^n$ such that the restriction of G^a to W is a diffeomorphism of W onto its image $G^a(W)$.

(iii) It is not difficult to prove that if $\{f^a,g_1^a,\ldots,g_m^a,h^a,x^a\}$ and $\{f^b,g_1^b,\ldots,g_m^b,h^b,x^b\}$ are two realizations of the same formal power series c, then, for all $0 < t_i < \varepsilon$, $1 \leq i \leq n$, with sufficiently small ε,

$$H^a \circ G^a(t_1,\ldots,t_n) = H^b \circ G^b(t_1,\ldots,t_n) \tag{6.4}$$

As a matter of fact, if ε is small then $G(t_1,\ldots,t_n)$ is a point of U_H , reached from x^o under the piecewise constant control defined by

$$u_j(t) = \bar{u}_j^i \quad \text{for } t \in [t_1+\ldots+t_{i-1}, t_1+\ldots+t_i)$$

Moreover, the values of the components of H (i.e. the values of functions $\lambda_1,\ldots,\lambda_n$) at a point were shown to coincide with the values of certain derivatives, at time $t = 0$, of some components of an output function $y(t)$ obtained under suitable piecewise constant controls (see proof of Theorem I.(7.8)). So, one may interpret the components of $H\circ G(t_1,\ldots,t_n)$ as the values at time $t = t_1+\ldots+t_n$ of certain derivatives of an output function $y(t)$ obtained under suitable piecewise constant controls.

Two minimal realizations of the same power series c characterize two systems which by definition display the same input-output behavior. These two systems, initialized respectively in x^a and x^b, under any piecewise constant control produce two identical output functions. Thus, the two sides of (6.4) must coincide.

(iv) Recall that, if the realization "a" is minimal, if $(t_1,\ldots,t_n)\in W$ and ε is sufficiently small, the mapping $H^a\circ G^a$ is a composition of diffeomorphisms. If also the realization "b" is minimal, H^b is indeed a diffeomorphism, but also G^b must be a diffeomorphism of W onto its image, because of the equality (6.4) and of the fact that the left-hand-side is itself a diffeomorphism. The following diagram

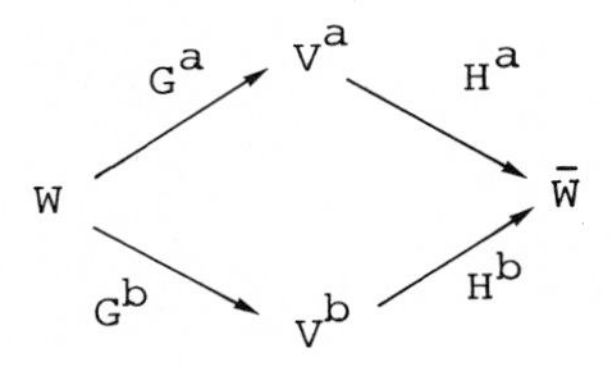

where $V^a = G^a(W)$, $V^b = G^b(W)$, $V^a \subset U^a_H$, $V^b \subset U^b_H$ and $\bar{W} = H^a\circ G^a(W) = H^b\circ G^b(W)$, is a commutative diagram of diffeomorphisms. Thus, we may define a diffeomorphism

$$F : V^a \to V^b$$

as

$$F = (H^b)^{-1}\circ H^a \tag{6.5a}$$

whose inverse may also be expressed as

$$F^{-1} = G^a\circ (G^b)^{-1} \tag{6.5b}$$

(v) By means of the same arguments as the ones already used in (iii) one may easily prove a more general version of (6.4). More precisely, setting

$$\theta^a = f^a + \sum_{i=1}^{m} g_i^a v_i \qquad \theta^b = f^b + \sum_{i=1}^{m} g_i^b v_i$$

one may deduce that, for sufficiently small t

$$H^a \circ \Phi_t^{\theta^a} \circ G^a(t_1,\dots,t_n) = H^b \circ \Phi_t^{\theta^b} \circ G^b(t_1,\dots,t_n)$$

Differentiating this one with respect to t and setting $t = 0$ one obtains

$$(H^a)_* \theta^a \circ G^a(t_1,\dots,t_n) = (H^b)_* \theta^b \circ G^b(t_1,\dots,t_n)$$

Because of the arbitrariness of $v_1,\dots,v_m$ one has then

$$(H^a)_* g_i^a \circ G^a(t_1,\dots,t_n) = (H^b)_* g_i^b G^b(t_1,\dots,t_n)$$

for all $0 \leq i \leq m$. But these ones, in view of the definitions (6.5), may be rewritten as

$$g_i^b(x) = F_* g_i^a \circ F^{-1}(x) \qquad 0 \leq i \leq m$$

for all $x \in V^b$, thus proving (6.2)

(vi) Again, using the same arguments already used in (iii) one may easily see that

$$h^a \circ G^a(t_1,\dots,t_n) = h^b \circ G^b(t_1,\dots,t_n)$$

i.e. that

$$h^b(x) = h^a \cdot F^{-1}(x)$$

for all $x \in V^b$, thus proving also (6.3). □

CHAPTER IV

DISTURBANCE DECOUPLING AND NON INTERACTING CONTROL

1. Nonlinear Feedback and Controlled Invariant Distributions

In this and in the following chapters, we assume that in the control system

$$\dot{x} = f(x) + \sum_{i=1}^{m} g_i(x)u_i \tag{1.1}$$

it is possible to assign the values of the inputs $u_1,\ldots,u_m$ at each time t as functions of the value at t of the state x and, possibly, of some other real-valued functions $v_1,\ldots,v_m$. This control mode is called a static state-feedback control. In order to preserve the structure of (1.1), we let u_i depend on x and $v_1,\ldots,v_m$ in the following form

$$u_i = \alpha_i(x) + \sum_{j=1}^{m} \beta_{ij}(x)v_j \tag{1.2}$$

where $\alpha_i(x)$ and $\beta_{ij}(x)$, $1 \leq i,j \leq m$, are real-valued smooth functions defined on the same open subset N of $\mathbb{R}^n$ on which (1.1) is defined.

In doing this we modify the original dynamics (1.1) and obtain the control system

$$\dot{x} = \tilde{f}(x) + \sum_{i=1}^{m} \tilde{g}_i(x)v_i \tag{1.3}$$

in which

$$\tilde{f}(x) = f(x) + \sum_{i=1}^{m} g_i(x)\alpha_i(x) \tag{1.4a}$$

$$\tilde{g}_i(x) = \sum_{j=1}^{m} g_j(x)\beta_{ji}(x) \tag{1.4b}$$

For reasons of notational simplicity, most of the times we consider $\alpha_i(x)$ as the i-th entry of an m-dimensional vector $\alpha(x)$, $\beta_{ij}(x)$ as the (i,j)-th entry of an m×m-dimensional matrix $\beta(x)$ and we consider the vector fields $g_j(x)$ and $\tilde{g}_j(x)$ as j-th columns of n×m-dimensional matrices $g(x)$ and $\tilde{g}(x)$.

In this way we may replace (1.4) with the shorter expressions

(1.5a) $$\tilde{f}(x) = f(x) + g(x)\alpha(x)$$

(1.5b) $$\tilde{g}(x) = g(x)\beta(x)$$

We also systematically assume that the $m\times m$ matrix $\beta(x)$ is invertible for all x. This makes it possible to invert the transformation (1.5), and to obtain

(1.6a) $$f(x) = \tilde{f}(x) - \tilde{g}(x)\beta^{-1}(x)\alpha(x)$$

(1.6b) $$g(x) = \tilde{g}(x)\beta^{-1}(x)$$

(1.8) *Remark*. Strictly speaking, only (1.5a) may be regarded as a "feedback", while (1.5b) should be regarded as a change of coordinates in the space of input values, depending on x. □

The purpose for which feedback is introduced is to obtain a dynamics with some nice properties that the original dynamics does not have. As we shall see later on, a typical situation is the one in which a modification is required in order to obtain the invariance of a given distribution Δ under the vector fields which characterize the new dynamics. This kind of problem is usually dealt with in the following way.

A distribution Δ is said to be *controlled invariant* on N if there exists a feedback pair (α,β) defined on N with the property that Δ is invariant under the vector fields $\tilde{f},\tilde{g}_1,\ldots,\tilde{g}_m$ (see (1.4)), i.e. if

(1.9a) $$[\tilde{f},\Delta](x) \subset \Delta(x)$$

(1.9b) $$[\tilde{g}_i,\Delta](x) \subset \Delta(x) \quad \text{for} \quad 1 \leq i \leq m$$

for all $x \in N$.

A distribution Δ is said to be *locally controlled invariant* if for each $x \in N$ there exists a neighborhood U of x with the property that Δ is controlled invariant on U. In view of the previous definition, this requires the existence of a feedback pair (α,β) defined on U such that (1.9) is true for all $x \in U$.

The notion of local controlled invariance lends itself to a simple geometric test. If we set

$$G = sp\{g_1, \ldots, g_m\}$$

we may express the test in question in the following terms.

(1.10) *Lemma*. Let Δ be an involutive distribution. Suppose Δ, G and $\Delta+G$ are nonsingular on N. Then Δ is locally controlled invariant if and only if

$$(1.11a) \qquad [f,\Delta] \subset \Delta + G$$

$$(1.11b) \qquad [g_i,\Delta] \subset \Delta + G \quad \text{for} \quad 1 \leq i \leq m$$

Proof. Necessity. Suppose Δ is locally controlled invariant. Let $x \in N$, U a neighborhood of x and (α,β) a feedback pair defined on U which makes (1.9) satisfied on U. Let τ be any vector field of Δ. Then we have

$$[\tilde{f},\tau] = [f+g\alpha,\tau] = [f,\tau] + \sum_{j=1}^{m} [g_j,\tau]\alpha_j + \sum_{j=1}^{m} (L_\tau \alpha_j) g_j$$

$$[\tilde{g}_i,\tau] = \sum_{j=1}^{m} [g_j \beta_{ji},\tau] = \sum_{j=1}^{m} [g_j,\tau]\beta_{ji} + \sum_{j=1}^{m} (L_\tau \beta_{ji}) g_j$$

for $1 \leq i \leq m$.

Since β is invertible, one may solve the last m equalities for $[g_j,\tau]$, obtaining

$$[g_j,\tau] \in \sum_{i=1}^{m} [\tilde{g}_i,\Delta] + G$$

for $1 \leq j \leq m$. Therefore, from (1.9b) we deduce (1.11b). Moreover, since

$$[f,\tau] \in [\tilde{f},\Delta] + \sum_{i=1}^{m} [g_i,\Delta] + G$$

again from (1.9) and (1.11b) we deduce (1.11a). □

In order to prove the sufficiency, we first need the following interesting result, which is a consequence of Frobenius Theorem.

(1.12) *Theorem*. Let U and V be open sets in $\mathbb{R}^m$ and $\mathbb{R}^n$ respectively. Let $x_1,\ldots,x_m$ denote coordinates of a point x in $\mathbb{R}^m$ and $y_1,\ldots,y_n$ coordinates of a point y in $\mathbb{R}^n$. Let $\Gamma^1,\ldots,\Gamma^m$ be smooth functions

$$\Gamma^i : U \to \mathbb{R}^{n\times n}$$

Consider the set of partial differential equations

$$(1.13) \qquad \frac{\partial y(x)}{\partial x_i} = \Gamma^i(x)y(x) \qquad 1 \le i \le m$$

where y denotes a function

$$y : U \to V$$

Given a point $(x^o,y^o) \in U \times V$ there exist a neighborhood U_o of x in U and a unique smooth function

$$y : U_o \to V$$

which satisfies the equations (1.13) and is such that $y(x^o) = y^o$ if and only if the functions $\Gamma^1,\ldots,\Gamma^n$ satisfy the conditions

$$(1.14) \qquad \frac{\partial \Gamma^i}{\partial x_k} - \frac{\partial \Gamma^k}{\partial x_i} + \Gamma^i\Gamma^k - \Gamma^k\Gamma^i = 0 \qquad 1 \le i,k \le m$$

for all $x \in U$.

Proof. Necessity. Suppose that for all (x^o,y^o) there is a function y which satisfies (1.13). Then from the property

$$\frac{\partial^2 y}{\partial x_i \partial x_k} = \frac{\partial^2 y}{\partial x_k \partial x_i}$$

one has

$$\frac{\partial}{\partial x_i}(\Gamma^k(x)y(x)) = \frac{\partial}{\partial x_k}(\Gamma^i(x)y(x))$$

Expanding the derivatives on both sides and evaluating them at $x = x^o$ one obtains

$$[\,(\frac{\partial \Gamma^k}{\partial x_i})_{x^o} + \Gamma^k(x^o)\Gamma^i(x^o)]\,y^o = [\,(\frac{\partial \Gamma^i}{\partial x_k})_{x^o} + \Gamma^i(x^o)\Gamma^k(x^o)]\,y^o$$

which, due to arbitrariness of x^o,y^o, yields the condition (1.14).

Sufficiency. The proof of this part consists of the following steps.

(i) It is shown that the fulfillment of (1.14) enables us to define on $U \times V$ a certain involutive distribution Δ, of dimension m.

(ii) Using Frobenius Theorem, one can find a neighborhood $U' \times V'$ of

(x^o, y^o) and a local coordinates tranformation

$$F : (x,y) \mapsto \xi$$

defined on $U' \times V'$, with the property that

$$\Delta(x,y) = \text{span}\{(\frac{\partial}{\partial \xi_1})_{(x,y)}, \ldots, (\frac{\partial}{\partial \xi_m})_{(x,y)}\}$$

for all $(x,y) \in U' \times V'$

(iii) From the transformation F one constructs a solution of (1.13).

As for the step (i), the distribution Δ is defined, at each $(x,y) \in U \times V$, by

$$\Delta(x,y) = \text{span}\{(\frac{\partial}{\partial x_i}) + \sum_{h=0}^{n} \sum_{k=0}^{n} \Gamma^i_{hk}(x) y_k (\frac{\partial}{\partial y_h}) : 1 \leq i \leq m\}$$

In other words, $\Delta(x,y)$ is spanned by m tangent vectors whose coordinates with respect to the canonical basis $\{(\frac{\partial}{\partial x_1}), \ldots, (\frac{\partial}{\partial x_m}), (\frac{\partial}{\partial y_1}), \ldots, (\frac{\partial}{\partial y_n})\}$ of the tangent space to $U \times V$ at (x,y) have the form

$$\begin{pmatrix} 1 \\ 0 \\ \vdots \\ 0 \\ \Gamma^1(x)y \end{pmatrix}, \begin{pmatrix} 0 \\ 1 \\ \vdots \\ 0 \\ \Gamma^2(x)y \end{pmatrix}, \ldots, \begin{pmatrix} 0 \\ 0 \\ \vdots \\ 1 \\ \Gamma^m(x)y \end{pmatrix}$$

These m vectors are linearly independent at all (x,y) and so the distribution Δ is nonsingular and of dimension m. Moreover, it is an easy computation to check that if the "integrability" condition (1.14) is satisfied, then Δ is involutive.

The possibility of constructing the coordinate transformation described in (ii) is a straightforward consequence of Frobenius theorem. The function F thus defined is such that if v is a vector in Δ, the last n components of $F_* v$ are vanishing. Since, moreover, the tangent vectors $(\frac{\partial}{\partial y_1}), \ldots, (\frac{\partial}{\partial y_n})$ span a subspace which is complementary to $\Delta(x,y)$ at all (x,y) and F is nonsingular, one may easily conclude that the function

$$\xi = F(x,y)$$

is such that the jacobian matrix

$$(1.15) \qquad \begin{pmatrix} \frac{\partial \xi_{m+1}}{\partial y_1} & \cdots & \frac{\partial \xi_{m+1}}{\partial y_n} \\ & \cdots & \\ \frac{\partial \xi_{m+n}}{\partial y_1} & \cdots & \frac{\partial \xi_{m+n}}{\partial y_n} \end{pmatrix}$$

is nonsingular at all $(x,y) \in U' \times V'$.

Without loss of generality we may assume that

$$\xi_i(x^o,y^o) = 0$$

for all $m+1 \le i \le m+n$. As a consequence, the integral submanifold of Δ passing through (x^o,y^o) is defined by the set of equations

$$\xi_{m+i}(x,y) = 0 \qquad 1 \le i \le n$$

Since the matrix (1.15) is nonsingular, thanks to the implicit function theorem the above equations may be solved for y, yielding a set of functions

$$(1.16) \qquad y_i = \eta_i(x) \qquad 1 \le i \le n$$

defined in a neighborhood $U_o \subset U'$ of x^o. Moreover

$$\eta_i(x^o) = y_i^o \qquad 1 \le i \le n$$

The functions (1.16) satisfy the differential equations (1.13) and therefore, are the required solutions. As a matter of fact, the functions

$$\varphi_i(x,y) = y_i - \eta_i(x) \qquad 1 \le i \le n$$

are constant on the integral submanifold of Δ passing through (x^o,y^o) and, therefore, if v is a vector in Δ,

$$d\varphi_i v = 0 \qquad 1 \le i \le n$$

at all pairs $(x,\eta(x))$. These equations, taking for v each one of the m vectors used to define Δ, yield exactly

$$\frac{\partial \eta_i}{\partial x_j} = (\Gamma^j(x)\eta(x))_i \qquad 1 \leq i \leq n, \quad 1 \leq j \leq m. \square$$

Proof. (of Lemma 1.10). Sufficiency. Recall that, by assumption, Δ, G and $\Delta+G$ are nonsingular; let d denote the dimension of Δ and let

$$p = \dim G - \dim \Delta \cap G$$

Given any $x^o \in N$ it is possible to find a neighborhood U of x^o and an $m\times m$ nonsingular matrix B, whose (i,j)-th element b_{ij} is a smooth real-valued function defined on U, such that, for

$$\hat{g}_i = \sum_{j=1}^{m} g_j b_{ji} \qquad 1 \leq i \leq m$$

the following is true

$$sp\{\hat{g}_{p+1},\ldots,\hat{g}_m\} \subset \Delta$$

$$(1.17) \qquad (\Delta+G) = \Delta \oplus sp\{\hat{g}_1,\ldots,\hat{g}_p\}$$

The tangent vectors $\hat{g}_1(x),\ldots,\hat{g}_p(x)$ are clearly linearly independent at all $x \in U$.

Now, observe that if the assumption (1.11b) is satisfied, then also

$$(1.18) \qquad [\hat{g}_i,\Delta] \subset \Delta + G$$

and let $\tau_1,\ldots,\tau_d$ be a set of vector fields which locally span Δ around x^o. From (1.17) and (1.18) we deduce the existence of a unique set of smooth real-valued functions c_{ji}^k, defined locally around x^o, and a vector field $\delta_i^k \in \Delta$ defined locally around x^o such that

$$(1.11b') \qquad [\hat{g}_i,\tau_k] = \sum_{j=1}^{p} c_{ji}^k \hat{g}_j + \delta_i^k$$

for all $1 \leq i \leq m$ and $1 \leq k \leq d$. Using the same arguments and setting

$$\hat{g}_0 = f$$

from (1.11a) and (1.18) we deduce the existence of a unique set of real-valued smooth functions c_{j0}^k and a vector field $\delta_0^k \in \Delta$, defined locally around x^o, such that

$$(1.11a') \qquad [\hat{g}_0, \tau_k] = \sum_{j=1}^{p} c^k_{j0}\hat{g}_j + \delta^k_0$$

Now, suppose there exists a nonsingular $m\times m$ matrix $\hat{B}$, whose (i,j)-th element $\hat{b}_{ij}$ is a smooth real-valued function defined locally around x, such that

$$(1.19) \qquad -L_{\tau_k}\hat{b}_{hi} + \sum_{j=1}^{m} c^k_{hj}\hat{b}_{ji} = 0$$

for $1 \leq k \leq d$, $1 \leq h \leq p$, $1 \leq i \leq m$. Then, it is easy to see that

$$(1.20) \qquad [\sum_{h=1}^{m} \hat{g}_h\hat{b}_{hi}, \tau_k] \in \Delta$$

for $1 \leq i \leq m$, $1 \leq k \leq d$. For,

$$[\sum_{h=1}^{m} \hat{g}_h\hat{b}_{hi}, \tau_k] = -\sum_{h=1}^{m} (L_{\tau_k}\hat{b}_{hi})\hat{g}_h + \sum_{j=1}^{m} \hat{b}_{ji}[\hat{g}_j, \tau_k]$$

$$= -\sum_{h=1}^{p} (L_{\tau_k}\hat{b}_{hi})\hat{g}_h + \sum_{j=1}^{m} \hat{b}_{ji} \sum_{h=1}^{p} c^k_{hj}\hat{g}_h + \bar{\delta}^k_i = \bar{\delta}^k_i$$

where $\bar{\delta}^k_i$ is a vector field in Δ. Since $\tau_1,\ldots,\tau_k$ locally span Δ,(1.20) implies that

$$[\sum_{h=1}^{m} \hat{g}_h\hat{b}_{hi}, \Delta] \subset \Delta$$

Therefore, the matrix

$$\beta = B\hat{B}$$

is such that (1.9b) is satisfied.

Using similar arguments, one can see that if there exists an $m\times 1$ vector $\hat{a}$, whose i-th element $\hat{a}_i$ is a smooth real-valued function defined locally around x^o, such that

$$(1.21) \qquad -L_{\tau_k}\hat{a}_h + \sum_{j=1}^{m} c^k_{hj}\hat{a}_j + c^k_{h0} = 0$$

for $1 \leq k \leq d$, $1 \leq h \leq p$, then

$$(1.22) \qquad [\hat{g}_0 + \sum_{h=1}^{m} \hat{g}_h\hat{a}_h, \tau_k] \in \Delta$$

for $1 \leq k \leq d$. For,

$$[\hat{g}_0 + \sum_{h=1}^{m} \hat{g}_h \hat{a}_h, \tau_k] = -\sum_{h=1}^{p} (L_{\tau_k} \hat{a}_h)\hat{g}_h + \sum_{j=1}^{m} \hat{a}_j \sum_{h=1}^{p} c_{hj}^k \hat{g}_h + \sum_{h=1}^{p} c_{h0}^k \hat{g}_h + \bar{\delta}^k = \bar{\delta}^k$$

where $\bar{\delta}^k$ is a vector field in Δ. From this one deduces that the vector

$$\alpha = B\hat{a}$$

is such that (1.9a) is satisfied.

Thus, we have seen that the possibility of finding $\hat{B}$ and $\hat{a}$ which satisfy (1.19) and (1.21) enables us to construct a pair of feedback functions that makes (1.9) satisfied. In order to complete the proof, we have to show that (1.19) and (1.21) can be solved for $\hat{B}$ and $\hat{a}$.

Since Δ is nonsingular and involutive, we may assume, without loss of generality, that our choice of local coordinates is such that

$$\tau_k = \frac{\partial}{\partial x_k} \qquad 1 \leq k \leq d.$$

The equations (1.19) and (1.21) may be rewritten as a set of partial differential equations of the form (1.13) by simply setting

$$\Gamma^k = \begin{pmatrix} c_{11}^k & \cdots & c_{1m}^k & c_{10}^k \\ \cdot & \cdots & \cdot & \cdot \\ c_{p1}^k & \cdots & c_{pm}^k & c_{p0}^k \\ 0 & \cdots & 0 & 0 \\ \cdot & \cdots & \cdot & \cdot \\ 0 & \cdots & 0 & 0 \end{pmatrix} \qquad 1 \leq k \leq d$$

As a matter of fact, for each fixed i, the equations (1.19) correspond to an equation for the i-th column of $\hat{B}$, of the form

$$\text{(1.23)} \qquad \frac{\partial}{\partial x_k}\begin{pmatrix} \hat{b}_i \\ 0 \end{pmatrix} = \Gamma^k \begin{pmatrix} \hat{b}_i \\ 0 \end{pmatrix} \qquad 1 \leq k \leq d$$

(where $\hat{b}_i$ stands for the i-th column of $\hat{B}$) and the equations (1.21) correspond to

$$\text{(1.24)} \qquad \frac{\partial}{\partial x_k}\begin{pmatrix} \hat{a} \\ 1 \end{pmatrix} = \Gamma^k \begin{pmatrix} \hat{a} \\ 1 \end{pmatrix} \qquad 1 \leq k \leq d$$

Both these equations have exactly the form

$$(1.25)\qquad \frac{\partial y}{\partial x_k} = \Gamma^k y \qquad 1 \le k \le d$$

the unknown vector y being m+1 dimensional. Since now the functions Γ^k depend also on the coordinates $x_{d+1},\ldots,x_n$ (with respect to which no derivative of y is considered), in order to achieve uniqueness, the value of y must be specified, for a given $x_1^o,\ldots,x_d^o$, at each $x_{d+1},\ldots,x_n$. For consistency, the last component of the initial value of the solution sought for the equations (1.23) must be set equal to zero, whereas the last component of the initial value of the solution sought for the equation (1.24) must be set equal to 1. In addition, the first m components of the initial values of the solutions sought for each of the equations (1.23) must be columns of a nonsingular m×m matrix, in order to let $\hat{B}$ be nonsingular.

The solvability of an equation of the form (1.25) depends, as we have seen, on the fulfillment of the integrability conditions (1.14). This, in turn, is implied by (1.11). Consider the Jacobi identity

$$-[[\hat{g}_i,\tau_k],\tau_h]+[[\hat{g}_i,\tau_h],\tau_k] = [\hat{g}_i,[\tau_h,\tau_k]]$$

for any $0 \le i \le m$. Using for $[\hat{g}_i,\tau_k]$ and $[\hat{g}_i,\tau_h]$ the expressions given by (1.11a') or (1.11b') and taking $\tau_k = \frac{\partial}{\partial x_k}$, $\tau_h = \frac{\partial}{\partial x_h}$ one easily obtains

$$[\sum_{j=1}^{p} c_{ji}^k \hat{g}_j + \delta_i^k, \frac{\partial}{\partial x_h}] - [\sum_{j=1}^{p} c_{ji}^h \hat{g}_j + \delta_i^h, \frac{\partial}{\partial x_k}] = 0$$

This yields

$$-\sum_{j=1}^{p} \frac{\partial c_{ji}^k}{\partial x_h}\hat{g}_j + \sum_{j=1}^{p} c_{ji}^k (\sum_{\ell=1}^{p} c_{\ell j}^h \hat{g}_\ell + \delta_j^h) + [\delta_i^k, \frac{\partial}{\partial x_h}]$$

$$+\sum_{j=1}^{p} \frac{\partial c_{ji}^h}{\partial x_k}\hat{g}_j - \sum_{j=1}^{p} c_{ji}^h (\sum_{\ell=1}^{p} c_{\ell j}^k \hat{g}_\ell + \delta_j^k) - [\delta_i^h, \frac{\partial}{\partial x_k}] = 0$$

Now, recall that $\frac{\partial}{\partial x_h}$ and $\frac{\partial}{\partial x_k}$ are both vector fields of Δ, which is involutive. Therefore, also $[\delta_i^k, \frac{\partial}{\partial x_h}]$ and $[\delta_i^h, \frac{\partial}{\partial x_k}]$ are in Δ. Since Δ and $sp\{\hat{g}_1,\ldots,\hat{g}_p\}$ are direct summands and $\hat{g}_1,\ldots,\hat{g}_p$ are linearly independent, the previous equality implies

$$-\frac{\partial c_{ji}^k}{\partial x_h} + \frac{\partial c_{ji}^h}{\partial x_k} + \sum_{\ell=1}^{p} c_{j\ell}^h c_{\ell i}^k - \sum_{\ell=1}^{p} c_{j\ell}^k c_{\ell i}^h = 0$$

for $1 \leq j \leq p$, $0 \leq i \leq m$, $1 \leq h,k \leq d$, which is easily seen to be identical to the condition (1.14). □

We see from this Lemma that, under reasonable assumptions (namely, the nonsingularity of Δ, G and $\Delta+G$) an involutive distribution is locally controlled invariant if and only if the conditions (1.11) are satisfied. These conditions are of special interest because they don't invoke the existence of feedback functions α and β, as the definition does, but are expressed only in terms of the vector fields $f,g_1,\ldots,g_m$ which characterize the given control system and of the distribution itself. The fulfillment of conditions (1.11) implies the existence of a pair of feedback functions which make Δ invariant under the new dynamics but the actual construction of such a feedback pair generally involves the solution of a set of partial differential equations, as we have seen in the proof of Lemma (1.10). There are cases, however, in which the solution of partial differential equations may be avoided and these, luckyly enough, include some situations of great importance in control theory. These will be examined later on in this chapter.

2. The Disturbance Decoupling Problem

The notion of locally controlled invariance will now be used in order to solve the following control problem. Consider a control system

$$\dot{x} = f(x) + \sum_{i=1}^{m} g_i(x)u_i + p(x)w \tag{2.1a}$$

$$y = h(x) \tag{2.1b}$$

where the additional input w represents an undesired perturbation, which influences the behavior of the system through the vector field p. The system is to be modified, via static state-feedback control on the inputs $u_1,\ldots,u_m$, in such a way that the disturbance w has no influence on the output y.

In view of some earlier results (Theorem III.(3.12) and Remark III.(3.13)) this problem consists in finding a feedback pair (α,β) and a distribution Δ which is invariant under $\tilde{f} = f+g\alpha$ and $\tilde{g}_i = (g\beta)_i$, $1 \leq i \leq m$, contains the vector field p and is contained in $(\mathrm{sp}\{dh_j\})^{\perp}$ for all $1 \leq j \leq \ell$.

According to the terminology introduced in the previous section, a distribution Δ which is invariant under $\tilde{f} = f+g\alpha$ and $\tilde{g}_i = (g\beta)_i$,

$1 \leq i \leq m$, for some feedback (α,β) is *controlled invariant*. If we set

$$H = \bigcap_{j=1}^{\ell} (\mathrm{sp}\{dh_j\})^{\perp} = (\mathrm{sp}\{dh_1,\ldots,dh_\ell\})^{\perp}$$

we may express the problem in question in the following terms.

Disturbance decoupling problem. Find a distribution Δ which

(i) is controlled invariant

(ii) is such that $p \in \Delta \subset H$. □

As we have seen in the previous section, the notion of *local* controlled invariance is sometimes easier to deal with than (global) controlled invariance. This motivates the consideration of the following problem.

Local disturbance decoupling problem. Find a distribution Δ which

(i) is locally controlled invariant

(ii) is such that $p \in \Delta \subset H$.

(2.2) *Remark.* Note that the distribution Δ is not required to be nonsingular, neither involutive. However, nonsingularity and involutivity may be needed in order to construct the pair of feedback functions (α,β) which make it possible to implement the disturbance-decoupling control mode. This typically happens when one has found a distribution Δ which satisfies (ii) and, instead of (i), satisfies the condition

$$\text{(i')} \qquad [f,\Delta] \subset \Delta + G$$

$$[g_i,\Delta] \subset \Delta + G \qquad 1 \leq i \leq m$$

In this case, we know from Lemma (1.10) that nonsingularity of Δ, G and $\Delta+G$ helps in finding at least locally a pair of feedback functions (α,β) with the desired properties.

If Δ is nonsingular and involutive, invariant under $\tilde{f}$ and $\tilde{g}_i$, $1 \leq i \leq m$, and satisfies (ii), then it is known from the analysis developed in chapter I that there exist local coordinate transformations which put the closed-loop system into the form

$$\begin{aligned}
\dot{x}_1 &= \tilde{f}_1(x_1,x_2) + \sum_{i=1}^{m} \tilde{g}_{i1}(x_1,x_2)u_i + p_1(x_1,x_2)w \\
\text{(2.3)} \qquad \dot{x}_2 &= \tilde{f}_2(x_2) + \sum_{i=1}^{m} \tilde{g}_{i2}(x_2)u_i \\
y &= h(x_2)
\end{aligned}$$

Here, once again, one sees that the disturbance w has no influence on the output y. □

A systematic way to deal with the Disturbance Decoupling Problem is to examine first whether or not the family of all controlled invariant distributions contained in H has a "maximal" element (an element which contains all other members of the family). For, if this is true, then the problem is solved if and only if this maximal element contains the vector field p.

If, rather than controlled invariant distributions, we look at *locally* controlled invariant distributions, then the existence of such a maximal element may be shown under rather mild assumptions. To this end, we introduce a notation and an algorithm. Let $\mathbb{I}(f,g;K)$ denote the collection of all smooth distributions which are contained in a given distribution K and satisfy the conditions (1.11). In view of Lemma (1.10), the maximal element of $\mathbb{I}(f,g;K)$ is the natural candidate for the maximal locally controlled invariant distribution in K. As a matter of fact, the maximal element of $\mathbb{I}(f,g;K)$ may be found by means of the following algorithm.

(2.4) *Lemma* (Controlled Invariant Distribution Algorithm). Let

$$\Omega_0 = K^{\perp} \tag{2.5}$$
$$\Omega_k = \Omega_{k-1} + L_f(G^{\perp} \cap \Omega_{k-1}) + \sum_{i=1}^{m} L_{g_i}(G^{\perp} \cap \Omega_{k-1})$$

Suppose there exists an integer k^* such that $\Omega_{k^*} = \Omega_{k^*+1}$. Then $\Omega_k = \Omega_{k^*}$ for all $k > k^*$.

If $\Omega_{k^*} \cap G^{\perp}$ and $\Omega_{k^*}^{\perp}$ are smooth, then $\Omega_{k^*}^{\perp}$ is the maximal element of $\mathbb{I}(f,g;K)$.

Proof. The first part of the statement is a trivial consequence of the definitions. As for the other, note first that from the equality $\Omega_{k^*+1} = \Omega_{k^*}$ we deduce

$$L_{g_i}(G^{\perp} \cap \Omega_{k^*}) \subset \Omega_{k^*}$$

for $1 \leq i \leq m$ and also for $i = 0$ if we set $f = g_0$, as sometimes we did before. Let ω be a one-form in $G^{\perp} \cap \Omega_{k^*}$, and τ a vector field in $\Omega_{k^*}^{\perp}$. In the expression

$$\langle L_{g_i}\omega,\tau\rangle = L_{g_i}\langle\omega,\tau\rangle - \langle\omega,[g_i,\tau]\rangle$$

we have

$$\langle L_{g_i}\omega,\tau\rangle = 0$$

because $L_{g_i}\omega \in \Omega_{k*}$ and

$$\langle \omega,\tau\rangle = 0$$

because $\tau \in \Omega_{k*}^{\perp} + G$. Thus

$$\langle \omega,[g_i,\tau]\rangle = 0$$

Since $G^{\perp} \cap \Omega_{k*}$ is smooth by assumption, $[g_i,\tau]$ annihilates every covector in $G^{\perp} \cap \Omega_{k*}$, i.e.

$$[g_i,\tau] \in \Omega_{k*}^{\perp} + G$$

for $0 \leq 1 \leq m$. Thus, $\Omega_{k*}^{\perp}$ is a member of $\mathbb{I}(f,g;K)$. Let $\bar{\Delta}$ be any other element of this collection. We will prove that $\bar{\Delta} \subset \Omega_{k*}^{\perp}$. First of all, note that if ω is a one-form in $\bar{\Delta}^{\perp} \cap G^{\perp}$ and τ a vector field in $\bar{\Delta}$ we have

$$\langle L_{g_i}\omega,\tau\rangle = 0$$

so that (recall that $\bar{\Delta}$ is a smooth distribution)

$$L_{g_i}(\bar{\Delta}^{\perp} \cap G^{\perp}) \subset \bar{\Delta}^{\perp}$$

Suppose

$$\bar{\Delta}^{\perp} \supset \Omega_k$$

for some $k \geq 0$. Then

$$\Omega_{k+1} \subset \Omega_k + L_f(\bar{\Delta}^{\perp} \cap G^{\perp}) + \sum_{i=1}^{m} L_{g_i}(\bar{\Delta}^{\perp} \cap G^{\perp}) \subset \bar{\Delta}^{\perp}$$

Thus, since $\Omega_0 = K^{\perp} \subset \bar{\Delta}^{\perp}$, we deduce that

$$\bar{\Delta} \subset \Omega_{k*}^{\perp}$$

and $\Omega_{k*}^{\perp}$ is the maximal element of $\mathbb{I}(f,g;K)$. □

For convenience, we introduce a terminology which is useful to remind both the convergence of the sequence (2.5) in a finite number of stages and the dependence of its final element on the distribution K. We set

$$(2.6) \qquad J(K) = (\Omega_0+\Omega_1+\ldots+\Omega_k+\ldots)^{\perp}$$

and we say that $J(K)$ is *finitely computable* if there exists an integer k^* such that, in the sequence (2.5), $\Omega_{k^*} = \Omega_{k^*+1}$. If this is the case, then obviously $J(K) = \Omega_{k^*}^{\perp}$.

In the Lemma (2.4) we have seen that if $J(K)$ is finitely computable and if $J(K)^{\perp} \cap G^{\perp}$ and $J(K)$ are smooth, then $J(K)$ is the maximal element of $\mathbb{I}(f,g;K)$. In order to let this distribution be locally controlled invariant all we need are the assumptions of Lemma (1.10), as stated below.

(2.7) *Lemma.* Suppose $J(K)$ is finitely computable. Suppose K is an involutive distribution and G, $J(K)$, $J(K)+G$ are nonsingular. Then $J(K)$ is involutive and is the largest locally controlled invariant distribution contained in K.

Proof. First, observe that the assumption of nonsingularity on G, $J(K)$, $J(K)+G$ indeed implies the smoothness of $J(K)^{\perp} \cap G^{\perp}$ and $J(K)$. So, in view of Lemma (1.10) we need only to show that $J(K)$ is involutive.

For, let d denote the dimension of $J(K)$. At any point x^o one may find a neighborhood U of x^o and vector fields $\tau_1,\ldots,\tau_d$ such that

$$J(K) = sp\{\tau_1,\ldots,\tau_d\}$$

on U. Consider the distribution

$$D = sp\{\tau_i\colon 1 \le i \le d\} + sp\{[\tau_i,\tau_j]\colon 1 \le i,j \le d\}$$

and suppose, for the moment, that D is nonsingular on U. Then, every vector field τ in D can be expressed as the sum of a vector field τ' in $J(K)$ and a vector field τ'' of the form

$$\tau'' = \sum_{i=1}^{d}\sum_{j=1}^{d} c_{ij}[\tau_i,\tau_j]$$

where c_{ij}, $1 \le i,j \le d$, are smooth real-valued functions defined on U.

We want to show that

$$[g_k, D] \subset D + G$$

for all $0 \leq k \leq m$. In view of the above decomposition of any vector field τ in D, this amounts to show that

$$[g_k, [\tau_i, \tau_j]] \subset D + G$$

The expression of the vector field on the left-hand-side via Jacobi identity yields

$$[g_k, [\tau_i, \tau_j]] = [\tau_i, [g_k, \tau_j]] - [\tau_j, [g_k, \tau_i]]$$

The vector field $[g_k, \tau_j]$ is in $J(K) + G$ and therefore, because of the nonsingularity of $J(K)$ and $J(K) + G$, it can be written as the sum of a vector field τ in $J(K)$ and a vector field g in G. Since, $[\tau_i, g] \in J(K) + G$ for any $g \in G$, we have

$$[\tau_i, [g_k, \tau_j]] = [\tau_i, \tau + g] \in D + J(K) + G = D + G$$

and we conclude that D is such that

$$[g_k, D] \subset D + G$$

for all $0 \leq k \leq m$.

Now, recall that K is involutive by assumption, and therefore that

$$D \subset K$$

From this and from the previous inclusions we deduce that D is an element of $\mathbb{I}(f,g;K)$. Since $D \supset J(K)$ by construction and $J(K)$ is the maximal element of $\mathbb{I}(f,g;K)$, we see that

$$D = J(K)$$

Thus, any Lie bracket of vector fields of $J(K)$, which is in D by construction, is still in $J(K)$ and the latter is an involutive distribution.

If we drop the assumption that D has constant dimension on U, we can still conclude that D coincides with $J(K)$ on the subset $\bar{U} \subset U$ consisting of all regular points of D. Then, using Lemma I.(2.11), we can as well prove that $D = J(K)$ on the whole of U. □

In the Local Disturbance Decoupling Problem one is interested in the largest locally controlled invariant distribution contained in H. Since this latter is involutive (see chapter I), in order to be able to use the previous Lemma, we need to assume that the distribution $J(H)$ is finitely computable and that G, $J(H)$, $J(H)+G$ are nonsingular. If this is the case, then, as we said before, the Local Disturbance Decoupling Problem is solvable if and only if

$$p \in J(H)$$

We conclude the section with a remark about the invariance of the algorithm (2.5) under feedback transformation.

(2.8) *Lemma*. Let $\tilde{f},\tilde{g}_1,\ldots,\tilde{g}_m$ be any set of vector fields deduced from $f,g_1,\ldots,g_m$ by setting $\tilde{f} = f + g\alpha$, $\tilde{g}_i = (g\beta)_i$, $1 \le i \le m$; then each codistribution Ω_k of the sequence (2.5) is such that

$$\Omega_k = \Omega_{k-1} + L_{\tilde{f}}(G^{\perp} \cap \Omega_{k-1}) + \sum_{i=1}^{m} L_{\tilde{g}_i}(G^{\perp} \cap \Omega_{k-1})$$

Proof. Recall that, given a covector field ω, a vector field τ and a scalar function γ,

$$L_{(\tau\gamma)}\omega = (L_{\tau}\omega)\gamma + \langle \omega,\tau \rangle d\gamma$$

If ω is a covector field in $G^{\perp} \cap \Omega_{k-1}$, then

$$L_{\tilde{f}}\omega = L_f\omega + \sum_{i=1}^{m} (L_{g_i}\omega)\alpha_i + \sum_{i=1}^{m} \langle \omega,g_i \rangle d\alpha_i$$

$$L_{\tilde{g}_i}\omega = \sum_{j=1}^{m} (L_{g_j}\omega)\beta_{ji} + \sum_{j=1}^{m} \langle \omega,g_j \rangle d\beta_{ji}$$

But $\langle \omega,g_j \rangle = 0$ because $\omega \in G^{\perp}$ and therefore

$$L_{\tilde{f}}(G^{\perp} \cap \Omega_{k-1}) + \sum_{i=1}^{m} L_{\tilde{g}_i}(G^{\perp} \cap \Omega_{k-1}) \subset L_f(G^{\perp} \cap \Omega_{k-1}) + \sum_{i=1}^{m} L_{g_i}(G^{\perp} \cap \Omega_{k-1})$$

Since β is invertible, one may also write $f = \tilde{f} - \tilde{g}\beta^{-1}\alpha$ and $g_i = (\tilde{g}\beta^{-1})_i$ and, using the same arguments, prove the reverse inclusion. The two sides of inclusion are thus equal and the Lemma is proved. □

3. Some Useful Algorithms

In this section we describe a practical implementation of the algorithm yielding the largest locally controlled invariant distribution contained in H. Moreover, we show that in some particular cases the construction of this distribution may be obtained with simpler methods.

We begin with the easiest situation, first. For each output function $h_i(x)$ we define an integer ρ_i , called the *characteristic number* of y_i , as the integer identified by the conditions

$$(3.1a) \qquad L_{g_j} L_f^k h_i(x) = 0$$

for all $k < \rho_i$, all $1 \leq j \leq m$, all $x \in N$ and

$$(3.1b) \qquad L_{g_j} L_f^{\rho_i} h_i(x) \neq 0$$

for some j and x.

Note that if for some output y_i the characteristic number is not defined (i.e. (3.1a) holds for all k, all j and all x), then the output y_j is in no way affected by any of the inputs $u_1,\dots,u_m$. The expansions described in chapter III show that if this is the case

$$y_i(t) = \sum_{k=0}^{\infty} L_f^k h_i(x^o)\frac{t^k}{k!} = h_i(\Phi_t^f(x^o))$$

Thus, it seems reasonable to assume that our control system is such that the characteristic numbers are defined for each output.

Once the characteristic numbers are known, we may define an $\ell\times m$ matrix $A(x)$ whose element $a_{ij}(x)$ on the i-th row and j-th column is

$$(3.2) \qquad a_{ij}(x) = L_{g_j} L_f^{\rho_i} h_i(x) = \langle dL_f^{\rho_i} h_i(x), g_j(x)\rangle$$

and an ℓ-vector $b(x)$ whose element $b_i(x)$ on the i-th row is

$$(3.3) \qquad b_i(x) = L_f^{\rho_i+1} h_i(x) = \langle dL_f^{\rho_i} h_i(x), f(x)\rangle$$

We point out first of all an interesting property of the objects defined so far

(3.4) *Lemma*. Let (α,β) be any pair of feedback functions and let $\tilde{f} = f + g\alpha$, $\tilde{g}_i = (g\beta)_i$. Then

$$L_{\tilde{f}}^{k}h_i(x) = L_f^k h_i(x)$$

for all $k \leq \rho_i$ and all $x \in N$. Moreover, let $\tilde{A}(x)$ be the $\ell \times m$ matrix whose (i,j)-th element a_{ij} is

$$\tilde{a}_{ij}(x) = L_{\tilde{g}_j} L_{\tilde{f}}^{\rho_i} h_i(x)$$

and $\tilde{b}(x)$ the ℓ-vector whose i-th element b_i is

$$\tilde{b}_i(x) = L_{\tilde{f}}^{\rho_i+1} h_i(x)$$

Then

$$\tilde{A}(x) = A(x)\beta(x)$$

$$\tilde{b}(x) = A(x)\alpha(x)+b(x)$$

Proof. The first equality is easily proved by induction. It is true for $k = 0$ and, if true for some $0 < k < \rho_i$, yields

$$L_{\tilde{f}}^{k+1}h_i(x) = L_{\tilde{f}}L_{\tilde{f}}^{k}h_i(x) = L_f^{k+1}h_i(x) + \sum_{j=1}^{m} L_{g_j} L_f^k h_i(x)\alpha_j(x) = L_f^{k+1}h_i(x)$$

The other equalities are straightforward consequences of the first one.

(3.5) *Remark*. Note that the invertibility of β implies the invariance of the integers $\rho_1,\ldots,\rho_\ell$ as well as that of rank of $A(x)$ under feedback transformations. □

From this one can deduce the following interesting result.

(3.6) *Lemma*. Every locally controlled invariant distribution contained in H is also contained in the distribution Δ_{sup} defined by

(3.7) $$\Delta_{sup} = \bigcap_{i=1}^{\ell} \bigcap_{k=0}^{\rho_i} (sp\{dL_f^k h_i\})^{\perp}$$

Suppose Δ_{sup} is a smooth distribution. A pair of feedback functions (α,β) is such that

(3.8a) $$[f + g\alpha, \Delta_{sup}] \subset \Delta_{sup}$$

(3.8b) $$[(g\beta)_i, \Delta_{sup}] \subset \Delta_{sup} \qquad 1 \leq i \leq m$$

if and only if the differentials of each entry of the column vector $A(x)\alpha(x) + b(x)$ and those of each entry of the matrix $A(x)\beta(x)$ belong to the codistribution $\Delta_{sup}^{\perp}$.

Proof. Let Δ be a locally controlled invariant distribution contained in H. Then, by definition, $\Delta \subset (sp\{dh_i\})^{\perp}$ for all $1 \leq i \leq \ell$. Moreover, for some local feedback α, $[\tilde{f},\Delta] \subset \Delta$. Suppose $\Delta \subset (sp\{dL_f^k h_i\})^{\perp}$ for some $k < \rho_i$; then using Lemma (3.4) we have for any vector field $\tau \in \Delta$

$$0 = \langle dL_f^k h_i, [\tilde{f},\tau]\rangle = L_{\tilde{f}}\langle dL_f^k h_i, \tau \rangle - \langle dL_{\tilde{f}} L_f^k h_i, \tau \rangle = \langle dL_f^{k+1} h_i, \tau \rangle$$

i.e. $\Delta \subset (sp\{dL_f^{k+1} h_i\})^{\perp}$. This proves that

$$\Delta \subset \bigcap_{i=1}^{\ell} \bigcap_{k=0}^{\rho_i} (sp\{dL_f^k h_i\})^{\perp}$$

and therefore the distribution (3.7) contains every locally controlled invariant distribution.

Now, suppose there exists a pair of feedback functions that makes (3.8) satisfied. Let τ be a vector field in Δ_{sup}. Then

$$\langle dL_f^k h_i, \tau \rangle = 0 \tag{3.9a}$$

$$\langle dL_f^k h_i, [\tilde{f},\tau]\rangle = 0 \tag{3.9b}$$

$$\langle dL_f^k h_i, [\tilde{g}_j,\tau]\rangle = 0 \tag{3.9c}$$

for all $1 \leq i \leq \ell$, $0 \leq k \leq \rho_i$, $1 \leq j \leq m$. From (3.9b) written for $k = \rho_i$, we deduce, using Lemma (3.4),

$$0 = L_{\tilde{f}}\langle dL_f^{\rho_i} h_i, \tau \rangle - \langle dL_{\tilde{f}} L_f^{\rho_i} h_i, \tau \rangle = \langle dL_{\tilde{f}}^{\rho_i+1} h_i, \tau \rangle = \langle d\tilde{b}_i, \tau \rangle$$

Similarly , for (3.9c) written for $k = \rho_i$ we deduce that

$$0 = \langle d\tilde{a}_{ij}, \tau \rangle$$

Therefore, the differentials of $\tilde{b}_i$ and $\tilde{a}_{ij}$ belong to the codistribution $\Delta_{sup}^{\perp}$. Conversely, if the differentials of $\tilde{b}_i$ and $\tilde{a}_{ij}$ belong to the codistribution $\Delta_{sup}^{\perp}$, we have that (3.9b) and (3.9c) hold for $k = \rho_i$. For values $k < \rho_i$ (3.9b) and (3.9c) hold for any feedback (α,β) because of Lemma (3.4) and, therefore, we deduce that Δ_{sup} is invariant under $\tilde{f}$ and $\tilde{g}_i$. □

From this result we see that there are cases in which the computation of the largest controlled invariant distribution contained in H is not terribly difficult. An interesting special case is the one in which the matrix A(x) has a rank equal to the number of its rows (i.e. the number of the output channels); this is explained in the following results.

(3.10) *Lemma*. Suppose that the matrix A(x) has rank ℓ at x^o. Then the covectors

$$dh_1(x^o),\dots,dL_f^{\rho_1}h_1(x^o),\dots,dh_\ell(x^o),\dots,dL_f^{\rho_\ell}h_\ell(x^o)$$

are linearly independent. As a consequence, the distribution Δ_{sup} is nonsingular in a neighborhood U of x^o and

$$\dim \Delta^{\perp}_{sup}(x) = \rho_1+\dots+\rho_\ell + \ell \le n \tag{3.11}$$

Proof. Suppose that the differentials are linearly dependent at x^o. Then there exist real numbers c_{ik}, $1 \le i \le \ell$, $0 \le k \le \rho_i$ such that

$$\sum_{i=1}^{\ell}\sum_{k=0}^{\rho_i} c_{ik}dL_f^k h_i(x^o) = 0 \tag{3.12}$$

Now consider the function

$$\lambda(x) = \sum_{i=0}^{\ell}\sum_{k=0}^{\rho_i} c_{ik}L_f^k h_i(x)$$

According to the definition of $\rho_1,\dots,\rho_m$, this function is such that

$$\langle d\lambda,g_j\rangle(x) = \sum_{i=1}^{\ell} c_{i\rho_i}\langle dL_f^{\rho_i}h_i,g_j\rangle(x) = \sum_{i=1}^{\ell} c_{i\rho_i}a_{ij}(x)$$

But, on the other hand, (3.12) shows that $d\lambda(x^o) = 0$ and therefore the above equality implies the linear dependence of the rows of the matrix $A(x^o)$, i.e. a contradiction. Therefore we conclude that if (3.12) holds, we must have $c_{1\rho_1} = \dots = c_{\ell\rho_\ell} = 0$.

Now consider the function

$$\gamma(x) = \sum_{i=0}^{\ell}\sum_{k=0}^{\rho_i-1} c_{ik}L_f^k h_i(x)$$

(with the understanding that the above sum is exendend over all non-negative k's) and observe that, if $0 \le k \le \rho_i-1$, then(*)

$$-\langle dL_f^k h_i,[f,g_j]\rangle = \langle dL_f^{k+1} h_i,g_j\rangle$$

Now, by the definition of $\rho_1,\ldots,\rho_m$ and from this formula, we have

$$\langle d\gamma,[f,g_j]\rangle(x) = -\sum_{i=0}^{\ell}\sum_{k=0}^{\rho_i-1} c_{ik}\langle dL_f^{k+1},g_j\rangle = -\sum_{i=1}^{\ell} c_{i,\rho_i-1}a_{ij}(x)$$

But since in the (3.12) the coefficients $c_{1\rho_1},\ldots,c_{\ell\rho_\ell}$ have already been proved being equal to 0, the function $\gamma(x)$ is such that $d\gamma(x^o)=0$ and the above equality implies again the linear dependence of the rows of the matrix $A(x^o)$, i.e. a contradiction. Therefore $c_{1,\rho_1-1} = \ldots = c_{\ell,\rho_\ell-1} = 0$ (for all c_{i,ρ_i-1} defined, i.e. such that $\rho_i \ge 1$).

By repeating the procedure one completes the proof.

(3.13) *Remark*. As a consequence of this Lemma, if the matrix $A(x^o)$ has rank ℓ, the functions $L_f^k h_i(x)$, $1 \le i \le \ell$, $0 \le k \le \rho_i$ are part of a coordinate system in a neighborhood U of x^o. This fact will be extensively used in the sequel. □

The assumption on the rank of A(x) identifies a special case in which the computation of the largest controlled invariant distribution contained in H is particularly simple.

(3.14) *Corollary*. Suppose the matrix A(x) has rank ℓ at x^o. Then in a neighborhood U of x^o the distribution Δ_{sup} coincides with the largest locally controlled invariant distribution contained in H.

Proof. If A(x) has rank ℓ at x^o, in a neighborhood U' of x^o the distribution Δ_{sup} is nonsingular and therefore smooth. Moreover, in a neighborhood $U \subset U'$ of x^o the equations

$$A(x)\alpha(x) + b(x) = \gamma(x) \tag{3.15a}$$

$$A(x)\beta(x) = \delta(x) \tag{3.15b}$$

where $\gamma(x)$ and $\delta(x)$ are an arbitrary ℓ-vector and respectively an

---

(*)

$$-\langle dL_f^k h_i,[f,g_j]\rangle = \langle dL_f^{k+1} h_i,g_j\rangle - L_f\langle dL_f^k h_i,g_j\rangle$$

and the last term is zero because $k \le \rho_i-1$.

arbitrary $\ell \times m$ matrix, have smooth solutions. If the entries of γ and δ are such that their differentials belong to $\Delta_{sup}^{\perp}$, then the feedback (α,β) is such that (3.8) are satisfied on U. In particular this is true if the entries of α and β are constants. Note that the matrix δ must have rank ℓ in order to let β be nonsingular.

(3.16) *Remark*. Recall that any pair of feedback functions α and β which makes Δ_{sup} invariant is a solution of (3.15), provided that γ and δ have entries with differentials in $\Delta_{sup}^{\perp}$ (see Lemma (3.6)). □

The procedures outlined so far are not always usable, because A(x) may fail to have rank ℓ or, more in general, Δ_{sup} may not be a locally controlled invariant distribution. In this case one may still use the general algorithm (2.4). A practical implementation of this algorithm can be obtained in the following way.

(3.17) *Algorithm* (Construction of the largest locally controlled invariant distribution contained in H).

Suppose that in a neighborhood of the point x^{o} the codistribution $sp\{dh_1,\ldots,dh_{\ell}\}$ has constant dimension, say s_0. Let $\lambda_0(x)$ be an s_0-vector whose entries $\lambda_{01},\ldots,\lambda_{0s_0}$ are entries of h, with the property that $d\lambda_{01},\ldots,d\lambda_{0s_0}$ are linearly independent at all x in a neighborhood of x^{o}.

The algorithm consists of a finite number of iterations, each one defined as follows.

Iteration (k). Consider the $s_k \times m$ matrix $A_k(x)$ whose (i,j)-th entry is $\langle d\lambda_{ki}(x), g_j(x)\rangle$. Suppose that in a neighborhood U_k of the point x^{o} the rank of $A_k(x)$ is constant and equal to r_k. Then it is possible to find r_k rows of $A_k(x)$ which, for all x in a neighborhood $U_k' \subset U_k$ of x^{o}, are linearly independent. Let

$$P_k = \begin{pmatrix} P_{k1} \\ P_{k2} \end{pmatrix}$$

be a $s_k \times s_k$ permutation matrix, chosen in such a way that the r_k rows of $P_{k1}A_k(x)$ are linearly independent at all $x \in U_k'$. Let $B_k(x)$ be an s_k-vector whose i-th element is $\langle d\lambda_{ki}, f\rangle(x)$. As a consequence of previous positions, the equations

$$P_{k1}A_k(x)\alpha(x) = -P_{k1}B_k(x) \tag{3.18a}$$

(3.18b) $$P_{k1}A_k(x)\beta(x) = K$$

(where K is a matrix of real numbers, of rank r_k) may be solved for α and β, an m-vector and an $m\times m$ invertible matrix whose entries are real-valued smooth functions defined in a neighborhood U_k'' of x^o.

Set $\tilde{g}_0 = f + g\alpha$ and $\tilde{g}_i = (g\beta)_i$, $1 \leq i \leq m$.

Consider the set of functions

$$\Lambda_k = \{\lambda = L_{\tilde{g}_i}\lambda_{kj} : 1 \leq j \leq s_k , 0 \leq i \leq m\}$$

and the codistributions

$$\Omega_{k1} = \sum_{j=1}^{s_k} sp\{d\lambda_{kj}\}$$

$$\Omega_{k2} = sp\{d\lambda: \lambda \in \Lambda_k\}$$

Suppose the codistribution $\Omega_{k1} + \Omega_{k2}$ has constant dimension, say s_{k+1} , in a neighborhood $U_k''' \subset U_k''$ of x^o. This integer s_{k+1} is necessarily larger than or equal to r_k because the r_k entries of $P_{k1}\lambda_k$ have linearly independent differentials at all $x \in U_k'$, otherwise $A_k(x)$ would not have rank r_k. Let $\lambda_{k+1,1},\ldots,\lambda_{k+1,s_{k+1}}$ be entries of λ_k and/or elements of Λ_k with the property that the differentials $d\lambda_{k+1,1},\ldots,d\lambda_{k+1,s_{k+1}}$ are linearly independent at all x in neighborhood $U_k'''' \subset U_k'''$ of x^o. Thus

$$\Omega_{k1} + \Omega_{k2} = \sum_{j=1}^{s_{k+1}} sp\{d\lambda_{k+1,j}\}$$

Define the s_{k+1}-vector λ_{k+1} whose i-th entry is the function $\lambda_{k+1,i}$.

This concludes the description of the algorithm. □

As a matter of fact, it is possible to show that the operations thus described are exactly the ones required in order to compute the codistribution Ω_k from codistribution Ω_{k-1} and therefore that, under suitable assumptions, the algorithm ends at a certain stage, yielding the required distribution. Since the possibility of completing the operations defined at the k-th stage depends on assumptions on the rank of A_k and on the dimension of $\Omega_{k1} + \Omega_{k2}$, we set for convenience all these assumptions in a suitable definition. We say that x^o is a *regular point* for the algorithm (3.17) if, for all $k \geq 0$, the matrix

A_k has constant rank in a neighborhood of x^o and the codistribution $\Omega_{k1} + \Omega_{k2}$ has constant dimension in a neighborhood of x.

In this case r_k, the rank of A_k, and s_{k+1}, the dimension of $\Omega_{k1} + \Omega_{k2}$ are well-defined quantities in a neighborhood of x^o. Note, however, that around a regular point x^1 other than x^o, r_k and s_{k+1} might be different.

The following statement shows that the algorithm in question provides the largest locally controlled invariant distribution contained in H.

(3.19) *Proposition.* Suppose x^o is a regular point for the algorithm (3.17). Then, there exists an integer k^* with the property that $s_{k^*+1} = s_{k^*}$ and, therefore, the algorithm terminates at the (k^*)-th iteration. Suppose also G is nonsingular. Then on a suitable neighborhood U of x^o distribution

$$\Delta^* = \bigcap_{i=0}^{s_{k^*}} (\mathrm{sp}\{d\lambda_{k^*,i}\})^{\perp}$$

coincides with the largest locally controlled invariant distribution contained in H. The pair of feedback functions that solve (3.18) for $k = k^*$ is such that

$$[f + g\alpha, \Delta^*] \subset \Delta^*$$

$$[(g\beta)_i, \Delta^*] \subset \Delta^* \qquad 1 \leq i \leq m$$

Proof. We shall prove by induction that the assumptions of Lemma (2.7) are satisfied and that

$$\Omega_k = \sum_{j=1}^{s_k} \mathrm{sp}\{d\lambda_{kj}\}$$

This is true for $k = 0$, by definition.

Suppose it is true for some k. To compute Ω_{k+1} we need to compute first $\Omega_k \cap G^{\perp}$. Note that Ω_k is nonsingular around x^o because the differentials $d\lambda_{kj}$, $1 \leq j \leq s_k$, are linearly independent at all $x \in U_k''''$. The intersection $\Omega_k \cap G^{\perp}$ at x is defined as the set of all linear combinations of the form

$$\sum_{i=1}^{s_k} c_i d\lambda_{kj}(x)$$

which annihilates $g_1(x),\dots,g_m(x)$. Therefore, it is easily seen that the coefficients $c_1,\dots,c_{s_k}$ of this combination must be solutions of the equation

$$(c_1 \dots c_{s_k})A_k(x) = 0$$

Since $A_k(x)$ has constant rank r_k in a neighborhood of x^o, $\Omega_k \cap G^{\perp}$ is nonsingular around x^o, has dimension $s_k - r_k$ and is spanned by covector fields which may be expressed as

$$\omega = (\gamma_1(x)P_{k2} + \gamma_2(x)P_{k1})d\lambda_k \tag{3.20}$$

$\gamma_1(x)$ being an arbitrary (s_k-r_k)-row vector of smooth functions. With $d\lambda_k$ we denote an s_k-column whose i-th entry is the covector field $d\lambda_{ki}$.

In computing Ω_{k+1} , we make also use of the fact that, if (α,β) is any feedback pair, then (see Lemma (2.8))

$$\Omega_k + \sum_{i=0}^{m} L_{g_i}(\Omega_k \cap G^{\perp}) = \Omega_k + \sum_{i=0}^{m} L_{\tilde{g}_i}(\Omega_k \cap G^{\perp})$$

Now, take the Lie derivative of (3.20) along $\tilde{g}_i$, with α and β solutions of (3.18). As a result one obtains

$$L_{\tilde{g}_i}\omega = ((L_{\tilde{g}_i}\gamma_1)P_{k2} + (L_{\tilde{g}_i}\gamma_2)P_{k1})d\lambda_k + \gamma_1 P_{k2} dL_{\tilde{g}_i}\lambda_k + \gamma_2 P_{k1} dL_{\tilde{g}_i}\lambda_k$$

But the way the $\tilde{g}_i$ are defined is such that

$$P_{k1}L_{\tilde{g}_0}\lambda_k = P_{k1}\langle d\lambda_k, \tilde{g}_0\rangle = 0$$

$$P_{k1}L_{\tilde{g}_i}\lambda_k = P_{k1}\langle d\lambda_k, \tilde{g}_i\rangle = \text{constant}$$

for all $1 \leq i \leq m$. Thus, in the above expression we may replace γ_2 with any arbitrary r_k-row vector $\bar{\gamma}_2$ of smooth functions. This makes it possible to express $L_{\tilde{g}_i}\omega$ in the form

$$L_{\tilde{g}_i}\omega = \gamma_3 d\lambda_k + \gamma_4 dL_{\tilde{g}_i}\lambda_k \tag{3.21}$$

where γ_3 is some s_k-row vector and γ_4 is an arbitrary s_k-row vector of smooth functions. The first term of this sum is already an element of Ω_k, by assumption, while the second, due to the arbitrariness of γ_4, spans the codistribution Ω_{k2} (see above). Thus, we may conclude that

$$\Omega_{k+1} = \Omega_k + \sum_{i=0}^{m} L_{\tilde{g}_i}(\Omega_k \cap G^\perp) = \Omega_k + \Omega_{k2} = \Omega_{k1} + \Omega_{k2} = \sum_{j=1}^{s_{k+1}} sp\{d\lambda_{k+1,j}\}$$

By assumption, the codistributions Ω_k, $k \geq 0$, are nonsingular around x^o (their dimension is s_k). Thus, there exists an integer k^* such that

$$\Omega_{k^*} = \Omega_{k^*+1}$$

This clearly implies the termination of the algorithm at the k^*-th step. We have also assumed that $\Omega_k \cap G^\perp$ are nonsingular around x^o (their dimension is r_k). So in particular $\Omega_{k^*}^\perp + G$ is nonsingular. If G is also nonsingular all the assumptions of Lemma (2.7) are satisfied and $\Omega_{k^*}^\perp$ is the required distribution.

In order to complete the proof, we have to show that the feedback pair which solves (3.18) for $k = k^*$ is such as to make $\Omega_{k^*}^\perp$ invariant under the new dynamics. To this end, consider again the expression (3.21) of the Lie derivative along $\tilde{g}_i$ of a covector field ω of $\Omega_k \cap G^\perp$. If the algorithm terminates at k^*, then

$$L_{\tilde{g}_i}(\Omega_{k^*} \cap G^\perp) \subset \Omega_{k^*}$$

and, therefore, we see from (3.21) that every entry of $dL_{\tilde{g}_i}\lambda_{k^*}$ (due to the arbitraryness of γ_4) is a covector field of Ω_{k^*}. But, since the entries of $d\lambda_{k^*}$ span Ω_{k^*}, this implies

$$L_{\tilde{g}_i}\Omega_{k^*} \subset \Omega_{k^*} \qquad 0 \leq i \leq m$$

and thus Ω_{k^*} is invariant under $\tilde{g}_0, \tilde{g}_1, \ldots, \tilde{g}_m$. Ω_{k^*} being nonsingular and therefore smooth, we may conclude that $\Omega_{k^*}^\perp$ is invariant under the new dynamics. □

This result is very important because it shows that, under suitable regularity assumptions, it is possible to find the largest locally controlled invariant distribution contained in H, and also a (locally defined) feedback pair α and β which makes it invariant under the new dynamics. The latter is particularly useful because we see that, as far as one is concerned with the maximal locally controlled invariant distribution contained in H, the computation of a such a feedback pair does not require solving partial differential equations (like in the general case, as seen from Lemma (1.10)) but may be carried out essentially solving x-dependent linear algebraic equations.

We conclude the section with some additional considerations about the properties of the algorithm (3.17). It is observed that, if the algorithm may be carried out until its final stage (i.e. if x^o is a regular point for the algorithm), as a by-product one obtains, for all $k \geq 0$, not only the dimension s_k of each codistribution Ω_k of the sequence (2.5) but also the dimension $s_k - r_k$ of the codistribution $\Omega_k \cap G^{\perp}$.

Thus the rank r_k of A_k may be interpreted as

$$r_k = \dim \frac{\Omega_k}{\Omega_k \cap G^{\perp}} \tag{3.22}$$

The integers $r_0, r_1, \ldots, r_{k^*}$ are rather important also for reasons not directly related to the construction of the distribution Δ^*. We will see in the next chapter, for instance, that the sequence of integers defined by setting

$$\begin{aligned} \delta_1 &= r_0 \\ \delta_2 &= r_1 - r_0 \\ &\cdots \\ \delta_{k^*+1} &= r_{k^*} - r_{k^*-1} \end{aligned} \tag{3.23}$$

may be, in a special case, directly evaluated starting from the coefficients of the functional expansion of the input-output behavior and plays an essential role in the problem of matching linear models.

It is also possible to relate the integers r_i, $0 \leq i \leq k^*$, to the characteristic numbers ρ_i, $1 \leq i \leq \ell$, as stated below.

(3.24) *Proposition.* Suppose that the outputs have been renumbered in such a way that the sequence of the characteristic numbers $\rho_1, \ldots, \rho_\ell$ is increasing. Let x^o be a regular point for the Algorithm (3.17). If rank $A(x^o) = \ell$, then the integers $\delta_1, \ldots, \delta_{k^*+1}$ defined by (3.23) are such that δ_i is equal to the number of outputs whose characteristic number is $(i-1)$ and, moreover, $\delta_1+\ldots+\delta_{k^*+1} = \ell$. □

The proof of this proposition is left as an exercise to the reader.

4. Noninteracting Control

Consider again a control system of the form

$$\dot{x} = f(x) + \sum_{i=1}^{m} g_i(x)u_i$$

$$y_i = h_i(x) \qquad 1 \leq i \leq \ell$$

and suppose $\ell \leq m$.

It is required to modify the system, via static state-feedback control, in order to obtain a closed loop system

$$\dot{x} = \tilde{f}(x) + \sum_{i=1}^{m} \tilde{g}_i(x)v_i$$

$$y_i = h_i(x) \qquad 1 \leq i \leq \ell$$

in which, for some suitable partition of the inputs $v_1,\ldots,v_m$ into ℓ disjoint sets, the i-th output is influenced only by the i-th set of inputs.

This control problem may easily be dealt with on the basis of the results discussed in chapter III (Theorem III.(3.12)) and its solution has interesting connections with the analysis developed so far in this chapter. In the present case, in order to ensure the independence of y_i from a set of inputs $v_{j_1},\ldots,v_{j_k}$ we have to find a distribution Δ_i which is invariant under $\tilde{f}$ and $\tilde{g}_j$, $1 \leq j \leq m$, is contained in $(\mathrm{sp}\{dh_i\})^{\perp}$, and contains the vector fields $\tilde{g}_{j_1},\ldots,\tilde{g}_{j_k}$. Since this is required to hold for each individual output, one has to find ℓ distributions $\Delta_1,\ldots,\Delta_\ell$ all invariant with respect to the same set of vector fields $\tilde{f},\tilde{g}_1,\ldots,\tilde{g}_m$.

A set of distributions $\Delta_1,\ldots,\Delta_\ell$ with the property that

(4.1a) $$[\tilde{f},\Delta_i] \subset \Delta_i$$

(4.1b) $$[\tilde{g}_j,\Delta_i] \subset \Delta_i \qquad 1 \leq j \leq m$$

for all $1 \leq i \leq \ell$ is called a set of *compatible* controlled invariant distributions. The feedback pair which makes (4.1) satisfied is called a compatible feedback. Obviously, in the very same way one can introduce the notion of compatible local controlled invariance.

Thus, the problem we face is the following one.

(Local) single-outputs noninteracting control problem. Find a set of distributions $\Delta_1,\ldots,\Delta_\ell$ which:

(i) are compatibly (locally) controlled invariant

(iia) satisfy the conditions $\Delta_i \subset (\mathrm{sp}\{dh_i\})^\perp$

(iib) for some partition $I_1 \cup I_2 \cup \ldots \cup I_\ell$ of the index set $\{1,\ldots,m\}$ and for some compatible feedback, satisfy the conditions

$$(g\beta)_j \in \Delta_i$$

for all $j \notin I_i$. □

The existence of a solution to this problem is characterized as follows

(4.2) *Theorem.* The Local Single-Outputs Noninteracting Control Problem is solvable if and only if the matrix $A(x)$ has rank ℓ for all x.

Proof. (Necessity). Suppose there exists a pair of feedback functions which solves the Single-Outputs Noninteracting Control Problem. Then, we know from the analysis of chapter III, section 3, that, in particular, for all k and all $1 \leq i \leq \ell$

$$L_{\tilde{g}_j} L_{\tilde{f}}^k h_i(x) = 0$$

whenever $j \notin I_i$. Without loss of generality we may assume the inputs $v_1,\ldots,v_m$ being renumbered in such a way that

$$I_i = \{m_{i-1}+1,\ldots,m_i\} \qquad 1 \leq i \leq \ell$$

with $m_0 = 1$ and $m_\ell = m$. The above condition, written for $k = \rho_i$ shows that the matrix $\tilde{A}(x)$ has a block-diagonal structure: on the i-th row only the elements whose indexes belong to the set I_i are nonzero. But we have also that

$$\tilde{A}(x) = A(x)\beta(x)$$

Thus, since the matrix β is nonsingular and each row of $A(x)$ is nonzero by construction (we assumed that all ρ_i's are defined), each row of $\tilde{A}(x)$ is nonzero. $\tilde{A}(x)$ being block-diagonal, this implies that the ℓ rows of $\tilde{A}(x)$ are linearly independent and so are the ℓ rows of $A(x)$.

(Sufficiency). It is known from the analysis given in the previous section that if the i-th row of $A(x)$ is nonzero for all x, the largest locally controlled invariant distribution contained in $(sp\{dh_i\})^{\perp}$ is nonsingular and given by

$$\Delta_i^* = \bigcap_{k=0}^{\rho_i} (sp\{dL_f^k h_i\})^{\perp}$$

A pair of feedback functions (α,β) such that

$$[f + g\alpha, \Delta_i^*] \subset \Delta_i^* \tag{4.3a}$$

$$[(g\beta)_j, \Delta_i^*] \subset \Delta_i^* \qquad 1 \leq j \leq m \tag{4.3b}$$

is a solution of equations of the form

$$A_i(x)\alpha(x) + b_i(x) = \gamma_i(x) \tag{4.4a}$$

$$A_i(x)\beta(x) = \delta_i(x) \tag{4.4b}$$

where $A_i(x)$ and $b_i(x)$ denote the i-th rows of $A(x)$ and $b(x)$. The scalar $\gamma_i(x)$ and the $1\times m$ row vector $\delta_i(x)$ are functions whose differentials belong to $(\Delta_i^*)^{\perp}$: in particular, real numbers.

Considering the equations (4.4) all together, for all $1 \leq i \leq \ell$, one sees immediately that, thanks to the assumption on the rank of $A(x)$, there exists a pair of feedback functions (α,β) that makes (4.3) satisfied simultaneously for all Δ_i^*, i.e. that $\Delta_1^*,\ldots,\Delta_\ell^*$ are compatible locally controlled invariant distributions. In particular if the right-hand-side of (4.4b) is chosen to be the i-th row of a block diagonal matrix, one has that in the i-th row of $A(x)\beta(x)$, i.e. in the i-th row of $\tilde{A}(x)$, the only elements whose indexes belong to the set I_i are nonzero. This proves that a compatible β exists with the property that

$$L_{\tilde{g}_j} L_{\tilde{f}}^{\rho_i} h_i = 0$$

for all $j \notin I_i$. But this, in view of Lemma (3.4) is equivalent to

$$L_{\tilde{g}_j} L_f^{\rho_i} h_i = 0$$

i.e., because by definition $L_{\tilde{g}_j} L_f^k h_i = 0$ for $0 \leq k < \rho_i$,

$$\tilde{g}_j \in \Delta_i^*$$

for all $j \notin I_i$.

This proves that the Local Single-Outputs Noninteracting Control Problem is solved. □

It may be interesting to look at the internal structure of the decoupled system obtained in the proof of this theorem. Suppose again that $A(x)$ has rank ℓ on some neighborhood U and let α and β be solutions of the equations (4.4) on U. One knows from Lemma (3.10) (see also Remark (3.13)) that the functions $L_f^k h_i(x)$, $1 \leq i \leq \ell$, $0 \leq k \leq \rho_i$, are part of a local coordinate system. Without loss of generality we may assume that they are coordinate functions exactly on the neighborhood U. We want to examine the special structure of the control system in the new coordinates, after the introduction of the decoupling feedback.

To this end, we set the new coordinates in the following way. Let

$$\xi_i(x) = \begin{pmatrix} z_{i0} \\ z_{i1} \\ \cdot \\ \cdot \\ \cdot \\ z_{i\rho_i} \end{pmatrix} = \begin{pmatrix} h_i(x) \\ L_f h_i(x) \\ \cdot \\ \cdot \\ \cdot \\ L_f^{\rho_i} h_i(x) \end{pmatrix}$$

for $1 \leq i \leq \ell$. If $\rho_1 + \ldots + \rho_\ell + \ell$ is strictly less than n, an extra set of coordinates, say $\xi_{\ell+1}$, is needed.

The computation of the form taken by the differential equations describing the system in the new coordinates is rather easy. For $1 \leq i \leq \ell$ and $k < \rho_i$

$$\begin{aligned} \dot{z}_{ik} &= \frac{\partial z_{ik}}{\partial x}(\tilde{f} + \sum_{j=1}^m \tilde{g}_j v_j) = L_{\tilde{f}} z_{ik} + \sum_{j=1}^m L_{\tilde{g}_j} z_{ik} v_j \\ (4.5) \qquad &= L_{\tilde{f}} L_f^k h_i + \sum_{j=1}^m L_{\tilde{g}_j} L_f^k h_i v_j = L_f^{k+1} h_i = z_{i,k+1} \end{aligned}$$

Whereas, for $k = \rho_i$ (see Lemma (3.4))

$$\dot{z}_{ik} = \gamma_i(x) + \sum_{j=1}^{m} \delta_{ij}(x) v_j$$

where $\gamma_i(x)$ is the right-hand-side of (4.4a) and $\delta_{ij}(x)$ is the j-th element of the right-hand-side of (4.4b). If this latter is chosen to be as the i-th row of a block-diagonal matrix,as in the proof of Theorem (4.2), then the above equation reduces to

$$(4.6) \qquad \dot{z}_{ik} = \gamma_i + \sum_{j \in I_i} \delta_{ij} v_j$$

Again from the proof of Theorem (4.2), it is seen that γ_i and δ_{ij} depend only on $z_{i0},\ldots,z_{i\rho_i}$(*). As a matter of fact, γ_i and δ_{ij} may be simply real numbers.

Finally, by definition, for all $1 \leq i \leq \ell$

$$(4.7) \qquad y_i = z_{i0}$$

As a result, we see that in the new coordinates the closed loop system may be described in the form

$$(4.8) \qquad \begin{aligned} \dot{\xi}_i &= \bar{f}_i(\xi_i) + \sum_{j \in I_i} \bar{g}_{ij}(\xi_i) v_j \qquad 1 \leq i \leq \ell \\ \dot{\xi}_{\ell+1} &= \bar{f}_{\ell+1}(\xi_1,\ldots,\xi_{\ell+1}) + \sum_{j=1}^{m} \bar{g}_{\ell+1,j}(\xi_1,\ldots,\xi_{\ell+1}) v_j \\ y_i &= \bar{h}_i(\xi_i) \end{aligned}$$

with

$$\bar{f}_i(\xi_i) = \begin{pmatrix} z_{i1} \\ \cdot \\ \cdot \\ \cdot \\ z_{i\rho_i} \\ \gamma_i(\xi_i) \end{pmatrix} \qquad \bar{g}_{ij}(\xi_i) = \begin{pmatrix} 0 \\ \cdot \\ \cdot \\ \cdot \\ 0 \\ \delta_{ij}(\xi_i) \end{pmatrix}$$

$$\bar{h}_i(\xi_i) = z_{i0}$$

(*) Let $\gamma_i(z) = \gamma_i \circ x(z)$. Then

$$\frac{\partial \gamma_i}{\partial z_{jk}} = d\gamma_i \frac{\partial x}{\partial z_{jk}} = \sum_{s=0}^{\rho_i} c_{is} \frac{\partial z_{is}}{\partial x} \frac{\partial x}{\partial z_{jk}} = 0$$

because $i \neq j$.

These equations clearly stress the decoupled structure of the closed loop system.

(4.9) *Remark.* The choice of $\gamma_i(\xi_i)$ linear in ξ_i , i.e. the choice

$$\gamma_i(x) = a_{i0}h_i(x) + a_{i1}L_f h_i(x) + \ldots + a_{i\rho_i} L_f^{\rho_i} h_i(x)$$

with $a_{i0},\ldots,a_{i\rho_i}$ real numbers, is admissible, because $d\gamma_i(x)$ in this case belongs to $sp\{dh_i,\ldots,dL_f^{\rho_i}h_i\}$. It is also possible to choose δ_{ij} constant, provided that, for some $j \in I_i$, δ_{ij} is nonzero because this is required for the solution β of the (4.4b) be nonsingular. The two facts show that a suitable choice of decoupling feedback makes linear the first ℓ subsystems of (4.8).

(4.10) *Remark.* Note that Δ_i^* , the largest locally controlled invariant distribution contained in $(sp\{dh_i\})^{\perp}$, in the coordinates is expressed as

$$\Delta_i^* = sp\{\frac{\partial}{\partial z_{jk}} : j \neq i,\ 0 \le k \le \rho_i\} + sp\{\frac{\partial}{\partial z_{\ell+1,k}} : 1 \le k \le d\}$$

where d denotes the dimension of $\xi_{\ell+1}$ (see chapter I, section 3). □

At the beginning of this section, we have formulated the Noninteracting Control Problem looking at the existence of a set of compatible controlled invariant distributions, each one contained in $(sp\{dh_i\})^{\perp}$ and containing the vector fields $\tilde{g}_j = (g\beta)_j$ for all $j \notin I_i$. One can also consider a complementary formulation in the following terms.

Local single-outputs noninteracting control problem. Find a set of distributions $\Delta_1,\ldots,\Delta_\ell$ which

(i) are compatibly locally controlled invariant

(iia) satisfy the conditions $\Delta_i \subset (sp\{dh_j\})^{\perp}$ for all $j \neq i$

(iib) for some partition $I_1 \cup I_2 \cup \ldots \cup I_\ell$ of the index set $\{1,\ldots,m\}$ and for some compatible feedback, satisfy the conditions

$$(g\beta)_j \in \Delta_i$$

for all $j \in I_i$. □

Also in this case, in fact, the output y_i of the closed-loop system will be affected only by the inputs whose index belongs to the set I_i.

Clearly the condition that the rank of $A(x)$ is equal to ℓ remains necessary and sufficient for the existence of a solution to the problem. If desired, one could directly prove the sufficiency in terms of the complementary formulation discussed above. As in Theorem (4.2), it is easy to prove that the assumption on $A(x)$ makes it possible to express the largest locally controlled invariant distribution contained in $\bigcap_{j\neq i} (sp\{dh_j\})^{\perp}$ as

$$K_i^* = \bigcap_{j\neq i} \bigcap_{k=0}^{\rho_j} (sp\{dL_f^k h_j\})^{\perp}$$

The distributions $K_1^*,\ldots,K_\ell^*$ are compatible and a compatible feedback is exactly the one that makes $\Delta_1^*,\ldots,\Delta_\ell^*$ compatible.

(4.11) *Remark*. Note that in the new coordinate system

$$K_i^* = sp\{\frac{\partial}{\partial z_{ik}} : 0 \leq k \leq \rho_i\} + sp\{\frac{\partial}{\partial z_{\ell+1,k}} : 1 \leq k \leq d\}$$

(4.12) *Remark*. Summarizing some of the above results, one may observe that if $A(x)$ has rank ℓ, there is a set of distributions $D_1,\ldots,D_{\ell+1}$, namely

$$D_i = sp\{\frac{\partial}{\partial z_{ik}} : 0 \leq k \leq \rho_i\} \qquad 1 \leq i \leq \ell$$

$$D_{\ell+1} = sp\{\frac{\partial}{\partial z_{\ell+1,k}} : 1 \leq k \leq d\}$$

which are independent, i.e. such that

$$D_i \cap (\sum_{j\neq i} D_j) = 0$$

and span the tangent space, i.e. are such that

$$D_1 + D_2 + \ldots + D_{\ell+1} = TM$$

Moreover,

$$\Delta_i^* = \bigoplus_{j\neq i} D_j$$

$$K_i^* = D_i + D_{\ell+1}$$

5. Controllability Distributions

The approach to the noninteracting control discussed at the end of the previous section, was the one of looking at a set of compatible locally controlled invariant distributions $\Delta_1,\ldots,\Delta_\ell$, such that

(5.1) $$\mathrm{sp}\{\tilde{g}_j : j \in I_i\} \subset \Delta_i \subset \bigcap_{j\neq i} (\mathrm{sp}\{dh_j\})^\perp$$

with $\tilde{g}_j$ obtained by means of a compatible feedback. It was shown that if $A(x)$ has rank ℓ (*), the largest locally controlled invariant distributions contained in $\bigcap_{j\neq i} (\mathrm{sp}\{dh_j\})^\perp$, denoted $K_1^*,\ldots,K_\ell^*$, are such as to satisfy these requirements. This approach essentially looks at the "maximal" Δ_i which satisfy (5.1); however, one could as well look at the "minimal" Δ_i which satisfy these conditions. This kind of approach yields the notion of a controllability distribution.

A distribution Δ is said to be a *controllability distribution* on N if it is involutive and there exist a feedback pair (α,β) defined on N and a subset I of the index set $\{1,\ldots,m\}$ with the property that $\Delta \cap G = \mathrm{sp}\{\tilde{g}_i : i \in I\}$, and Δ is the smallest distribution which is invariant under the vector fields $\tilde{f},\tilde{g}_1,\ldots,\tilde{g}_m$ and contains $\tilde{g}_i$ for all $i \in I$.

A distribution Δ is said to be a *local controllability distribution* if for each $x \in N$ there exists a neighborhood U of x with the property that Δ is a controllability distribution on U.

It is clear that, by definition, a (local) controllability distribution is (locally) controlled invariant. Therefore, according to the result of Lemma (1.10), such a distribution must satisfy (1.11) (note that the necessity of (1.11) is not dependent on the assumptions made in Lemma (1.10) but only on the controlled invariance and the nonsingularity of β). Therefore it is interesting to look for the extra condition to be added to (1.11) in order to let a given controlled invariant distribution become a local controllability distribution. To this purpose, it is useful to introduce the following algorithm.

(5.2) *Lemma* (Controllability Distribution Algorithm). Let Δ be a fixed distribution. Define a sequence of distributions S_i setting

(5.3) $$S_0 = \Delta \cap G$$
$$S_k = \Delta \cap ([f,S_{k-1}] + \sum_{j=1}^{m} [g_j,S_{k-1}] + G)$$

(*) This condition is indeed necessary in the Single-Outputs Noninteracting Control Problem if β is nonsingular and all ρ_i's are defined.

This sequence is nondecreasing. If there exists an integer k^* such that $S_{k^*} = S_{k^*+1}$, then $S_k = S_{k^*}$ for all $k > k^*$.

Proof. We need only to prove that $S_k \supset S_{k-1}$. This is clearly true for $k = 1$. If true for some k, then

$$([f,S_k] + \sum_{j=1}^{m}[g_j,S_k]) \supset ([f,S_{k-1}] + \sum_{j=1}^{m}[g_j,S_{k-1}])$$

and therefore

$$S_{k+1} \supset S_k$$

(5.4) *Remark.* Note that we may as well represent S_k as

$$S_k = \Delta \cap ([f,S_{k-1}] + \sum_{j=1}^{m}[g_j,S_{k-1}] + G) + S_{k-1}$$

or as

$$S_k = \Delta \cap ([f,S_{k-1}] + \sum_{j=1}^{m}[g_j,S_{k-1}] + S_{k-1} + G)$$

The last one comes from the first and from the modular distributive rule, which holds because $S_{k-1} \subset \Delta$. □

As we did for the algorithm (2.5) we introduce now a terminology which will be used in order to remind both the convergence of the sequence (5.3) in a finite number of stages and the dependence of its final element on the distribution Δ. We set

(5.6) $$S(\Delta) = (S_0 + S_1 + \ldots + S_k + \ldots)$$

and we say that $S(\Delta)$ is *finitely computable* if there exists an integer k^* such that, in the sequence (5.3), $S_{k^*} = S_{k^*+1}$. If this is the case, then obviously $S(\Delta) = S_{k^*}$.

An interesting property of the algorithm (5.3) is the following one.

(5.7) *Lemma.* Let $\tilde{f},\tilde{g}_1,\ldots,\tilde{g}_m$ be any set of vector fields deduced from $f,g_1,\ldots,g_m$ by setting $\tilde{f} = f+g\alpha$ and $\tilde{g}_i = (g\beta)_i$, $1 \leq i \leq m$; then each distribution S_k of the sequence (5.3) is such that

$$S_k = \Delta \cap ([\tilde{f},S_{k-1}] + \sum_{j=1}^{m}[\tilde{g}_j,S_{k-1}] + G)$$

Proof. Let τ be a vector field of S_{k-1}. Then, we have

$$[\tilde{f},\tau] = [f+g\alpha,\tau] = [f,\tau] + \sum_{j=1}^{m} ([g_j,\tau]\alpha_j - (L_\tau\alpha_j)g_j)$$

$$[\tilde{g}_i,\tau] = [(g\beta)_i,\tau] = \sum_{j=1}^{m} ([g_j,\tau]\beta_{ji} - (L_\tau\beta_{ji})g_j)$$

Therefore

$$[\tilde{f},S_{k-1}] + \sum_{j=1}^{m}[\tilde{g}_j,S_{k-1}] + G \subset [f,S_{k-1}] + \sum_{j=1}^{m}[g_j,S_{k-1}] + G$$

But, since β is invertible, then $f = \tilde{f}-g\beta^{-1}\alpha$ and $g_i = (\tilde{g}\beta^{-1})_i$ so that, by doing the same computations, it is found that the reverse inclusion holds. The two sides are thus equal and the lemma is proved. □

From this it is now possible to deduce the desired "intrinsic" characterization of a local controllability distribution.

(5.8) *Lemma*. Let Δ be an involutive distribution. Suppose Δ, G, $\Delta + G$ are nonsingular and that $S(\Delta)$ is finitely computable. Then Δ is a local controllability distribution if and only if

(5.9a) $$[f,\Delta] \subset \Delta + G$$

(5.9b) $$[g_i,\Delta] \subset \Delta + G \qquad 1 \leq i \leq m$$

(5.10) $$S(\Delta) = \Delta$$

Proof. Necessity. Suppose Δ is a local controllability distribution. Then it is locally controlled invariant and (5.9) are satisfied. Moreover, locally around each x there exists a feedback (α,β) with the property that $\Delta \cap G = \mathrm{sp}\{\tilde{g}_i, i \in I\}$, where I is a subset of $\{1,\ldots,m\}$, and Δ is the smallest distribution which is invariant under $\tilde{f},\tilde{g}_1,\ldots,\tilde{g}_m$ and contains $\tilde{g}_i$ for all $i \in I$. Consider the sequence of distributions defined by setting

(5.11a) $$\Delta_0 = \Delta \cap G$$

(5.11b) $$\Delta_k = [\tilde{f},\Delta_{k-1}] + \sum_{i=1}^{m}[\tilde{g}_i,\Delta_{k-1}] + \Delta_0$$

It is easily seen, by induction, that

$$\Delta_k \subset \Delta$$

for all k. This is true for k = 0 and, if true for some k = 0, the invariance of Δ under $\tilde{f},\tilde{g}_1,\ldots,\tilde{g}_m$ shows that $\Delta_{k+1} \subset \Delta$. Therefore, one has

$$\Delta_k = \Delta \cap ([\tilde{f},\Delta_{k-1}] + \sum_{i=1}^{m}[\tilde{g}_i,\Delta_{k-1}] + G)$$

i.e., from Lemma (5.7)

$$\Delta_k = S_k \tag{5.12}$$

It is also seen that, by definition, $\Delta_0 = \mathrm{sp}\{\tilde{g}_i : i \in I\}$ and that, by construction, $\Delta_{k-1} \subset \Delta_k$ for all $1 \leq k$. Thus, the sequence of distributions generated by the algorithm (5.11) is exactly the same as the one yielding $\langle \tilde{f},\tilde{g}_1,\ldots,\tilde{g}_m | \mathrm{sp}\{\tilde{g}_i : i \in I\}\rangle$, the smallest distribution invariant under $\tilde{f},\tilde{g}_1,\ldots,\tilde{g}_m$ and containing $\mathrm{sp}\{\tilde{g}_i : i \in I\}$. From (5.12) and from the assumption that $S(\Delta)$ is finitely computable we know that there is an integer k^* such that $\Delta_{k^*} = \Delta_{k^*+1}$. Therefore, in view of Lemma I.(6.3), the largest distribution in the sequence (5.11) is exactly $\langle \tilde{f},\tilde{g}_1,\ldots,\tilde{g}_m | \mathrm{sp}\{\tilde{g}_i : i \in I\}\rangle$. From this, one concludes that the largest distribution in the sequence (5.11) must coincide with Δ, i.e. again from (5.12), that the condition (5.10) is satisfied.

Sufficiency. We know from Lemma (1.10) that if Δ is involutive, if G, Δ and $G+\Delta$ are nonsingular and if the conditions (5.9) are satisfied, then locally around each x there exists a pair of feedback functions (α,β) with the property that Δ is invariant under $\tilde{f},\tilde{g}_1,\ldots,\tilde{g}_m$. From this fact one may deduce that

$$\begin{aligned}
\Delta \cap ([\tilde{f},S_{k-1}] + \sum_{i=1}^{m}[\tilde{g}_i,S_{k-1}] + G) + S_{k-1} &= \\
= [\tilde{f},S_{k-1}] + \sum_{i=1}^{m}[\tilde{g}_i,S_{k-1}] + \Delta \cap G + S_{k-1} &= \\
= [\tilde{f},S_{k-1}] + \sum_{i=1}^{m}[\tilde{g}_i,S_{k-1}] + S_{k-1}&
\end{aligned}$$

In view of Lemma (5.7) and Remark (5.4), this shows that

$$S_k = [\tilde{f},S_{k-1}] + \sum_{i=1}^{m}[\tilde{g}_i,S_{k-1}] + S_{k-1} \tag{5.13}$$

Without loss of generality, we may assume that $\tilde{g}_1,\ldots,\tilde{g}_m$ are such that $\Delta \cap G = \mathrm{sp}\{\tilde{g}_i : i \in I\}$ for some index set I. In fact, $\Delta \cap G$ is nonzero because, otherwise $S(\Delta)$ would be zero, thus contradicting (5.10). Since $\Delta \cap G$ is nonsingular, one may find a new feedback function $\bar{\beta}$ and construct new vector fields $\bar{g}_i = (\tilde{g}\bar{\beta})_i$, $1 \leq i \leq m$, such that, for some index set I, $\mathrm{sp}\{\bar{g}_i : i \in I\} = \Delta \cap G$ and $\bar{g}_i = \tilde{g}_i$ for $i \notin I$. This new set of vector fields still keeps Δ invariant because $\bar{g}_i \in \Delta$ for $i \in I$ and Δ is involutive.

So $S_0 = G \cap \Delta = \mathrm{sp}\{\tilde{g}_i : i \in I\}$, and the sequence of distributions S_k coincides with the sequence of distributions yielding $\langle \tilde{f}, \tilde{g}_1, \ldots, \tilde{g}_m | \mathrm{sp}\{\tilde{g}_i : i \in I\}\rangle$. Since, by assumption, for some k^*, $S_{k^*} = S_{k^*+1}$ we deduce from Lemma I.(6.3) that S_{k^*} is the smallest distribution which is invariant under $\tilde{f}, \tilde{g}_1, \ldots, \tilde{g}_m$ and contains $\mathrm{sp}\{\tilde{g}_i : i \in I\}$. But (5.10) says that S_{k^*} coincides with Δ and this completes the proof. □

In view of the use of the notion of local controllability distribution in problems of decoupling or noninteracting control, it is useful to be able to construct a "maximal" local controllability distribution contained in a given distribution K. To this end one may use the following result.

(5.14) *Lemma*. Let Δ be an involutive distribution. Suppose $G, \Delta, G+\Delta$ are nonsingular and

$$[f,\Delta] \subset \Delta + G$$

$$[g_i,\Delta] \subset \Delta + G \qquad 1 \leq i \leq m$$

Moreover, suppose $S(\Delta)$ is finitely computable and nonsingular. Then $S(\Delta)$ is the largest local controllability distribution contained in Δ.

Proof. As in the proof of Lemma (5.8) (sufficiency) it is easily seen that the assumptions imply that locally around each x there exists a pair of feedback functions with the property that $\Delta \cap G = \mathrm{sp}\{\tilde{g}_i : i \in I\}$ and $S(\Delta)$ is the smallest distribution which is invariant under $\tilde{f}, \tilde{g}_1, \ldots, \tilde{g}_m$ and contains $\mathrm{sp}\{\tilde{g}_i : i \in I\}$. Moreover, since

$$\mathrm{sp}\{\tilde{g}_i : i \in I\} \subset S(\Delta) \subset \Delta$$

and $\Delta \cap G = \mathrm{sp}\{\tilde{g}_i : i \in I\}$, it is seen that

$$S(\Delta) \cap G = \mathrm{sp}\{\tilde{g}_i : i \in I\}$$

Thus $S(\Delta)$ is a local controllability distribution.

Let $\bar{\Delta}$ be another local controllability distribution contained in Δ. Then, by definition, in a neighborhood U of each x there exists a feedback $(\bar{\alpha},\bar{\beta})$ with the property that $\bar{\Delta} \cap G = sp\{\bar{g}_i : i \in \bar{I}\}$ for some subset $\bar{I}$ of $\{1,\ldots,m\}$, and $\bar{\Delta}$ is invariant under $\bar{f},\bar{g}_1,\ldots,\bar{g}_m$, where $\bar{f} = f+g\bar{\alpha}$ and $\bar{g}_i = (g\bar{\beta})_i$ for $1 \leq i \leq m$. Consider the sequence of distributions

$$\bar{\Delta}_0 = sp\{\bar{g}_i : i \in \bar{I}\}$$

$$\bar{\Delta}_k = [\bar{f},\bar{\Delta}_{k-1}] + \sum_{i=1}^{m} [\bar{g}_i,\bar{\Delta}_{k-1}] + \bar{\Delta}_{k-1}$$

Note that $\bar{\Delta}_k \subset \bar{\Delta} \subset \Delta$. Thus

$$\bar{\Delta}_k \subset \Delta \cap ([\bar{f},\bar{\Delta}_{k-1}] + \sum_{i=1}^{m} [\bar{g}_i,\bar{\Delta}_{k-1}] + \bar{\Delta}_{k-1} + G)$$

Since $\bar{\Delta}_0 = \bar{\Delta} \cap G \subset \Delta \cap G = S_0$, it is easy to show, by induction, by means of Lemma (5.7) and Remark (5.4) that $\bar{\Delta}_k \subset S_k$ for all $k \geq 0$, i.e.

$$\bar{\Delta}_k \subset S(\Delta)$$

Now recall (see Lemma I.(6.4)) that there exists a dense subset U with the property that at each $x \in U$, $\bar{\Delta}(x) = \bar{\Delta}_k(x)$ for some integer k. Thus, we have that

$$\bar{\Delta}(x) \subset S(\Delta)(x)$$

for all x in a dense subset. Since $\bar{\Delta}$ is smooth and $S(\Delta)$ is nonsingular, this implies $\bar{\Delta} \subset S(\Delta)$. □

If the distribution K in which one seeks the maximal controllability distribution does not satisfy the above conditions, one may proceed finding first the largest locally controlled invariant distribution contained in K. From Lemma (2.7) we know that this one is given by $J(K)$, provided that this distribution is finitely computable, K is involutive and $G, J(K), J(K)+G$ are nonsingular. If $S(J(K))$ is finitely computable and nonsingular, then $S(J(K))$ itself is the required distribution. In fact, we know from Lemma (5.14) that $S(J(K))$ is not only the largest local controllability distribution contained in $J(K)$, but also the largest local controllability distribution contained in K, because any controllability distribution contained in K, being locally controlled invariant, must be also contained in $J(K)$.

(5.15) *Remark.* From (5.13) it is also seen that the distribution $S(\Delta)$ is left invariant by any set of vector fields $\tilde{f},\tilde{g}_1,\ldots,\tilde{g}_m$ which leaves Δ invariant. As a matter of fact, the condition

$$S_{k^*} = S_{k^*+1}$$

implies $[\tilde{f},S_{k^*}] \subset S_{k^*}$ and $[\tilde{g}_i,S_{k^*}] \subset S_{k^*}$, $1 \leq i \leq m$.

6. More on Noninteracting Control

In this section we shall see that the notion of controllability distribution makes it possible to analyze under a different perspective the kind of problems dealt with in the section 4. Consider again the Local Single-Outputs Noninteracting Control Problem, that we know is solvable if and only if the matrix $A(x)$ has rank ℓ. In order to avoid unessential notational complications, we may assume that the number of input channels is equal to that of the output channels, i.e. $\ell = m$, so that each decoupled channel is single-input and single-output. In section 4 we have seen that a pair of feedback functions which solves the problem may be found as a solution of the equations (4.4) (where, in particular, γ_i may be zero and δ_i the i-th row of the identity matrix). We have also observed that this solution provides a feedback which makes the following simultaneously invariant:

- Δ^*, the largest locally controlled invariant distribution contained in H,
- Δ_i^* , the largest locally controlled invariant distribution contained $(sp\{dh_i\})^{\perp}$, $1 \leq i \leq \ell$,
- K_i^* , the largest locally controlled invariant distribution contained $\bigcap_{j\neq i} (sp\{dh_j\})^{\perp}$, $1 \leq i \leq \ell$.

We have also investigated the internal structure of the system thus obtained, and found a local state-space description of the form (4.8), i.e.

$$\dot{\xi}_i = \bar{f}_i(\xi_i) + \bar{g}_i(\xi_i)v_i$$

$$\dot{\xi}_{\ell+1} = \bar{f}_{\ell+1}(\xi_1,\ldots,\xi_{\ell+1}) + \sum_{j=1}^{\ell} \bar{g}_{\ell+1,j}(\xi_1,\ldots,\xi_{\ell+1})v_j \quad (6.1)$$

$$y_i = \bar{h}_i(\xi_i)$$

The approach to the noninteracting control problem via the solution of (4.4) makes simultaneously invariant a set of distributions which generally are not independent. For instance, the set $K_1^*,\ldots,K_\ell^*$ is indeed a set of compatibly locally controlled invariant distributions which satisfy the conditions

$$(g\beta)_i \subset K_i^* \subset \bigcap_{i\neq j} (\mathrm{sp}\{dh_j\})^\perp$$

for some compatible feedback but, as we have seen before (Remark (4.12)), if

$$d = n - (\rho_1+\rho_2+\ldots+\rho_\ell+\ell) > 0$$

then for any pair $1 \leq i \leq \ell$

$$K_i^* \cap (\sum_{k\neq i} K_k^*) = K_j^* \cap (\sum_{k\neq j} K_k^*) = K_1^* \cap K_2^* \ldots \cap K_\ell^* \neq \{0\}$$

The existence of such a nonzero intersection corresponds to the presence of the set of coordinates $\xi_{\ell+1} = (z_{\ell+1,1},\ldots,z_{\ell+1,d})$ which characterizes the $(\ell+1)$-th subsystem of (6.1).

Motovated by this consideration, we want to investigate in this section a slightly different version of the noninteracting control problem, defined as follows.

Local, single-outputs, strong noninteracting control problem. Find a set of distributions $\Delta_1,\ldots,\Delta_\ell$ which:

(i) are compatibly locally controlled invariant

(ii) for some compatible feedback satisfy the conditions

$$(g\beta)_i \subset \Delta_i \subset \bigcap_{j\neq i} (\mathrm{sp}\{dh_j\})^\perp\ , \quad 1 \leq i \leq \ell$$

(iii) are nonsingular, independent and span the tangent space

(iv) are simultaneously integrable. □

In view of Theorem I.(3.12), one may replace the requirement (iv) with the requirement

(iv') for each $i = 1,\ldots,\ell$ the distribution $D_i = \sum_{j\neq i} \Delta_j$ is involutive.

Note that, for instance, the distributions $K_1^*,\ldots,K_\ell^*$ were already nonsingular and spanned the tangent space, so that the real new constraint added in (iii) is the one of the independence of the set of distributions in question. On the other hand, simultaneous integrability, introduced in (iv), is useful because it makes it possible to find local coordinates in which the system, once decoupled, appears as the aggregate of ℓ independent single-input single-output subsystems. We discover such a decomposition as an intermediate step in the proof of the following result.

(6.2) *Lemma*. Let $\ell = m$. The Local Single-Outputs Strong Noninteracting Control Problem has a solution if and only if there exists a set of distributions $\Delta_1,\ldots,\Delta_\ell$ which:

(i) are locally controlled invariant

(ii) satisfy the conditions $\Delta_i \subset \bigcap_{j\neq i} (\mathrm{sp}\{dh_j\})^{\perp}$, $1 \le i \le \ell$

(iii) are nonsingular, independent and span the tangent space

(iv) are simultaneously integrable

(v) are such that $\Delta_i \cap G$ is nonsingular and one-dimensional, for all $1 \le i \le \ell$

(6.3) *Remark*. In other words, this Lemma shows that the simpler statement "$\Delta_i \cap G$ is nonsingular and one-dimensional for all $1 \le i \le \ell$" essentially replaces the statement "$\Delta_1,\ldots,\Delta_\ell$ are compatible and, for some compatible feedback, $(g\beta)_i \in \Delta_i$ for all $1 \le i \le \ell$".

Proof. Necessity. All we have to show is that (v) is true. Recall that the matrix $A(x)$ has necessarily rank ℓ for all x. Since

$$A(x) = \begin{pmatrix} dL_f^{\rho_1} h_1(x) \\ \cdot \\ \cdot \\ \cdot \\ dL_f^{\rho_\ell} h_\ell(x) \end{pmatrix} (g_1(x)\ldots g_m(x))$$

and $\ell = m$, we deduce that $\dim G(x) = \ell$ for all x.

On the other hand, from the condition $(g\beta)_i \in \Delta_i$ we have also

$$\sum_{i=1}^{\ell} (\Delta_i \cap G) \subset G = \mathrm{sp}\{(g\beta)_1,\ldots,(g\beta)_\ell\} \subset \sum_{i=1}^{\ell} (\Delta_i \cap G)$$

i.e.

$$G = \sum_{i=1}^{\ell} (\Delta_i \cap G)$$

Since the distributions $\Delta_1,\ldots,\Delta_\ell$ are independent the latter is a direct sum and this yields

$$\ell = \sum_{i=1}^{\ell} \dim(\Delta_i \cap G)$$

i.e. the condition (v) because $\dim(\Delta_i \cap G) > 0$.

Sufficiency. Suppose there exists involutive distributions $\Delta_1,\ldots,\Delta_\ell$ which are nonsingular, independent, span the tangent space and are simultaneously integrable. So, around every point, there exist local coordinates of the form

$$\xi = \mathrm{col}(\xi_1,\ldots,\xi_\ell)$$

with

$$\xi_i = \mathrm{col}(\xi_{i1},\ldots,\xi_{in_i})$$

such that

$$\Delta_i = \mathrm{sp}\{\frac{\partial}{\partial \xi_{ij}} : 1 \leq j \leq n_i\}, \quad 1 \leq i \leq \ell \tag{6.4}$$

From (v) one also deduces the existence of a (locally defined) $\ell\times\ell$ nonsingular matrix $\hat{\beta}$ of smooth functions with the property that

$$\hat{g}_i = (g\hat{\beta})_i$$

spans the one-dimensional distribution $\Delta_i \cap G$.

Moreover, from the fact that the distributions Δ_i are locally controlled invariant and from Lemma (1.10) (necessity) we have

$$[f,\Delta_j] \subset \Delta_j + \mathrm{sp}\{\hat{g}_1,\ldots,\hat{g}_\ell\}$$

$$[\hat{g}_i,\Delta_j] \subset \Delta_j + \mathrm{sp}\{\hat{g}_1,\ldots,\hat{g}_\ell\}, \quad 1 \leq i \leq \ell$$

for all $1 \leq j \leq \ell$. From these we get, in particular

$$[f,\Delta_2+\ldots+\Delta_\ell] \subset \Delta_2+\ldots+\Delta_\ell+\mathrm{sp}\{\hat{g}_1,\ldots,\hat{g}_\ell\} =$$

$$= \Delta_2+\ldots+\Delta_\ell + \mathrm{sp}\{\hat{g}_1\}$$

$$[\hat{g}_1,\Delta_2+\ldots+\Delta_\ell] \subset \Delta_2+\ldots+\Delta_\ell + \mathrm{sp}\{\hat{g}_1\}$$

These two conditions have the form (1.11). Thus, since $\Delta_2+\ldots+\Delta_\ell$ is involutive (see Theorem I.(3.12)) and constant dimensional, $\dim \mathrm{sp}\{\hat{g}_1\} = 1$ and $\mathrm{sp}\{\hat{g}_1\} \cap (\Delta_2+\ldots+\Delta_\ell) = 0$, from Lemma (1.10) (sufficiency) we deduce the existence of a locally defined pair of scalar functions, α_1 and β_1 such that

$$[f+\hat{g}_1\alpha_1, \Delta_2+\ldots+\Delta_\ell] \subset \Delta_2+\ldots+\Delta_\ell$$

$$[\hat{g}_1\beta_1, \Delta_2+\ldots+\Delta_\ell] \subset \Delta_2+\ldots+\Delta_\ell$$

One can proceed in the same way and find other pairs of functions α_2 and $\beta_2,\ldots,$ up to α_ℓ and β_ℓ which makes conditions like

(6.5a) $$[f+\hat{g}_i\alpha_i \, , \, \bigoplus_{j\neq i} \Delta_j] \subset \bigoplus_{j\neq i} \Delta_j$$

(6.5b) $$[\hat{g}_i\beta_i \, , \, \bigoplus_{j\neq i} \Delta_j] \subset \bigoplus_{j\neq i} \Delta_j$$

satisfied for $1 \leq i \leq \ell$.

From $\hat{\beta}$, $\alpha_1,\ldots,\alpha_\ell$, $\beta_1,\ldots,\beta_\ell$ we construct an overall feedback pair α and β setting

(6.6) $$\alpha = \hat{\beta}\begin{pmatrix} \alpha_1 \\ \alpha_2 \\ \cdot \\ \alpha_\ell \end{pmatrix} \qquad \beta = \hat{\beta}\begin{pmatrix} \beta_1 & 0 & \ldots 0 \\ 0 & \beta_2 & \ldots 0 \\ \cdot & \cdot & \ldots \\ 0 & 0 & \ldots \beta_\ell \end{pmatrix}$$

This feedback is clearly such that

$$\tilde{f} = f + g\alpha = f + \hat{g}_1\alpha_1+\ldots+\hat{g}_\ell\alpha_\ell$$

$$\tilde{g}_i = \hat{g}_i\beta_i \qquad\qquad 1 \leq i \leq \ell$$

We show now that this is a compatible feedback for $\Delta_1,\ldots,\Delta_\ell$. The check of this property is particularly easy in the local coordinate chosen to satisfy (6.4). Since $\hat{g}_i \in \Delta_i$, we deduce that the i-th group of components of $\tilde{f}$ coincides with the i-th of components of $f+\hat{g}_i\alpha_i$. Moreover, from (6.5a), using the same kind of arguments employed in the proof of Lemma I.(4.3), it is easily seen that the i-th group of components of $f+\hat{g}_i\alpha_i$ depends only on the local coordinates ξ_i. For similar reasons it is also seen that in $\hat{g}_i\beta_i$ the only nonzero group of components is the i-th one, which depends only on the local coordinates ξ_i. Thus, in the local coordinates $\xi = \text{col}(\xi_1,\ldots,\xi_\ell)$, the vector fields $\tilde{f}$ and $\tilde{g}_i$, $1 \leq i \leq \ell$, are represented in the form

$$\tilde{f}(\xi) = \begin{pmatrix} \tilde{f}_1(\xi_1) \\ \tilde{f}_2(\xi_2) \\ \cdot \\ \cdot \\ \cdot \\ \tilde{f}_\ell(\xi_\ell) \end{pmatrix}, \quad \tilde{g}_1(\xi) = \begin{pmatrix} \tilde{g}_1(\xi_1) \\ 0 \\ \cdot \\ \cdot \\ \cdot \\ 0 \end{pmatrix}, \ldots, \tilde{g}_\ell(\xi) = \begin{pmatrix} 0 \\ 0 \\ \cdot \\ \cdot \\ \cdot \\ \tilde{g}_\ell(\xi_\ell) \end{pmatrix}$$

This clearly shows that the feedback (6.6) is a compatible feedback and completes the proof.

(6.7) *Remark*. In the coordinates $\xi = \text{col}(\xi_1,\ldots,\xi_\ell)$, the i-th output depends only on ξ_i (because of (ii)). Therefore, the decoupled system is described as a set of independent single-input single-output subsystems of the form

$$\dot{\xi}_i = f_i(\xi_i) + g_i(\xi_i)v_i$$
(6.8)
$$y_i = h_i(\xi_i)$$

(6.9) *Remark*. The distributions $K_1^*,\ldots,K_\ell^*$ satisfy all the requirements (i) to (v) if and only if $\rho_1+\ldots+\rho_\ell + \ell = n$, i.e. if and only if Δ^* has dimension 0. If this is not the case, then, in order to be able to solve the Local Single-Outputs Strong Noninteracting Control Problem, one has to try with smaller controlled invariant distributions. □

If the set $K_1^*,\ldots,K_\ell^*$ is not suited, a reasonable alternative for the solution of this control problem is the set $S(K_1^*),\ldots,S(K_\ell^*)$. As a matter of fact, it is possible to prove that, if the matrix $A(x)$ has rank ℓ (a condition which is indeed necessary for the solvability of the problem), the only extra condition needed to let this set of distributions solve the problem in question is simply the condition (iii) of Lemma (6.2).

(6.10) *Theorem*. Let $\ell = m$. Suppose the Local Single-Outputs Noninteracting Control Problem is solvable. Suppose also that, for each $1 \leq i \leq \ell$, $S(K_i^*)$ is finitely computable and nonsingular. If the set $S(K_1^*),\ldots,S(K_\ell^*)$ is independent and spans the tangent space then the Local Single-Outputs Strong Noninteracting Control Problem is also solvable.

Proof. If the matrix $A(x)$ has rank ℓ for all x, then G also has rank ℓ for all x (see proof of Lemma (6.3)), K_i^* is nonsingular (see Remark (4.11)) and $K_i^* \cap G$ also is nonsingular. For, the intersection $K_i^* \cap G$

at x is given by the set of all linear combinations of the form

$$\sum_{i=1}^{\ell} g_i(x)c_i$$

which annihilate $dh_j(x),\dots,dL_f^{\rho_j}h_j(x)$, for all $j \neq i$. The coefficients $c_1,\dots,c_\ell$ of this combination must be solution of the equation

$$\begin{pmatrix} a_{11}(x) & \dots a_{1\ell}(x) \\ \cdot & \dots \cdot \\ a_{i-1,1}(x) & \dots a_{i-1,\ell}(x) \\ a_{i+1,1}(x) & \dots a_{i+1,\ell}(x) \\ \cdot & \dots \cdot \\ a_{\ell 1}(x) & \dots a_{\ell\ell}(x) \end{pmatrix} \begin{pmatrix} c_1 \\ \cdot \\ \cdot \\ \cdot \\ c_\ell \end{pmatrix} = \begin{pmatrix} 0 \\ \cdot \\ \cdot \\ \cdot \\ 0 \end{pmatrix}$$

The matrix on the left-hand-side of this equation has rank $\ell-1$ and therefore, at each x, the set of vectors in G which are also in K_i^* is exactly one-dimensional.

From these properties, using Lemma (5.14), we deduce that if $S(K_i^*)$ is finitely computable and nonsingular, then it is the largest local controllability distribution contained in K_i^*.

Moreover, it is known that $K_1^*,\dots,K_\ell^*$ are compatible, i.e. invariant under the same set of vector fields $\tilde{f},\tilde{g}_1,\dots,\tilde{g}_\ell$. Therefore, from Remark (5.15), it is deduced that also $S(K_1^*),\dots,S(K_\ell^*)$ are invariant under $\tilde{f},\tilde{g}_1,\dots,\tilde{g}_\ell$. Without loss of generality, one may assume that

$$\tilde{g}_i \in K_i^* \qquad 1 \leq i \leq \ell$$

so that

$$K_i^* \cap G = \mathrm{sp}\{\tilde{g}_i\} \tag{6.11}$$

because $K_i^* \cap G$ is one-dimensional.

By definition

$$G \cap K_i^* \subset S(K_i^*) \subset K_i^*$$

so that

$$G \cap S(K_i^*) = G \cap K_i^*$$

and

$$G \cap S(K_i^*) = \mathrm{sp}\{\tilde{g}_i\} \tag{6.12}$$

Consider now the distribution

$$D_i = \bigoplus_{j \neq i} S(K_j^*)$$

It is easy to see that this distribution is also invariant under $\tilde{f}, \tilde{g}_1, \ldots, \tilde{g}_\ell$ and that

$$D_i \supset \mathrm{sp}\{\tilde{g}_j : j \neq i\}$$

Therefore,

$$D_i \supset \langle \tilde{f}, \tilde{g}_1, \ldots, \tilde{g}_\ell | \mathrm{sp}\{\tilde{g}_j : j \neq i\} \rangle$$

We will show now that also the reverse inclusion holds, so that D_i is actually the smallest distribution invariant under $\tilde{f}, \tilde{g}_1, \ldots, \tilde{g}_\ell$ which contains $\mathrm{sp}\{\tilde{g}_j : j \neq i\}$. As a matter of fact, consider the sequence of distributions

$$S_{i0} = \mathrm{sp}\{\tilde{g}_i\}$$

$$S_{ik} = [\tilde{f}, S_{i,k-1}] + \sum_{j=1}^{\ell} [\tilde{g}_j, S_{i,k-1}] + S_{i,k-1}$$

From (5.13), and (6.11), we deduce that for some k^*

$$S_{i,k^*+1} = S_{ik^*} = S(K_i^*)$$

and therefore, from Lemma I.(6.3), that

$$S(K_i^*) = \langle \tilde{f}, \tilde{g}_1, \ldots, \tilde{g}_\ell | \mathrm{sp}\{\tilde{g}_i\} \rangle$$

This shows that

$$D_i \subset \langle \tilde{f}, \tilde{g}_1, \ldots, \tilde{g}_\ell | \mathrm{sp}\{\tilde{g}_j : j \neq i\} \rangle$$

Using Lemma I.(6.6) we have that D_i is involutive and this, in view of Theorem I.(3.12), shows that the set $S(K_1^*), \ldots, S(K_\ell^*)$ is simultaneously integrable, i.e. that the condition (iv) of Lemma (6.2) is satisfied. Conditions (i) and (ii) are satisfied by definition and condition (iii) by assumption. Moreover, the fulfillment of (v) derives from (6.12). This completes the proof. □

(6.13) *Remark*. The interest in the set $S(K_1^*),\ldots,S(K_\ell^*)$ is also motivated by the fact that there exists a well defined algorithm which produces each $S(K_i^*)$.

(6.14) *Remark*. Unfortunately, the condition expressed by the above Theorem is not generally necessary for the solution of this noninteracting control problem.

(6.15) *Remark*. From the proof of Theorem (6.10) it is seen that, when rank $A(x) = \ell$ and $S(K_1^*),\ldots,S(K_\ell^*)$ are independent and span the tangent space, then any feedback solving the Local Single-Outputs Noninteracting Control Problem also solves the strong version of this problem. □

We conclude the section with an example which illustrates the difference between the approach taken in section 4 and the one discussed here.

(6.16) *Example*. Suppose

$$f(x) = \begin{pmatrix} x_2 + x_1x_3 \\ x_1x_2 + x_2^2x_3 \\ x_1^2 - x_3 \end{pmatrix}, g_1(x) = \begin{pmatrix} x_1x_3 \\ x_2^2x_3 \\ 1 \end{pmatrix}, g_2(x) = \begin{pmatrix} x_1x_3 \\ x_2^2x_3 \\ x_2 \end{pmatrix}$$

$$h_1(x) = x_1$$

$$h_2(x) = x_3$$

and consider first the Local Single-Outputs Noninteracting Control Problem.

Since

$$dh_1 = (1 \quad 0 \quad 0)$$

$$dh_2 = (0 \quad 0 \quad 1)$$

we have

$$L_{g_1}h_1 = \langle dh_1, g_1 \rangle = x_1x_3$$

$$L_{g_2}h_1 = x_1x_3$$

$$L_{g_1}h_2 = 1$$

$$L_{g_2}h_2 = x_2$$

Then $\rho_1 = 0$, $\rho_2 = 0$ and

$$A(x) = \begin{pmatrix} x_1x_3 & x_1x_3 \\ 1 & x_2 \end{pmatrix}$$

Since

$$\det A(x) = x_1x_3(x_2 - 1)$$

is nonzero at all x in the dense subset of $\mathbb{R}^3$

$$U = \{x \in \mathbb{R}^3 : x_1 \neq 0,\ x_2 \neq 0,\ x_2 \neq 1\}$$

the problem in question can be solved on U.

A feedback solving the problem is found via the equations (4.4). Taking $\gamma_i = 0$ and δ_i = i-th row of the 2×2 identity matrix, these become

$$A(x)\alpha(x) = -b(x)$$

$$A(x)\beta(x) = I$$

where

$$b(x) = \begin{pmatrix} L_fh_1(x) \\ L_fh_2(x) \end{pmatrix} = \begin{pmatrix} x_2 + x_1x_3 \\ x_1^2 - x_3 \end{pmatrix}$$

This yields

$$\alpha(x) = \frac{-1}{x_1x_3(x_2-1)} \begin{pmatrix} x_2^2 + x_1x_2x_3 + x_1x_3^2 - x_1^3x_3 \\ -x_2 - x_1x_3 + x_1^3x_3 - x_1x_3^2 \end{pmatrix}$$

$$\beta(x) = \frac{1}{x_1x_3(x_2-1)}\begin{pmatrix} x_2 & -x_1x_3 \\ & \\ -1 & x_1x_3 \end{pmatrix}$$

One may wish to examine the form taken by $\tilde{f},\tilde{g}_1,\tilde{g}_2$ in the new local coordinates $\xi_1,\ldots,\xi_{\ell+1}$. In this case we have

$$\xi_1 = h_1(x) = x_1$$

$$\xi_2 = h_2(x) = x_3$$

and ξ_3 may be chosen as x_2. The equations (4.8) became

$$\dot{\xi}_1 = v_1$$

$$\dot{\xi}_2 = v_2$$

$$\dot{\xi}_3 = f(\xi) + \sum_{i=1}^{2} g_i(\xi)v_i$$

$$y_1 = \xi_1$$

$$y_2 = \xi_2$$

Let us see now how $K_1^*,K_2^*,S(K_1^*),S(K_2^*)$ look like.

Computation of $S(K_1^*)$. We need first K_1^*, the largest locally controlled invariant distribution contained in $(\mathrm{sp}\{dh_2\})^{\perp}$. By Corollary (3.14), since $A(x)$ has rank 2 (on U),

$$K_1^* = (\mathrm{sp}\{dh_2\})^{\perp} = \mathrm{sp}\{\frac{\partial}{\partial x_1}, \frac{\partial}{\partial x_2}\}$$

and we may proceed to compute $S(K_1^*)$ via the algorithm (5.2). In this case, in order to find $S_0 = K_1^* \cap G$ we have to solve a set of equations of the form

$$\begin{pmatrix} x_1x_3 & x_1x_3 \\ x_2^2x_3 & x_2^2x_3 \\ 1 & x_2 \end{pmatrix}\begin{pmatrix} c_1 \\ c_2 \end{pmatrix} = \begin{pmatrix} * \\ * \\ 0 \end{pmatrix}$$

for c_1,c_2. From this it is seen that $K_1^* \cap G$ is a one-dimensional distribution, spanned by the vector field

$$\tau = \begin{pmatrix} x_1 \\ x_2^2 \\ 0 \end{pmatrix}$$

Now, note that

$$[g_1,\tau] \in G, \ [g_2,\tau] \in G$$

So that

$$[f,S_0]+[g_1,S_0]+[g_2,S_0]+G = [f,S_0]+G = sp\{[f,\tau],g_1,g_2\}$$

Since

$$[f,\tau] = \begin{pmatrix} x_2(1-x_2) \\ x_2x_1(x_2-1) \\ -2x_1^2 \end{pmatrix}$$

then

$$sp\{[f,\tau],g_1,g_2\} = T_x\mathbb{R}^3$$

From this, it is seen that on U

$$S(K_1^*) = K_1^* = sp\{\frac{\partial}{\partial x_1}, \frac{\partial}{\partial x_2}\}$$

Computation of $S(K_2^*)$. In this case K_2^*, the largest locally controlled invariant distribution contained in $(sp\{dh_1\})^{\perp}$, is given by

$$K_2^* = (sp\{dh_1\})^{\perp} = sp\{\frac{\partial}{\partial x_2}, \frac{\partial}{\partial x_3}\}$$

The algorithm (5.2) now yields

$$S_0 = K_2^* \cap G = sp\{\frac{\partial}{\partial x_3}\}$$

Moreover,

$$[f,\frac{\partial}{\partial x_3}] \in G, \ [g_1,\frac{\partial}{\partial x_3}] \in G, \ [g_3,\frac{\partial}{\partial x_3}] \in G$$

so that

$$S_1 = K_2^* \cap ([f,S_0] + [g_1,S_0] + [g_2,S_0] + G) = K_2^* \cap G = S_0$$

From this, it is seen that on U

$$S(K_2^*) = K_2^* \cap G = \mathrm{sp}\{\frac{\partial}{\partial x_3}\}$$

The distributions thus found are such that

$$K_1^* \cap K_2^* = \mathrm{sp}\{\frac{\partial}{\partial x_2}\}$$

whereas

$$S(K_1^*) \cap S(K_2^*) = 0.$$

CHAPTER V

EXACT LINEARIZATION METHODS

1. Linearization of the Input-Output Response

Throughout this chapter we consider again a control system described by equations of the form

(1.1a) $$\dot{x} = f(x) + \sum_{i=1}^{m} g_i(x)u_i$$

(1.1b) $$y = h(x)$$

and we want to examine to what extent the behavior of such a system could be made "linear" under the effect of an appropriate feedback control law. In the first five sections we concentrate our analysis on the input-output response, whereas in the last two ones the input-state and the state-output behavior will be considered. We shall refer to all of these subjects as to "exact" linearization problems, as opposite to "approximate" linearization, which generally indicates the approximation of the behavior of a nonlinear system by means of its first-order truncated power series expansion.

The first problem we deal with is the one of finding a static state-feedback, i.e. a feedback of the form

(1.2) $$u_i = \alpha_i(x) + \sum_{j=1}^{m} \beta_{ij}(x)v_j$$

under which the input-output behavior of the system (1.1) becomes the same as the one of a linear system. To this end, we shall first deduce a set of conditions which express in simple terms the property, for a nonlinear system of the form (1.1), of displaying an essentially linear input-output response.

Consider the Volterra series expansion of the input-output response of (1.1) (see III.(2.4), where the individual kernels have,e.g., the expressions III.(2.8)). Suppose the first order kernels $w_i(t,\tau_1,x)$, $1 \leq i \leq m$, depend only on the difference $(t-\tau_1)$ and do not depend on x, in a neighborhood U of the initial point x^o. If this is the case we see from III.(2.8") that, because of the independence of $w_i(t,\tau_1,x)$ of x, all kernels of order higher than one are vanishing on U. Thus the whole expansion III.(2.4) reduces to an expansion of the form

(1.3) $$y(t) = Q(t,x^o) + \sum_{i=1}^{m} \int_0^t k_i(t-\tau)u_i(\tau)d\tau$$

with

$$k_i(t-\tau) = w_i(t,\tau,x)$$

The response (1.3) is very much close to the one of a linear system. Indeed, it is exactly the one of a linear system if one neglects the effect of the *zero-input* term $Q(t,x^o)$. Anyhow the *input-dependent* part of the response (1.3) is linear in the input. Since in most practical situations one is essentially interested in getting linearity only between input and output, the achievement of a response of the form (1.3) will be considered as satisfactory.

(1.4) *Remark.* Suppose, for instance, that the initial state x^o is an equilibrium state. In this case, it is readily seen from III.(2.6) that $Q(t,x^o) = h(x^o)$ and, therefore, by subtracting from $y(t)$ the constant term $h(x^o)$, one obtains in (1.3) exactly the zero-state behavior of a linear system. □

Note that, if a Volterra series expansion takes the particular form (1.3), then necessarily the first order kernels $w_i(t,\tau_1,x)$ are independent of x and depend only on the difference $t-\tau_1$, so that this particular property of the first order kernels becomes a necessary and sufficient condition for (1.3) to hold.

If, instead of the expression III.(2.8), one considers the Taylor series expansion III.(2.12b) of $w_i(t,\tau_1,x)$, it is found that a necessary and sufficient condition for this kernel to be independent of x and dependent only on $t-\tau_1$, or - in other words - for (1.3) to hold is that

(1.5) $$L_{g_i} L_f^k h_j(x) = \text{independent of } x$$

for all $k \geq 0$ and all $1 \leq j \leq \ell$, $1 \leq i \leq m$. We may summarize this by saying that the input-dependent part of the response of a nonlinear system of the form (1.1) is linear in the input if and only if the conditions (1.5) are satisfied.

In general, the conditions (1.5) will not be satisfied for a specific nonlinear system. If this is the case, we may wish to have them satisfied via feedback, thus setting a rather interesting synthesis problem. As usual, we could look at a global problem, in which a globally defined feedback is sought which solves the problem for all $x \in N$ or, more simply, a local problem in which a point x^o is given and one wishes to find a feedback defined in a neighborhood U of x^o. The latter, which is easier, will be dealt with in the sequel. For the sake of completeness we state this as follows.

Input-Output Linearization Problem. Given (f,g,h) and an initial state x^o, find (if possible) a neighborhood U of x^o and a pair feedback functions α and β, with invertible β, defined on U, such that for all $k \geq 0$ and for all $1 \leq j \leq \ell$, $1 \leq i \leq m$ (*)

$$(1.6) \qquad L_{\tilde{g}_i} L_{\tilde{f}}^k h_j(x) = \text{independent of } x \text{ on } U. \; \Box$$

The possibility of solving this problem may be expressed as a property of the functions $L_{g_j} L_f^k h_i(x)$ which characterize the Taylor series expansions of the kernels $w_j(t,0,x)$ around $t = 0$. For convenience, we arrange these data into $\ell \times m$ matrices and let $T_k(x)$ denote the matrix whose entry $t_{ij}(x)$ on the i-th row and j-th column is $L_{g_j} L_f^k h_i(x)$. As a matter of fact, the possibility of solving the problem in question may be expressed in different forms, each one being related to a different way in which the data $T_k(x)$, $k \geq 0$, are arranged.

One way of arranging these data is to consider a formal power series $T(s,x)$ in the indeterminate s, defined as

$$(1.7) \qquad T(s,x) = \sum_{k=0}^{\infty} T_k(x) s^{-k-1}$$

We will see below that the problem in question may be solved if and only if $T(s,x)$ satisfies a suitable separation condition.

Another equivalent condition for the existence of solutions is based on the construction of a sequence of Toeplitz matrices, denoted $M_k(x)$, $k \geq 0$, and defined as

$$(1.8) \qquad M_k(x) = \begin{pmatrix} T_0(x) & T_1(x) & \cdots & T_k(x) \\ 0 & T_0(x) & \cdots & T_{k-1}(x) \\ \cdot & \cdot & \cdots & \cdot \\ 0 & 0 & \cdots & T_0(x) \end{pmatrix}$$

(*) Recall that $\tilde{f} = f + g\alpha$ and $\tilde{g}_i = (g\beta)_i$ (see IV.(1.4) and IV.(1.5)).

In this case, one is interested in the special situation in which linear dependence between rows may be tested by taking linear combinations with constant coefficients only.

In view of the relevance of this particular property throughout all the subsequent analysis, we discuss the point with a little more detail. Let $M(x)$ be an $\ell \times m$ matrix whose entries are smooth real-valued functions. We say that x^o is a *regular* point of M if there exists a neighborhood U of x^o with the property that

$$\text{rank } M(x) = \text{rank } M(x^o) \tag{1.9}$$

for all $x \in U$. If this is the case, the integer rank $M(x^o)$ is denoted $r_{\mathbb{K}}(M)$; clearly $r_{\mathbb{K}}(M)$ depends on the point x^o, because on a neighborhood V of another point x^1, rank $M(x^1)$ may be different.

With the matrix M we will associate another notion of "rank", in the following way. Let x^o be a regular point of M, U an open set on which (1.9) holds, and $\bar{M}$ a matrix whose entries are the restrictions to U of the corresponding entries of M. We consider the vector space defined by taking *linear combinations of rows* of $\bar{M}$ *over the field* $\mathbb{R}$, the set of real numbers, and denote $r_{\mathbb{R}}(M)$ its dimension (note that again $r_{\mathbb{R}}(M)$ may depend on x^o). Clearly, the two integers $r_{\mathbb{R}}(M)$ and $r_{\mathbb{K}}(M)$ are such that

$$r_{\mathbb{R}}(M) \geq r_{\mathbb{K}}(M) \tag{1.10}$$

The equality of these two integers may easily be tested in the following way. Note that both remain unchanged if M is multiplied on the left by a nonsingular matrix of real numbers. Let us call a *row-reduction* of M the process of multiplying M on the left by a nonsingular matrix V of real numbers with the purpose of annihilating the maximal number of rows in VM (here also the row-reduction process may depend on the point x^o). Then, it is trivially seen that the two-sides of (1.10) are equal if and only if any process of row-reduction of M leaves a number of nonzero rows in VM which is equal to $r_{\mathbb{K}}(M)$.

We may now return to the original synthesis problem and prove the main result.

(1.11) *Theorem.* There exists a solution at x^o to the Input-Output Linearization Problem if and only if either one of the following equivalent conditions is satisfied:

(a) there exist a formal power series

$$K(s) = \sum_{k=0}^{\infty} K_k s^{-k-1}$$

whose coefficients are $\ell \times m$ matrices of real numbers, and a formal power series

$$R(s,x) = R_{-1}(x) + \sum_{k=0}^{\infty} R_k(x) s^{-k-1}$$

whose coefficients are $m \times m$ matrices of smooth functions defined on a neighborhood U of x^o, with invertible $R_{-1}(x)$, which factorize the formal power series $T(s,x)$ as follows:

(1.12) $$T(s,x) = K(s) \cdot R(s,x)$$

(b) for all $i \geq 0$, the point x^o is a regular point of the Toeplitz matrix M_i and

(1.13) $$r_{\mathbb{R}}(M_i) = r_{\mathbb{K}}(M_i). \ \square$$

The proof of this Theorem consists in the following steps. First we introduce a recursive algorithm, known as the Structure Algorithm, which operates on the sequence of matrices $T_k(x)$, $k \geq 0$. Then, we prove the sufficiency of (b), essentially by showing that this assumption makes it possible to continue the Structure Algorithm at each stage and that from the data thus extracted one may construct a feedback solving the problem. Then, we complete the proof that (a) is necessary and that (a) implies (b).

(1.14) *Remark*. For the sake of notational compactness, from this point on we make systematic use of the following notation. Let γ be an $s \times 1$ vector of smooth functions and $\{g_1,\dots,g_m\}$ a set of vector fields. We let $L_g\gamma$ denote the $s \times m$ matrix whose i-th column is the vector $L_{g_i}\gamma$, i.e.

$$L_g\gamma = [L_{g_1}\gamma \ \dots \ L_{g_m}\gamma]. \ \square$$

(1.15) *Algorithm* (Structure Algorithm).
Step 1. Let x^o be a regular point of T_0 and suppose $r_{\mathbb{R}}(T_0) = r_{\mathbb{K}}(T_0)$. Then, there exists a nonsingular matrix of real numbers, denoted by

$$V_1 = \begin{bmatrix} P_1 \\ K_1^1 \end{bmatrix}$$

where P_1 performs row permutations, such that

$$V_1 T_0(x) = \begin{bmatrix} S_1(x) \\ 0 \end{bmatrix}$$

where $S_1(x)$ is an $r_0 \times m$ matrix and rank $S_1(x^o) = r_0$. Set

$$\delta_1 = r_0$$

$$\gamma_1(x) = P_1 h(x)$$

$$\bar{\gamma}_1(x) = K_1^1 h(x)$$

and note that

$$L_g \gamma_1(x) = S_1(x)$$

$$L_g \bar{\gamma}_1(x) = 0$$

If $T_0(x) = 0$, then P_1 must be considered as a matrix with no rows and K_1^1 is the identity matrix.

Step i. Consider the matrix

$$\begin{bmatrix} L_g\gamma_1(x) \\ \cdot \\ \cdot \\ \cdot \\ L_g\gamma_{i-1}(x) \\ L_g L_f \bar{\gamma}_{i-1}(x) \end{bmatrix} = \begin{bmatrix} S_{i-1}(x) \\ L_g L_f \bar{\gamma}_{i-1}(x) \end{bmatrix}$$

and let x^o be a regular point of this matrix. Suppose

$$(1.16) \qquad r_{\mathbb{R}} \begin{bmatrix} S_{i-1} \\ L_g L_f \bar{\gamma}_{i-1} \end{bmatrix} = r_{\mathbb{K}} \begin{bmatrix} S_{i-1} \\ L_g L_f \bar{\gamma}_{i-1} \end{bmatrix}$$

Then, there exists a nonsingular matrix of real numbers, denoted by

$$V_i = \begin{bmatrix} I_{\delta_1} & \dots & 0 & 0 \\ \cdot & \dots & \cdot & \cdot \\ \cdot & \dots & \cdot & \cdot \\ \cdot & \dots & \cdot & \cdot \\ 0 & \dots & I_{\delta_{i-1}} & 0 \\ 0 & \dots & 0 & P_i \\ K_1^i & \dots & K_{i-1}^i & K_i^i \end{bmatrix}$$

where P_i performs row permutations, such that

$$V_i \begin{bmatrix} L_g\gamma_1(x) \\ \vdots \\ L_g\gamma_{i-1}(x) \\ L_gL_f\bar{\gamma}_{i-1}(x) \end{bmatrix} = \begin{bmatrix} S_i(x) \\ \\ 0 \end{bmatrix}$$

where $S_i(x)$ is an $r_{i-1} \times m$ matrix and rank $S_i(x^o) = r_{i-1}$. Set

$$\delta_i = r_{i-1} - r_{i-2}$$

$$\gamma_i(x) = P_iL_f\bar{\gamma}_{i-1}(x)$$

$$\bar{\gamma}_i(x) = K_1^i\gamma_1(x)+\dots+K_{i-1}^i\gamma_{i-1}(x)+K_i^iL_f\bar{\gamma}_{i-1}(x)$$

and note that

$$\begin{bmatrix} L_g\gamma_1(x) \\ \vdots \\ L_g\gamma_i(x) \end{bmatrix} = S_i(x)$$

$$L_g\bar{\gamma}_i(x) = 0$$

If the condition (1.16) is satisfied but the last $\ell-r_{i-2}$ rows of the matrix depend on the first r_{i-2}, then the step degenerates, P_i

must be considered as a matrix with no rows, K_i^i is the identity matrix, $\delta_i = 0$ and $S_i(x) = S_{i-1}(x)$. □

As we said before, this algorithm may be continued at each stage if and only if the assumption (b) is satisfied, because of the following fact.

(1.17) *Lemma*. Let x^o be a regular point of T_0 and suppose $r_{\mathbb{R}}(T_0)=r_{\mathbb{K}}(T_0)$. Then x^o is a regular point of

$$\begin{bmatrix} S_{i-1} \\ \\ L_g L_f \bar{\gamma}_{i-1} \end{bmatrix}$$

and the condition (1.16) holds for all $2 \le i \le k$ if and only if x^o is a regular point of T_i and the condition (1.13) holds for all $1 \le i \le k-1$.

Proof. We sketch the proof for the case k = 2. Recall that

$$M_1 = \begin{pmatrix} T_0 & T_1 \\ \\ 0 & T_0 \end{pmatrix} = \begin{pmatrix} L_g h & L_g L_f h \\ \\ 0 & L_g h \end{pmatrix}$$

Moreover, let V_1, γ_1 and $\bar{\gamma}_1$ be defined as in the first step of the algorithm. Multiply M_1 on the left by

$$V = \begin{pmatrix} V_1 & 0 \\ \\ 0 & V_1 \end{pmatrix}$$

As a result, one obtains

$$VM_1 = \begin{pmatrix} V_1 L_g h & L_g V_1 L_f h \\ \\ 0 & V_1 L_g h \end{pmatrix} = \begin{pmatrix} L_g P_1 h & L_g L_f P_1 h \\ 0 & L_g L_f K_1^1 h \\ 0 & L_g P_1 h \\ 0 & 0 \end{pmatrix} =$$

$$= \begin{pmatrix} S_1 & L_g L_f \gamma_1 \\ 0 & L_g L_f \bar{\gamma}_1 \\ 0 & S_1 \\ 0 & 0 \end{pmatrix}$$

Note that $r_{\mathbb{R}}(S_1) = r_{\mathbb{K}}(S_1)$. Thus, because of the special structure of VM_1, x^o is a regular point of M_1 and the condition $r_{\mathbb{R}}(M_1) = r_{\mathbb{K}}(M_1)$ is satisfied if and only if x^o is a regular point of

$$\begin{pmatrix} L_g L_f \bar{\gamma}_1 \\ S_1 \end{pmatrix}$$

and

$$r_{\mathbb{R}} \begin{pmatrix} L_g L_f \bar{\gamma}_1 \\ S_1 \end{pmatrix} = r_{\mathbb{K}} \begin{pmatrix} L_g L_f \bar{\gamma}_1 \\ S_1 \end{pmatrix}$$

i.e. the condition (1.16) holds for $i = 2$. For higher values of k one may proceed by induction. □

From this, we see that the Structure Algorithm may be continued up to the k-th step if and only if the condition (1.13) holds for all i up to k-1. The Structure Algorithm may be indefinitely continued if and only if the assumption (b) is satisfied.

Proof of Theorem (1.11). Sufficiency of (b): construction of the linearizing feedback. If the Structure Algorithm may be continued indefinitely, two possibilities may occur. Either there is a step q such that the matrix

$$\begin{bmatrix} L_g \gamma_1(x) \\ \vdots \\ L_g \gamma_{q-1}(x) \\ L_g L_f \bar{\gamma}_{q-1}(x) \end{bmatrix}$$

has rank ℓ at x^o. Then the algorithm terminates. Formally, one can still set P_q = identity, V_q = identity

$$\gamma_q = P_q L_f \bar{\gamma}_{q-1}(x)$$

and

$$\begin{bmatrix} S_{q-1}(x) \\ L_g \gamma_q(x) \end{bmatrix} = S_q(x)$$

and consider $K_1^q,\ldots,K_q^q$ as matrices with no rows. Or,else, from a certain step on all further steps are degenerate. In this case, let q denote the index of the last nondegenerate step. Then, for all $j > q$, P_j will be a matrix with no rows, K_j^j the identity and $\delta_j = 0$.

From the functions $\gamma_1,\ldots,\gamma_q$ generated by the Structure Algorithm, one may construct a linearizing feedback in the following way. Set

$$\Gamma(x) = \begin{bmatrix} \gamma_1(x) \\ \cdot \\ \cdot \\ \cdot \\ \gamma_q(x) \end{bmatrix}$$

and recall that $S_q = L_g\Gamma$ is an $r_{q-1} \times m$ matrix, of rank r_{q-1} at x^o. Then the equations

$$[L_g\Gamma(x)]\alpha(x) = -L_f\Gamma(x) \tag{1.18a}$$

$$[L_g\Gamma(x)]\beta(x) = [I_{r_{q-1}} \quad 0] \tag{1.18b}$$

on a suitable neighborhood U of x^o are solved by a pair of smooth functions α and β.

Sufficiency of (b): proof that the above feedback solves the problem. We show first that

$$P_1 L_{\tilde g} L_{\tilde f}^k h(x) = \text{independent of } x \tag{1.19a}$$

$$P_i K_{i-1}^{i-1} \ldots K_1^1 L_{\tilde g} L_{\tilde f}^k h(x) = \text{independent of } x \tag{1.19b}$$

for all $2 \leq i \leq q$ and that

$$K_q^q K_{q-1}^{q-1} \ldots K_1^1 L_{\tilde g} L_{\tilde f}^k h(x) = \text{independent of } x \tag{1.19c}$$

To this end, note that (1.18) imply

$$L_{\tilde f}\gamma_i = 0 \tag{1.20a}$$

$$L_{\tilde g}\gamma_i = \text{independent of } x \tag{1.20b}$$

for all $1 \leq i \leq q$. Moreover, since $L_g\bar{\gamma}_i = 0$ for all $i \geq 1$, also

(1.20c) $$L_{\tilde{f}}\bar{\gamma}_i = L_f\bar{\gamma}_i$$

(1.20d) $$L_{\tilde{g}}\bar{\gamma}_i = 0$$

for all $1 \leq i$. Using (1.20) repeatedly, it is easy to see that, if $k \geq i$

(1.21)
$$\begin{aligned} K_i^i\ldots K_1^1 L_{\tilde{f}}^k h &= K_i^i\ldots K_2^2 L_{\tilde{f}}^{k}\bar{\gamma}_1 \\ &= K_i^i\ldots K_3^3 L_{\tilde{f}}^{k-1}\bar{\gamma}_2 = \ldots \\ &= K_i^i L_{\tilde{f}}^{k-i+2}\bar{\gamma}_{i-1} \\ &= L_{\tilde{f}}^{k-i+1}\bar{\gamma}_i \end{aligned}$$

If $k < i$

(1.22) $$K_i^i\ldots K_1^1\, L_{\tilde{f}}^k h = K_i^i\ldots K_{k+1}^{k+1} L_{\tilde{f}}\bar{\gamma}_k$$

These expressions hold for every $i \geq 1$ (recall that, if $i > q$, K_i^i is an identity matrix).

Thus, if $i \leq q$ and $k \geq i-1$ we get from (1.21)

$$P_iK_{i-1}^{i-1}\ldots K_1^1 L_{\tilde{g}}L_{\tilde{f}}^k h = L_{\tilde{g}}P_iL_{\tilde{f}}^{k-i+2}\bar{\gamma}_{i-1} = L_{\tilde{g}}L_{\tilde{f}}^{k-i+1}\gamma_i$$

which is either independent of x (if $k = i-1$) or zero, while for $i \leq q$ and $k < i-1$ we get from (1.22)

$$P_iK_{i-1}^{i-1}\ldots K_1^1 L_{\tilde{g}}L_{\tilde{f}}^k h = P_i\ldots K_{k+2}^{k+2}L_{\tilde{g}}(\bar{\gamma}_{k+1} - \sum_{j=1}^{k} K_j^{k+1}\gamma_j)$$

The right-hand-side of this expression is again independent of x and this complete the proof of (1.19b).

Moreover, if $k \geq q$, (1.21) yields

$$\begin{aligned} K_q^q\ldots K_1^1 L_{\tilde{g}}L_{\tilde{f}}^k h &= K_k^k\ldots K_1^1 L_{\tilde{g}}L_{\tilde{f}}^k h = L_{\tilde{g}}L_{\tilde{f}}\bar{\gamma}_k = \\ &= L_{\tilde{g}}K_{k+1}^{k+1}L_{\tilde{f}}\bar{\gamma}_k = L_{\tilde{g}}(\bar{\gamma}_{k+1} - \sum_{j=1}^{q} K_j^{k+1}\gamma_j) \end{aligned}$$

and this, together with (1.22) written for $i = q$, which holds for $k < q$, shows that also (1.19c) is true. Finally, (1.19a) is also true, because $P_1 L_{\tilde{g}} L_{\tilde{f}}^k h = L_{\tilde{g}} L_{\tilde{f}}^k \gamma_1$ and the latter is either independent of x (if k = 0) or zero.

In order to complete the proof of the sufficiency of (b), we need only to prove that the matrix

$$(1.23) \qquad H = \begin{bmatrix} P_1 \\ P_2 K_1^1 \\ \cdot \\ \cdot \\ \cdot \\ P_q K_{q-1}^{q-1} \ldots K_1^1 \\ K_q^q K_{q-1}^{q-1} \ldots K_1^1 \end{bmatrix}$$

is square and nonsingular. This, together with the (1.19) already proved, shows in fact that

$$L_{\tilde{g}} L_{\tilde{f}}^k h(x) = \text{independent of } x$$

for all $k \geq 0$. But the nonsingularity of (1.23) is a straightforward consequence of the fact that this matrix may be deduced from the matrix $V_q \ldots V_2 V_1$ by means of elementary row operations.

Necessity of (a). Let

$$\hat{\beta}(x) = \beta^{-1}(x)$$

$$\hat{\alpha}(x) = -\beta^{-1}(x)\alpha(x)$$

and let

$$\tilde{T}_k(x) = L_{\tilde{g}} L_{\tilde{f}}^k h(x)$$

If the feedback pair α and β is such as to make $\tilde{T}_k(x)$ independent of x for all k (i.e. to solve the problem), then

$$(1.24) \qquad L_f^k h = L_{\tilde{f}}^k h + \tilde{T}_{k-1}\hat{\alpha} + \tilde{T}_{k-2} L_f \hat{\alpha} + \ldots + \tilde{T}_0 L_f^{k-1}\hat{\alpha}$$

This expression may be easily proved by induction. In fact, one has

$$L_f^{k+1} h = L_{(\tilde{f}+\tilde{g}\hat{\alpha})} L_{\tilde{f}}^k h + L_f(\tilde{T}_{k-1}\hat{\alpha} + \ldots + \tilde{T}_0 L_f^{k-1}\hat{\alpha}) =$$

$$= L_{\tilde{f}}^{k+1} h + L_{\tilde{g}} L_{\tilde{f}}^k h\hat{\alpha} + \tilde{T}_{k-1} L_f \hat{\alpha} + \ldots + \tilde{T}_0 L_f^k \hat{\alpha}$$

From (1.24) one then deduces

$$L_g L_f^k h = (L_{\tilde{g}} L_{\tilde{f}}^k h)\hat{\beta} + \tilde{T}_{k-1} L_g \hat{\alpha} + \tilde{T}_{k-2} L_g L_f \hat{\alpha} + \ldots + \tilde{T}_0 L_g L_f^{k-1} \hat{\alpha}$$

or,

$$(1.25) \qquad T_k(x) = \tilde{T}_k \hat{\beta}(x) + \tilde{T}_{k-1} L_g \hat{\alpha}(x) + \tilde{T}_{k-2} L_g L_f \hat{\alpha}(x) + \ldots + \tilde{T}_0 L_g L_f^{k-1} \hat{\alpha}(x)$$

Now, consider the formal power series

$$K(s) = \sum_{k=0}^{\infty} \tilde{T}_k s^{-k-1}$$

$$R(s,x) = \hat{\beta}(x) + \sum_{k=0}^{\infty} (L_g L_f^k \hat{\alpha}(x)) s^{-k-1}$$

and note that the latter is invertible (i.e. the coefficient of the 0-th power of s is an invertible matrix). At this point, the expression (1.25) tells us exactly that the Cauchy product of the two series thus defined is equal to the series (1.7), thus proving the necessity of (a)

(a) ⇒ (b). If (1.7) is true, we may write

$$M_k(x) = \begin{pmatrix} K_0 & K_1 & \ldots & K_k \\ 0 & K_0 & \ldots & K_{k-1} \\ \cdot & \cdot & \ldots & \cdot \\ 0 & 0 & \ldots & K_0 \end{pmatrix} \begin{pmatrix} R_{-1}(x) & R_0(x) & R_1(x) \ldots R_{k-1}(x) \\ 0 & R_{-1}(x) & R_0(x) \ldots R_{k-2}(x) \\ \cdot & \cdot & \cdot \quad \ldots \cdot \\ 0 & 0 & 0 \quad \ldots R_{-1}(x) \end{pmatrix}$$

The factor on the left of this matrix is a matrix of real numbers, whereas the factor on the right is nonsingular at x^o, as a consequence of the nonsingularity of $R_{-1}(x)$. Thus x^o is a regular point of M_k and the condition (1.13) holds. □

(1.26) *Remark*. We stress again the importance of the Structure Algorithm as a test for the fulfillment of the conditions (a) (or (b)) as well as a procedure for the construction of a linearizing feedback.

(1.27) *Remark*. An obvious sufficient condition for the existence of a solution to the Input-Output Linearization Problem is that the rank of the matrix A(x) is equal to ℓ, i.e. that there exists a solution to the Local Single-Outputs Noninteracting Control Problem. If this condition holds, the Structure Algorithms terminates at a finite stage q, yielding $S_q(x) = A(x)$.

2. The Internal Structure of the Linearized System

In this section we analyze some interesting features of the linearization procedure discussed so far. First of all, we examine some simple properties relating the Structure Algorithm with the Algorithm IV.(2.5), the one yielding the largest locally controlled invariant distribution contained in H.

We begin with a simple remark, which will be recalled several times later on, and then we give two lemmas which establish the required relation between the Algorithm (1.15) and Algorithm IV.(2.5)

(2.1) *Remark.* The submatrix

$$\begin{pmatrix} P_i \\ K_i^i \end{pmatrix}$$

of the matrix V_i introduced at the i-th stage of the algorithm (1.15) is nonsingular by definition. This makes it possible to express $L_f\bar{\gamma}_{i-1}$ as a linear combination of $\gamma_1,\ldots,\gamma_i$ and $\bar{\gamma}_i$. For, let Q_i' and Q_i'' be the two matrices of real numbers defined by

$$(Q_i' \quad Q_i'')\begin{pmatrix} P_i \\ K_i^i \end{pmatrix} = Q_i'P_i + Q_i''K_i^i = I$$

Then, one has

$$L_f\bar{\gamma}_{i-1} = Q_i'\gamma_i + Q_i''(\bar{\gamma}_i - K_1^i\gamma_1 - \ldots - K_{i-1}^i\gamma_{i-1})$$

If the i-th stage is trivial, Q_i' is a matrix with no columns and $Q_i'' = I$. If the algorithm terminates at the q-th stage, then Q_q'' is a matrix with no columns and $Q_q' = I$. □

In what follows, in order to simplify the notation, whenever we have an s×1 vector γ of real-valued functions and we want to consider the codistribution $\mathrm{sp}\{d\gamma_1,\ldots,d\gamma_s\}$, we denote the latter by $\mathrm{sp}\{d\gamma\}$.

(2.2) *Lemma.* Suppose the Input-Output Linearization Problem is solvable at x^o. Suppose G is nonsingular around x^o, and the codistributions Ω_k generated via the Algorithm IV.(2.5), initialized with $\Omega_0 = \mathrm{sp}\{dh\}$, are nonsingular around x^o. Then, for all $k \geq 0$

$$\Omega_k = \left(\sum_{i=1}^{k+1} \mathrm{sp}\{d\gamma_i\} + \sum_{i=1}^{k+1} \mathrm{sp}\{d\bar{\gamma}_i\}\right) \tag{2.3}$$

$$\left(\sum_{i=1}^{k+1} \mathrm{sp}\{d\gamma_i\} \cap \left(\sum_{i=1}^{k+1} \mathrm{sp}\{d\bar{\gamma}_i\}\right) = 0 \tag{2.4}$$

$$\Omega_k \cap G^{\perp} = \sum_{i=1}^{k+1} \mathrm{sp}\{d\bar{\gamma}_i\} \tag{2.5}$$

and $\Omega_k \cap G^{\perp}$ is nonsingular around x^o.

Proof. The proof proceeds by induction. For $k = 0$, (2.3) reduces to

$$\Omega_0 = \mathrm{sp}\{d\gamma_1\} + \mathrm{sp}\{d\bar{\gamma}_1\}$$

which is clearly true because

$$\begin{pmatrix} \gamma_1 \\ \\ \bar{\gamma}_1 \end{pmatrix} = V_1 h$$

and V_1 is a nonsingular matrix. Moreover, by definition $L_g\bar{\gamma}_1 = 0$, i.e.

$$\mathrm{sp}\{d\bar{\gamma}_1\} \subset G^{\perp}$$

which implies

$$\Omega_0 \cap G^{\perp} = (\mathrm{sp}\{d\gamma_1\} + \mathrm{sp}\{d\bar{\gamma}_1\}) \cap G^{\perp} = \mathrm{sp}\{d\gamma_1\} \cap G^{\perp} + \mathrm{sp}\{d\bar{\gamma}_1\}$$

But

$$\mathrm{sp}\{d\gamma_1\} \cap G^{\perp} = 0$$

because, if this were not true at x^o, then there would exist a $1\times\delta_1$ row vector of real numbers λ such that

$$\lambda L_g\gamma_1(x^o) = \lambda S_1(x^o) = 0$$

thus contradicting the linear independence of the δ_1 rows of $S_1(x)$ at x^o. Therefore, we conclude, that

$$\Omega_0 \cap G^{\perp} = \mathrm{sp}\{d\bar{\gamma}_1\}$$

i.e. (2.5) for $k = 0$. Moreover, this argument also shows that

$$\mathrm{sp}\{d\gamma_1\} \cap \mathrm{sp}\{d\bar{\gamma}_1\} = 0$$

i.e. (2.4) for $k = 0$.

The codistribution $\mathrm{sp}\{d\gamma_1\}$ has constant dimension δ_1 (because, otherwise, the matrix $S_1(x)$ would not have rank δ_1 at each x in a neighborhood of x^o). Ω_0 has constant dimension by assumption and therefore also $\mathrm{sp}\{d\bar{\gamma}_1\}$, i.e. $\Omega_0 \cap G^{\perp}$, has constant dimension.

Suppose now (2.3),(2.4),(2.5) are true for some k and $\Omega_k \cap G^{\perp}$ has constant dimension around x^o. From (2.5) we see that

$$\sum_{i=1}^{m} L_{g_i}(\Omega_k \cap G^{\perp}) \subset \sum_{j=1}^{k+1} \mathrm{sp}\{d\bar{\gamma}_j\} \subset \Omega_k$$

(because $L_{g_i}\bar{\gamma}_j = 0$) and, therefore, that

$$\Omega_{k+1} = \Omega_k + L_f(\Omega_k \cap G^{\perp}) \tag{2.6}$$

This, in turn, yields

$$\Omega_{k+1} = \sum_{i=1}^{k+1} \mathrm{sp}\{d\gamma_i\} + \sum_{i=1}^{k+1} \mathrm{sp}\{d\bar{\gamma}_i\} + L_f(\sum_{i=1}^{k+1} \mathrm{sp}\{d\bar{\gamma}_i\}$$

$$= \sum_{i=1}^{k+1} \mathrm{sp}\{d\gamma_i\} + \sum_{i=1}^{k+1} \mathrm{sp}\{d\bar{\gamma}_i\} + \sum_{i=1}^{k+1} \mathrm{sp}\{dL_f\bar{\gamma}_i\}$$

$$= \sum_{i=1}^{k+1} \mathrm{sp}\{d\gamma_i\} + \sum_{i=1}^{k+1} \mathrm{sp}\{d\bar{\gamma}_i\} + \mathrm{sp}\{d\gamma_{k+2}\} + \mathrm{sp}\{d\bar{\gamma}_{k+2}\}$$

(the last equality being a consequence of the Remark (2.1)), and this proves (2.3) for $k+1$.

Moreover, it is easily seen that

$$\sum_{i=1}^{k+2} \mathrm{sp}\{d\bar{\gamma}_i\} \subset G^{\perp}$$

(because $L_g\bar{\gamma}_i = 0$) and that

$$\sum_{i=1}^{k+2} \mathrm{sp}\{d\gamma_i\} \cap G^{\perp} = 0$$

(because otherwise the linear independence of the rows of S_{k+2} would be contradicted). The two conditions together prove (2.4) for k+1 and also that

$$\Omega_{k+1} \cap G^{\perp} = (\sum_{i=1}^{k+2} sp\{d\gamma_i\} + \sum_{i=1}^{k+2} sp\{d\bar{\gamma}_i\}) \cap G^{\perp} = \sum_{i=1}^{k+2} sp\{d\bar{\gamma}_i\}$$

i.e. (2.5) for k+1.

The codistribution $\sum_{i=1}^{k+2} sp\{d\gamma_i\}$ has constant dimension $\delta_1+\ldots+\delta_{k+2}$ (because otherwise the linear independence of the rows of S_{k+2} would be contradicted) and this, together with the assumption that Ω_{k+1} has constant dimension, proves that $\Omega_{k+1} \cap G^{\perp}$ has constant dimension around x^o. □

(2.7) *Remark*. Note, from the proof of Lemma (2.2), that an obvious necessary condition for the existence of a solution to the Input-Output Linearization Problem is that the sequence of codistributions Ω_k generated by means of the Algorithm IV.(2.5) coincides with the one generated by means of the (simpler) algorithm (2.6).

(2.8) *Lemma*. For all $k \geq 0$

$$\dim \frac{\Omega_k}{\Omega_k \cap G^{\perp}} = \dim(\sum_{i=1}^{k+1} sp\{d\gamma_i\}) = \sum_{i=1}^{k+1} \delta_i \tag{2.9}$$

Proof. The first equality follows directly from (2.3),(2.4) and (2.5). The second one is a consequence of the fact that the $r_k = \delta_1+\ldots+\delta_{k+1}$ rows of $S_{k+1}(x)$ are linearly independent at each x in a neighborhood of x^o. □

From these Lemmas one may deduce a series of interesting conclusions. First of all, the comparison of (2.9) with IV.(3.22) shows the coincidence of the δ_i's defined by means of IV.(3.23) with the δ_k's defined by means of the Structure Algorithm. Since the latter operates on data associated with the input-output behavior (the matrices $T_k(x)$), it follows that at least in the case of systems in which the Input-Output Linearization Problem has solutions, the integers IV.(3.23) have an interpretation in terms of input-output data. As a matter of fact, there is an explicit formula relating the δ_k's to the matrices $T_k(x)$'s. Following a procedure similar to the one suggested in the proof of Lemma (1.17), one may arrive at the conclusion that

$$r_K(M_k) = (k+1)\delta_1 + k\delta_2+\ldots+\delta_{k+1}$$

or, in other words, that

$$\delta_1 = r_{\mathbb{K}}(M_0)$$

$$\sum_{i=1}^{k+1} \delta_i = r_{\mathbb{K}}(M_k) - r_{\mathbb{K}}(M_{k-1}) \qquad k \geq 1$$

Since by definition $\delta_i = 0$ for $i > q$ (the last nondegenerate stage of the Structure Algorithm) and $\delta_q \neq 0$, one deduces from (2.9) that $\Omega_{q-1} \supsetneq \Omega_{q-2}$ and, therefore, that the integer k^* (which characterizes the last meaningful stage of the Algorithm IV.(2.5)) is related to q by the inequality

$$k^* \geq q-1 \tag{2.10}$$

A sufficient condition for (2.10) to become an equality is the following one.

(2.11) *Lemma.* If the number of rows of S_q is equal to ℓ, then $k^* = q-1$.

Proof. Suppose the number of rows of S_q is equal to ℓ. Then the algorithm (1.15) terminates at the q-th stage. From Lemma (2.2) we deduce that

$$\Omega_q = \Omega_{q-1}$$

i.e. that $k^* = q-1$. □

The case in which the assumption of this Lemma holds (namely, the case in which the algorithm (1.15) terminates at a finite stage) deserves a special attention, because of some interesting properties that will be pointed out hereafter.

(2.12) *Lemma.* If the number of rows of S_q is equal to ℓ, then, in a neighborhood of x^o, the distribution

$$\Delta^* = \bigcap_{i=1}^{q} sp\{d\gamma_i\}^{\perp} \cap \bigcap_{i=1}^{q-1} sp\{d\bar{\gamma}_i\}^{\perp}$$

coincides with the largest locally controlled invariant distribution contained in H and any pair of feedback functions α and β which solves the equations (1.18) is such as to make Δ^* invariant.

Proof. The first part of the statement is a consequence of (2.3) and Lemma (2.11). The second part may be proved exactly as done in the last part of the proof of Proposition IV.(3.19). □

(2.13) *Lemma*. If the number of rows of S_q is equal to ℓ, then the differentials of the entries of the vectors γ_i, $1 \le i \le q$, and $\bar{\gamma}_i$, $1 \le i \le q-1$, are linearly independent at x^o.

Proof. Let $n_i = \ell - r_{i-1}$ denote the number of entries of $\bar{\gamma}_i$. We prove that if

$$\dim(\sum_{i=1}^{p} sp\{d\bar{\gamma}_i(x^o)\}) < \sum_{i=1}^{p} n_i \tag{2.14}$$

for some p, then all $\bar{\gamma}_i$'s with $i \ge p+1$ are nontrivial.

We know from Lemma (2.2) that the codistribution

$$\Omega_{p-1} \cap G^{\perp} = \sum_{i=1}^{p} sp\{d\bar{\gamma}_i\}$$

has constant dimension around x^o. Thus, if (2.14) holds, then there exist $k \le p$ row vectors $\lambda_1,\ldots,\lambda_k$ of smooth real-valued functions (whose dimensions are respectively $1\times n_1,\ldots,1\times n_k$) defined in a neighborhood of x^o, with $\lambda_k \ne 0$, such that

$$\lambda_k(x)d\bar{\gamma}_k(x) = \lambda_1(x)d\bar{\gamma}_1(x)+\ldots+\lambda_{k-1}(x)d\bar{\gamma}_{k-1}(x) \tag{2.15}$$

for all x around x^o.

Differentiating (2.15) along f yields

$$(L_f\lambda_k)d\bar{\gamma}_k + \lambda_k(dL_f\bar{\gamma}_k) = \sum_{i=1}^{k-1}((L_f\lambda_i)d\bar{\gamma}_i + \lambda_i(dL_f\bar{\gamma}_i))$$

and also (see Remark (2.1))

$$(L_f\lambda_k)d\bar{\gamma}_k + \lambda_k(Q'_{k+1}d\gamma_{k+1}+Q''_{k+1}d\bar{\gamma}_{k+1}-Q''_{k+1}K_1^{k+1}d\gamma_1-\ldots-Q''_{k+1}K_k^{k+1}d\gamma_k) =$$

$$= \sum_{i=1}^{k-1}((L_f\lambda_i)d\bar{\gamma}_i+\lambda_i(Q'_{i+1}d\gamma_{i+1}+Q''_{i+1}d\bar{\gamma}_{i+1}-Q''_{i+1}K_1^{i+1}d\gamma_1-\ldots-Q''_{i+1}K_i^{i+1}d\gamma_i))$$

This may be rewritten as

$$\lambda_k(Q'_{k+1}d\gamma_{k+1} + Q''_{k+1}d\bar{\gamma}_{k+1}) = \sum_{i=1}^{k}(\mu_i d\gamma_i + \bar{\mu}_i d\bar{\gamma}_i) \tag{2.16}$$

for suitable μ_i's and $\bar{\mu}_i$'s.

We will see now that $\bar{\gamma}_{k+1}$ is nontrivial and that there exist k+1 row vectors $\lambda'_1,\ldots,\lambda'_{k+1}$, of smooth real-valued functions defined in a neighborhood of x^o, with $\lambda'_{k+1} \ne 0$, such that

$$(2.17) \qquad \lambda'_{k+1}(x)d\bar{\gamma}_{k+1}(x) = \lambda'_1(x)d\bar{\gamma}_1(x)+\ldots+\lambda'_k(x)d\bar{\gamma}_k(x)$$

To this end note that, bearing in mind (2.4) and (2.9),(2.16) yields

$$\lambda_k Q'_{k+1} = 0$$

$$\mu_i = 0 \qquad 1 \leq i \leq k$$

$$\lambda_k Q''_{k+1} \, d\bar{\gamma}_{k+1} = \bar{\mu}_1 d\bar{\gamma}_1+\ldots+\bar{\mu}_k d\bar{\gamma}_k$$

If $\bar{\gamma}_{k+1}$ were trivial, then $Q'_{k+1} = I$ and $\lambda_k = 0$, i.e. a contradiction. Thus $\bar{\gamma}_{k+1}$ is nontrivial and, also, $\lambda_k Q''_{k+1} \neq 0$ because otherwise the equality

$$\lambda_k (Q'_{k+1} \quad Q''_{k+1}) = 0$$

would contradict the nonsingularity of $(Q'_{k+1} \quad Q''_{k+1})$. This shows that (2.17) holds, with $\lambda'_{k+1} = \lambda_k Q''_{k+1}$ and $\lambda'_i = \bar{\mu}_i$ for $1 \leq i \leq k$.

We can iterate this argument and conclude that all $\bar{\gamma}'_i$s with $i \geq k+1$ are nontrivial. If the algorithm terminates at some step q, then (2.14) is contradicted and the differentials of the entries of $\bar{\gamma}_1,\ldots,\bar{\gamma}_{q-1}$ are linearly independent at x^o. □

The above results enable us to investigate the effect of the linearizing feedback on the state-space description of the system. From the last Lemma it is seen that the entries of $\gamma_i(x)$ and $\bar{\gamma}_i(x)$ are part of a local coordinate system. Thus, one may set

$$\xi_i = \gamma_i(x) \qquad 1 \leq i \leq q$$

$$\bar{\xi}_i = \bar{\gamma}_i(x) \qquad 1 \leq i \leq q-1$$

and find a suitable vector-valued function η with the property that the mapping

$$x \longmapsto (\xi_1,\ldots,\xi_q,\bar{\xi}_1,\ldots,\bar{\xi}_{q-1},\eta)$$

is a local coordinate transformation.

The description of the system

$$\dot{x} = \tilde{f}(x) + \tilde{g}(x)v$$

$$y = h(x)$$

in the new coordinates may be easily obtained in the following way. Consider the right-hand-side of (1.18b) and let $E_1,\ldots,E_q$ be $\delta_i \times m$ matrices which partition $[I_\ell \;\; 0]$ as

(2.18) $$[I_\ell \;\; 0] = \begin{bmatrix} E_1 \\ \vdots \\ E_q \end{bmatrix}$$

Then, if $\alpha(x)$ and $\beta(x)$ are solutions of (1.18), one has

$$L_{\tilde{f}}\gamma_i = 0$$

$$L_{\tilde{g}}\gamma_i = E_i$$

for $1 \leq i \leq q$. These yield for $\xi_1,\ldots,\xi_q$ the equations

(2.19a) $$\dot{\xi}_i = \dot{\gamma}_i = L_{\tilde{f}}\gamma_i + L_{\tilde{g}}\gamma_i v = E_i v$$

Moreover (see Remark (2.1)),

$$L_{\tilde{f}}\bar{\gamma}_i = L_f\bar{\gamma}_i = Q'_{i+1}\gamma_{i+1} + Q''_{i+1}(\bar{\gamma}_{i+1} - K_1^{i+1}\gamma_1 \ldots - K_i^{i+1}\gamma_i)$$

$$L_{\tilde{g}}\bar{\gamma}_i = 0$$

for $1 \leq i \leq q-2$. For $i = q-1$

$$L_{\tilde{f}}\bar{\gamma}_{q-1} = \gamma_q$$

$$L_{\tilde{g}}\bar{\gamma}_{q-1} = 0$$

From these one gets

(2.19b) $$\dot{\bar{\xi}}_i = Q'_{i+1}\xi_{i+1} + Q''_{i+1}\bar{\xi}_{i+1} - Q''_{i+1}K_1^{i+1}\xi_1 - \ldots - Q''_{i+1}K_i^{i+1}\xi_i$$

for $1 \leq i \leq q-2$ and

(2.19c) $$\dot{\bar{\xi}}_{q-1} = \xi_q.$$

The output y is related to ξ_1 and $\bar{\xi}_1$ in the following way

(2.19d) $$y = Q_1'\xi_1 + Q_1''\bar{\xi}_1$$

Combining the (2.19)'s, one finds in the new coordinates a state space description of the form

$$\dot{z} = Fz + Gv$$

$$\dot{\eta} = f(z,\eta) + g(z,\eta)v$$

$$y = Hz$$

with

$$z = \mathrm{col}(\xi_1,\ldots,\xi_q,\bar{\xi}_1,\ldots,\bar{\xi}_{q-1})$$

and

$$F = \begin{bmatrix} 0 & 0 & 0 & \cdots & 0 & 0 & 0 & 0 & 0 & \cdots & 0 \\ \cdot & \cdot & \cdot & \cdots & \cdot & \cdot & \cdot & \cdot & \cdot & \cdots & \cdot \\ 0 & 0 & 0 & \cdots & 0 & 0 & 0 & 0 & 0 & \cdots & 0 \\ -Q_2''K_1^2 & Q_2' & 0 & \cdots & 0 & 0 & 0 & Q_2'' & 0 & \cdots & 0 \\ -Q_3''K_1^3 & -Q_3''K_2^3 & Q_3' & \cdots & 0 & 0 & 0 & 0 & Q_3'' & \cdots & 0 \\ \cdot & \cdot & \cdot & \cdots & \cdot & \cdot & \cdot & \cdot & \cdot & \cdots & \cdot \\ -Q_{q-1}''K_1^{q-1} & -Q_{q-1}''K_2^{q-1} & -Q_{q-1}''K_3^{q-1} & \cdots & Q_{q-1}' & 0 & 0 & 0 & 0 & \cdots & Q_{q-1}'' \\ 0 & 0 & 0 & \cdots & 0 & I & 0 & 0 & 0 & \cdots & 0 \end{bmatrix}$$

$$G = \begin{bmatrix} E_1 \\ \cdot \\ E_q \\ 0 \\ \cdot \\ 0 \end{bmatrix}$$

$$H = [Q_1' \ 0 \ \ldots \ 0 \ Q_1'' \ 0 \ \ldots \ 0]$$

The particular choice of feedback makes Δ^* invariant (see Lemma (2.11)) and this is the reason for the presence of an unobservable (and nonlinear) subsystem. The other subsystem, which is the only one contributing to the input-output response, is fully linear.

We conclude the section with two remarks, which are consequences of the above result.

(2.20) *Remark.* From the above equations, we see that if the Algorithm (1.15) terminates at the q-th stage (i.e. if the number of rows of S_q is equal to the number ℓ of outputs), the response of the closed loop system becomes

$$y(t) = He^{Ft}z^o + \int_0^t He^{F(t-\tau)}Gu(\tau)d\tau$$

The input-dependent part is linear in the input, as expected, but also the zero-input term is linear in the initial state z^o.

(2.21) *Remark.* The structure of the matrix F which characterizes the linear part of the closed-loop system shows that all its eigenvalues are vanishing. Thus, one might wish to add an additional feedback in order to achieve not only a linear input-output behavior, but a linear and *stable* input-output behavior, if possible. As a matter of fact, the pair of matrices (F,G) turns out to be a *reachable* pair and so a matrix K may always be found which assigns the spectrum to F+GK. In order to obtain a linear input-output behavior with prescribed spectral properties, instead of the feedback $\alpha(x)$ and $\beta(x)$ proposed so far, one has to consider the feedback

$$\alpha'(x) = \alpha(x) + \beta(x)Kz(x)$$

$$\beta'(x) = \beta(x)$$

The reachability of the pair (F,G) may be checked by direct computation of the rank of $(G \;\; FG \;\; F^2G \; \ldots)$. At each stage, it is suggested to take advantage of the nonsingularity of $[Q_i' \;\; Q_i'']$ in order to prove that new linearly independent columns are added. □

3. Some Algebraic Properties

In this section we analyze the structure of the formal power series (1.7) with some detail, and show that the integers $\delta_1,\ldots,\delta_q$ are related to the behavior of $T(s,x)$ for $s \to \infty$.

In the proof of Theorem (1.11), we have shown that the existence of a solution of the Input-Output Linearization Problem at x^o makes it possible to separate $T(s,x)$ as a product of two formal power series as in (1.12). In particular, it was shown that, if α and β are a linearizing feedback, then

$$K(s) = \sum_{k=0}^{\infty} (L_{\tilde g} L_{\tilde f}^k h) s^{-k-1}$$

($L_{\tilde g} L_{\tilde f}^k h$ being independent of x for all $k \geq 0$) and

$$R(s,x) = \beta^{-1}(x) - \sum_{k=0}^{\infty} (L_g L_f^k \beta^{-1} \alpha(x)) s^{-k-1}$$

Clearly, (1.12) holds in the neighborhood U of x^o where the feedback α and β is defined.

An explicit expression for $L_{\tilde g} L_{\tilde f}^k h$, that is for $K(s)$, is not difficult to obtain. For, consider again the matrix H defined in (1.23) and let α and β be any solution of (1.18). Simple computations, based on appropriate use of the properties (1.20), yield

$$Hh = \begin{pmatrix} \gamma_1 \\ P_2 \bar{\gamma}_1 \\ P_3 K_2^2 \bar{\gamma}_1 \\ \vdots \\ K_q^q \dots K_2^2 \bar{\gamma}_1 \end{pmatrix}$$

$$L_{\tilde f} Hh = \begin{pmatrix} 0 \\ \gamma_2 \\ P_3(\bar{\gamma}_2 - K_1^2 \gamma_1) \\ \vdots \\ K_q^q \; \dots \; K_3^3(\bar{\gamma}_2 - K_1^2 \gamma_1) \end{pmatrix}$$

$$L_{\tilde f}^2 Hh = \begin{pmatrix} 0 \\ 0 \\ \gamma_3 \\ \vdots \\ K_q^q \; \dots \; K_4^4(\bar{\gamma}_3 - K_1^3 \gamma_1 - K_2^3 \gamma_2) \end{pmatrix}$$

and so on, until

$$L_{\tilde{f}}^{q-1}Hh = \begin{pmatrix} 0 \\ \cdot \\ \cdot \\ 0 \\ \gamma_q \\ \bar{\gamma}_q - K_1^q\gamma_1 - \ldots - K_{q-1}^q\gamma_{q-1} \end{pmatrix}$$

and

$$L_{\tilde{f}}^{q+i}Hh = \begin{pmatrix} 0 \\ \cdot \\ \cdot \\ 0 \\ 0 \\ \bar{\gamma}_{q+i+1} - K_1^{q+i+1}\gamma_1 - \ldots - K_q^{q+i+1}\gamma_q \end{pmatrix}$$

which holds for all $i \geq 0$.

Differentiation of these along $\tilde{g}_1, \ldots, \tilde{g}_m$ enables us to obtain the expression of $H(L_{\tilde{g}}L_{\tilde{f}}^k h)$ for all $k \geq 0$. Use of the partition (2.18) makes it possible to get

$$HL_{\tilde{g}}h = \begin{pmatrix} E_1 \\ 0 \\ 0 \\ \vdots \\ 0 \end{pmatrix}$$

$$HL_{\tilde{g}}L_{\tilde{f}}h = \begin{pmatrix} 0 \\ E_2 \\ -P_3K_1^2E_1 \\ \cdot \\ \cdot \\ -K_q^q \ldots K_3^3K_1^2E_1 \end{pmatrix}$$

and so on, until

$$HL_{\tilde g}L_{\tilde f}^{q-1}h = \begin{pmatrix} 0 \\ \cdot \\ \cdot \\ 0 \\ E_q \\ -K_1^q E_1 - \ldots - K_{q-1}^q E_{q-1} \end{pmatrix}$$

and, for all $i \geq 0$,

$$HL_{\tilde g}L_{\tilde f}^{q+i}h = \begin{pmatrix} 0 \\ \cdot \\ \cdot \\ 0 \\ 0 \\ -K_1^{q+i+1} E_1 - \ldots - K_q^{q+i+1} E_q \end{pmatrix}$$

Since $E_1, \ldots, E_q$ are rows of the matrix $[I_q \;\; 0]$, one easily understands that the formal power series

$$W(s) = H \sum_{k=0}^{\infty} (L_{\tilde g}L_{\tilde f}^{k}h) s^{-k-1} \tag{3.1}$$

displays the following pattern of elements

$$W(s) = \begin{bmatrix} I_{\delta_1} s^{-1} & 0 & \ldots & & 0 & 0 \\ 0 & I_{\delta_2} s^{-2} & \ldots & & 0 & 0 \\ W_{31}(s) & 0 & \ldots & & 0 & 0 \\ W_{41}(s) & W_{42}(s) & \ldots & & 0 & 0 \\ \cdot & \cdot & \ldots & & \cdot & \cdot \\ W_{q,1}(s) & W_{q,2}(s) & \ldots & & I_{\delta_q} s^{-q} & 0 \\ W_{q+1,1}(s) & W_{q+1,2}(s) & \ldots & W_{q+1,q-1}(s) & W_{q+1,q}(s) & 0 \end{bmatrix} \tag{3.2}$$

We recall that the partition for the rows corresponds to a partition of the output vector into q+1 blocks of dimensions

$\delta_1,\ldots,\delta_q,(\ell-r_q-1)$, while the one for the columns corresponds to a partition of the input vector into q+1 blocks of dimensions $\delta_1,\ldots,\delta_q,(m-r_q-1)$.

(3.3) *Remark*. Note that, if the Algorithm (1.15) terminates at the q-th stage (i.e. if the number of rows of S_q is equal to the number ℓ of outputs), the (q+1)-th block-row of the matrix W(s) does not exist, and the matrix itself is *right-invertible*. □

From the previous expression for $HL_{\tilde{g}}L_{\tilde{f}}^k h$, one also sees that in the j-th block column of (3.1), $1 \leq j \leq q$, the largest power of s appearing in any off-diagonal element is $-(j+1)$. As a consequence,one may conclude that $W_{ij}(s)s^j$ is a strictly proper formal power series. This property will be immediately used in the following way. Set

$$(3.4)\qquad P_1(s) = \begin{bmatrix} I & 0 & \ldots & 0 \\ 0 & I & \ldots & 0 \\ -W_{31}(s)s & 0 & \ldots & 0 \\ -W_{41}(s)s & 0 & \ldots & 0 \\ \cdot & \cdot & \ldots & \cdot \\ -W_{q+1,1}(s)s & 0 & \ldots & I \end{bmatrix}$$

and note that

$$P_1(s)W(s) = \begin{bmatrix} I_{\delta_1}s^{-1} & 0 \\ 0 & \hat{W}_{22}(s) \end{bmatrix}$$

$\hat{W}_{22}(s)$ being the lower-right-hand $(\ell-\delta_1)\times(m-\delta_1)$ submatrix of W(s).

The power series $P_1(s)$ is proper (because of the aforementioned property of $-W_{31}(s)s,\ldots,-W_{q+1,1}(s)s$) and its inverse too. A proper formal power series whose inverse is also proper is called a *biproper* power series. Thus, we have that the power series (3.4) is biproper.

Continuing this process, one can find biproper formal power series $P_2(s),\ldots,P_q(s)$ which reduce W(s) to a purely diagonal form, and prove the following interesting result.

(3.5) *Theorem*. Suppose the system (1.1) is such that the Input-Output Linearization Problem has a solution at x^o. Then there exist a biproper formal power series

$$R(s) = R_{-1}(x) + \sum_{k=0}^{\infty} R_k(x)s^{-k-1}$$

whose coefficients are m×m matrices of smooth functions defined on a neighborhood U of x^o, and a biproper formal power series

$$L(s) = L_{-1} + \sum_{k=0}^{\infty} L_k s^{-k-1}$$

whose coefficients are $\ell\times\ell$ matrices of real numbers such that

$$T(s,x) = L(s)\Lambda(s)R(s,x) \tag{3.6}$$

where

$$\Lambda(s) = \mathrm{diag}\{I_{\delta_1}\frac{1}{s}, I_{\delta_2}\frac{1}{s^2}, \ldots, I_{\delta_q}\frac{1}{s^q}, 0\} \tag{3.7}$$

Proof. The formal power series

$$P(s) = P_q(s)\ldots P_1(s)H$$

is biproper, because each $P_i(s)$ is and H is invertible. On the other hand,

$$P_q(s)\ldots P_1(s)W(s) = \Lambda(s)$$

and thus (3.6) follows from

$$L(s) = P^{-1}(s). \square$$

A factorization of the form (3.6) reveals the behavior of T(s,x) as $s \to \infty$. As a matter of fact, the limits of L(s) and R(s,x) for $s \to \infty$ are nonsingular; in i-th set of diagonal elements of $\Lambda(s)$, each function has *a zero of multiplicity* i at the infinity. For this reason, the string $\{\delta_1, \delta_2, \ldots\}$ is known as the *structure at the infinity* of the formal power series T(s,x), or of the system (1.1). Note that the string $\{\delta_1, \delta_2, \ldots\}$ is uniquely associated with T(s,x) and does not depend on the particular procedure chosen to obtain a factorization of the form (3.6).

(3.8) *Remark*. We have seen before that the integers $\delta_1, \delta_2, \ldots$ are related to the dimensions of the codistributions $\Omega_0, \Omega_1, \ldots$ generated by means of the Controlled Invariant Distribution Algorithm (see Lemma (2.8)). In other words, we have

$$\delta_1 = \dim \frac{\Omega_0}{\Omega_0 \cap G^{\perp}} \tag{3.9a}$$

$$\delta_{i+1} = \dim \frac{\Omega_i}{\Omega_i \cap G^{\perp}} - \dim \frac{\Omega_{i-1}}{\Omega_{i-1} \cap G^{\perp}} \qquad i \geq 1 \tag{3.9b}$$

Since $\Omega_0, \Omega_1, \ldots$ and G are invariant under feedback transformations (see Lemma IV.(2.8)), it turns out that the structure at the infinity of a system is invariant under feedback transformations.

4. Linear model matching

In the first section of this Chapter we have seen that, under suitable conditions, it is possible to synthesize a feedback under which the input-dependent part of the response of a given nonlinear system becomes the same as that of a linear system. Our aim was the one of achieving a response of the form (1.3), without any particular prescription on the first order kernels $k_i(t)$, $1 \leq i \leq m$. As a matter of fact, the transfer function $K(s)$ obtained for the linearized part of the response, whose form was analized in the previous section, happens to depend on the particular choice of feedback, i.e. on the particular matrices $P_i, K_1^i, \ldots, K_{i-1}^i$ selected at each stage of the Structure Algorithm.

The purpose of the present section is to discuss a more demanding problem, the one in which a *prescribed* linear input-output behavior rather than *some* linear input-output behavior is sought. We tackle this new synthesis problem in a more general setting than before, letting the state-feedback to be *dynamic* rather than static. This means that we let u_i to be related to the state x and, possibly, to other input variables $v_1, \ldots, v_\mu$ by means of equations of the form

$$\dot{\xi} = a(\xi, x) + \sum_{j=1}^{\mu} b_j(\xi, x) v_j \tag{4.1a}$$

$$u_i = c_i(\xi, x) + \sum_{i=1}^{\mu} d_{ij}(\xi, x) v_j \tag{4.1b}$$

These equations characterize a new dynamical system, whose state ξ evolves on an open subset of $\mathbb{R}^\nu$. As usual, we assume that all functions which characterize these equations are smooth functions, defined now on a subset of $\mathbb{R}^\nu \times \mathbb{R}^n$. Most of the times, we shall consider $b_j(\xi,x)$ as the j-th column of a $\nu \times \mu$ matrix $b(\xi,x)$, $c_i(\xi,x)$ as the i-th row of an $m\times 1$ vector $c(\xi,x)$ and $d_{ij}(\xi,x)$ the (i,j)-th entry of a matrix $d(\xi,x)$. Note that the number μ of new inputs may be different from m.

The composition of (4.1) with (1.1) defines a new dynamical system, with input $v = \mathrm{col}(v_1,\ldots,v_\mu)$, output $y = \mathrm{col}(y_1,\ldots,y_\ell)$ described by equations of the form

$$\begin{pmatrix}\dot{\xi}\\ \dot{x}\end{pmatrix} = \hat{f}(\xi,x) + \sum_{i=1}^{\mu} \hat{g}_i(\xi,x)v_i \tag{4.2a}$$

$$y_i = \hat{h}_i(\xi,x) \tag{4.2b}$$

in which

$$\hat{f}(\xi,x) = \begin{pmatrix} a(\xi,x) \\ f(x) + \sum_{i=1}^{m} g_i(x)c_i(\xi,x) \end{pmatrix}$$

$$\hat{g}_i(\xi,x) = \begin{pmatrix} b_i(\xi,x) \\ \sum_{j=1}^{m} g_j(x)d_{ji}(\xi,x) \end{pmatrix}$$

$$\hat{h}_i(\xi,x) = h_i(x)$$

The integer ν, which characterizes the dimension of the dynamical system (4.1), and the quadruplet (a,b,c,d) are to be chosen in such a way as to obtain, for the closed loop system (4.2), an input-output response of the form (see (1.3))

$$y(t) = Q(t,(\xi^o,x^o)) + \int_0^t W_M(t-\tau)v(\tau)d\tau \tag{4.3}$$

$W_M(t)$ being a *fixed* $\ell\times\mu$ matrix of functions of t, the impulse-response matrix of a prespecified linear model. As before, we seek local solutions defined in a neighborhood of the initial state. In view of our earlier discussions, this yields the following formal statement.

Linear Model Matching Problem. Given (f,g,h), an initial state x^o, and a linear model (A,B,C), find (if possible) an integer ν, an initial state $\xi^o \in \mathbb{R}^\nu$, a quadruplet of smooth functions (a,b,c,d) defined in a neighborhood U of (ξ^o, x^o) such that for all $k \geq 0$

$$(4.4) \qquad L_{\hat{g}} L_{\hat{f}}^{k} \hat{h}(\xi, x) = CA^k B. \ \square$$

If the system (1.1) is such that a solution to the Input-Output Linearization Problem exists, then it is quite simple to find the extra conditions needed for the existence of a solution to the Linear Model Matching Problem and to construct such a solution. The main tool is again the Structure Algorithm described in the first section.

The data of a Linear Model Matching Problem are, besides the initial point x^o, the triplet (f,g,h) which characterizes the system to be controlled and the triplet (A,B,C) which characterizes the model to be reproduced. These data will be used in order to define an *extended* system, described by the following set differential equations

$$(4.5) \qquad \begin{aligned} \dot{x} &= f(x) + g(x)u \\ \dot{z} &= Az + Bv \\ w &= h(x) - Cz \end{aligned}$$

The output w of this system is actually the difference between the output of the system (1.1) and that of the model. For convenience, we represent (4.5) in the form

$$\begin{aligned} \dot{x}^E &= f^E(x^E) + g^E(x^E)u^E \\ w &= h^E(x^E) \end{aligned}$$

letting $x^E = \text{col}(x,z)$, $u^E = \text{col}(u,v)$ and

$$(4.6a) \qquad f^E(x,z) = \begin{pmatrix} f(x) \\ Az \end{pmatrix} \qquad g^E(x,z) = \begin{pmatrix} g(x) & 0 \\ 0 & B \end{pmatrix}$$

$$(4.6b) \qquad h^E(x,z) = h(x) - Cz$$

The conditions for the existence of a solution to the Linear Model Matching Problem may easily be expressed in terms of properties of the system thus defined, as we will see hereafter.

Suppose the system (1.1) is such that the Input-Output Linearization Problem has a solution at x^o. Then, the triplet (f,g,h) is such as to fulfill the condition (a) of Theorem (1.11). It is easily seen that, for any z^o, also the triplet (f^E,g^E,h^E) is such as to fulfill a similar condition at (x^o,z^o). For, let

$$T_k^E(x,z) = L_{g^E}L_{f^E}^k h^E(x,z) = [\,L_gL_f^k h(x) \quad CA^kB\,]$$

Then

$$\sum_{k=0}^{\infty} T_k^E(x,z)s^{-k-1} = [\,T(s,x) \quad \sum_{k=0}^{\infty} CA^kBs^{-k-1}\,] =$$

$$= [\,K(s) \quad W_M(s)\,]\begin{bmatrix} R(s,x) & 0 \\ 0 & I \end{bmatrix}$$

where

$$W_M(s) = \sum_{k=0}^{\infty} CA^kBs^{-k-1}$$

denotes the transfer function of the model.

As a consequence of this, the Algorithm (1.15) may also be performed on the triplet (f^E,g^E,h^E), around the point (x^o,z^o), and one may define on the formal power series $T^E(s,x)$ a structure at the infinity, characterized by a string of integers $\{\delta_1^E,\delta_2^E,\ldots\}$.

The coincidence between the structure at infinity of the formal power series $T(s,x)$ and that of the formal power series $T^E(s,x)$ is exactly the condition that characterizes the possibility of solving a Linear Model Matching Problem. In order to be able to prove this result and give an explicit construction of the required feedback, we need a little more notation.

Let $P_i,K_1^i,\ldots,K_i^i$ be the set of matrices determined at the i-th stage of the Structure Algorithm, when operating on the triplet (f,g,h). Let the triplet (A,B,C) characterize the model to be followed. We set

$$C_1 = P_1C \tag{4.7a}$$

(4.7b) $$\bar{C}_1 = K_1^1 C$$

and, for $i \geq 2$

(4.7c) $$C_i = P_i \bar{C}_{i-1} A$$

(4.7d) $$\bar{C}_i = K_1^i C_1 + \ldots + K_{i-1}^i C_{i-1} + K_i^i \bar{C}_{i-1} A$$

With the functions $\gamma_1(x),\ldots,\gamma_q(x)$ determined at each nondegenerate stage of the Algorithm we associate, as before, a matrix

$$\Gamma(x) = \begin{bmatrix} \gamma_1(x) \\ \vdots \\ \gamma_q(x) \end{bmatrix}$$

and with the matrices $C_1,\ldots,C_q$ defined above we associate the matrix

(4.8) $$D = \begin{bmatrix} C_1 \\ \vdots \\ C_q \end{bmatrix}$$

The constructions defined above are helpful in finding a solution to the problem in question.

(4.9) *Theorem*. Suppose the system (1.1) is such that the Input-Output Linearization Problem is solvable at x^o. The Linear Model Matching Problem is solvable at x^o if and only if either one of the following equivalent conditions is satisfied

(a) $\bar{C}_i B = 0$ for all $i \geq 1$

(b) the system (1.1) and the extended system (4.5) are characterized by the same structure at the infinity.

A dynamical state-feedback which solves the problem is the one described by the following equations

(4.10a) $$\dot{\xi} = A\xi + Bv$$

(4.10b) $$u = \alpha(x) - \beta(x)DA\xi - \beta(x)DBv$$

in which $\alpha(x)$ and $\beta(x)$ are solutions of

$$L_g\Gamma(x)\alpha(x) = -L_f\Gamma(x) \tag{4.11a}$$

$$L_g\Gamma(x)\beta(x) = I_{r_{q-1}} \tag{4.11b}$$

with D defined as in (4.8). The initial state ξ^o of (4.10) may be set arbitrarily.

Proof. (a) ⇔ (b). Consider the sequence of functions thus defined

$$\gamma_1^E(x) = P_1 h^E(x)$$

$$\bar{\gamma}_1^E(x) = K_1^1 h^E(x)$$

and, for $i \geq 2$,

$$\gamma_i^E(x) = P_i L_{f^E}\bar{\gamma}_{i-1}^E(x)$$

$$\bar{\gamma}_i^E(x) = K_1^i\gamma_1^E(x)+\ldots+K_{i-1}^i\gamma_{i-1}^E(x)+K_i^i L_{f^E}\bar{\gamma}_{i-1}^E(x)$$

Note, also, that for all $i \geq 1$

$$\gamma_i^E(x) = \gamma_i(x) + C_i z$$

$$\bar{\gamma}_i^E(x) = \bar{\gamma}_i(x) + \bar{C}_i z$$

Suppose that, for all $1 \leq k \leq i$,

$$\bar{C}_k B = 0$$

Then, for all $1 \leq k \leq i$,

$$L_{g^E}\bar{\gamma}_k^E(x) = 0$$

and the matrix

$$\begin{bmatrix} L_{g^E}\gamma_1^E(x) \\ \vdots \\ L_{g^E}\gamma_k^E(x) \end{bmatrix} = \begin{bmatrix} L_g\gamma_1(x) & C_1B \\ \vdots & \vdots \\ L_g\gamma_k(x) & C_kB \end{bmatrix}$$

has a rank equal to the number $\delta_1+\ldots+\delta_k$ of its rows, at (x^o,z^o) (note that z^o is irrelevant). As a consequence, one may conclude that the first i steps of the Structure Algorithm on the triplet(f^E,g^E,h^E) may be performed exactly in the same way as on the triplet(f,g,h). At each of these steps, the same set of matrices $P_k,K_1^k,\ldots,K_k^k$ makes it possible to perform the required operations. In particular, since the integers which characterize the structure at the infinity do not depend on the choice of matrices in the Structure Algorithm, we see that the first i entries in the structure at the infinity of (f,g,h) and (f^E,g^E,h^E) coincide. Now, let V_{i+1} be the matrix determined at the (i+1)-th stage of the Algorithm (1.15) and observe that

$$V_{i+1}\begin{bmatrix} L_{g^E}\gamma_1^E(x^E) \\ \vdots \\ L_{g^E}\gamma_i^E(x^E) \\ L_{g^E}L_{f^E}\bar{\gamma}_i^E(x^E) \end{bmatrix} = V_{i+1}\begin{bmatrix} L_g\gamma_1(x) & C_1B \\ \vdots & \vdots \\ L_g\gamma_i(x) & C_iB \\ L_gL_f\bar{\gamma}_i(x) & \bar{C}_iAB \end{bmatrix} = \begin{bmatrix} L_g\gamma_1(x) & C_1B \\ \vdots & \vdots \\ L_g\gamma_{i+1} & C_{i+1}B \\ 0 & \bar{C}_{i+1}B \end{bmatrix}$$

From this we see that the (i+1)-th entries in the structure at the infinity of (f,g,h) and (f^E,g^E,h^E) coincide if and only if

$$\bar{C}_{i+1}B = 0$$

Sufficiency of (b). At the last nondegenerate step of the algorithm one ends up with a matrix

$$\Gamma^E(x^E) = \begin{pmatrix} \gamma_1^E(x^E) \\ \vdots \\ \gamma_q^E(x^E) \end{pmatrix} = \begin{pmatrix} \gamma_1(x) & C_1z \\ \vdots & \vdots \\ \gamma_q(x) & C_qz \end{pmatrix} = (\Gamma(x) \quad Dz)$$

of rank $r_{q-1}=\delta_1+\ldots+\delta_q$ at (x^o,z^o). Consider now the two equations

$$L_{g^E}\Gamma^E(x^E)\alpha^E(x^E) = -L_{f^E}\Gamma^E(x^E) \tag{4.12a}$$

$$L_{g^E}\Gamma^E(x^E)\beta^E(x^E) = (I_{r_{q-1}} \quad 0) \tag{4.12b}$$

which correspond to the equations (1.18). If $\alpha(x)$ and $\beta(x)$ are solutions of (4.11), solutions $\alpha^E(x^E)$ and $\beta^E(x^E)$ of (4.12) may be found as

$$\alpha^E(x,z) = \begin{pmatrix} \alpha(x)-\beta(x)DAz \\ 0 \end{pmatrix}$$

$$\beta^E(x,z) = \begin{pmatrix} \beta(x) & -\beta(x)DB \\ 0 & I_\mu \end{pmatrix}$$

Note that β is $m\times r_{q-1}$ and that β^E is $(m+\mu)\times(r_{q-1}+\mu)$.

Now, suppose the functions α^E and β^E thus defined are used in a static state-feedback loop on the extended system (4.5). As a consequence of all previous discussions, we get

$$(4.13) \qquad L_{g^E\beta^E} L^k_{f^E+g^E\alpha^E} h^E(x^E) = \text{independent of } x^E$$

for all $k \geq 0$ around (x^o,z^o), for any z^o. In particular (see e.g. the structure of (3.2)), the last μ columns of these matrices vanish for all $k \geq 0$.

The extended system (4.5) subject to the feedback thus defined is described by equations of the form

$$\dot{x} = f(x)+g(x)\alpha(x)-g(x)\beta(x)DAz+g(x)\beta(x)\bar{u}-g(x)\beta(x)DB\bar{v}$$

$$(4.14) \quad \dot{z} = Az + B\bar{v}$$

$$w = h(x) - Cz$$

where $\bar{u}$ and $\bar{v}$ represent new inputs. The response of this system consists of a "zero-input" term $w_0(t,(x^o,z^o))$ and of a linear term in $\bar{u}$ alone, because, as we observed, the last μ columns of (4.13) are vanishing. This means that, if $\bar{u} = 0$, the response of such a system consists of $w_0(t,(x^o,z^o))$ alone. Equations (4.14) with $\bar{u} = 0$ may be interpreted as the composition of the original system (1.1) with the dynamic feedback

$$\dot{z} = Az + B\bar{v}$$

$$u = \alpha(x)-\beta(x)D(Az+B\bar{v})$$

together with the a new output map

$$w = y - Cz$$

For all initial states around (x^o, z^o) and all inputs $\bar{v}$

$$w(t) = w_0(t,(x^o,z^o))$$

and, therefore,

$$y(t) = w_0(t,(x^o,z^o)) + Ce^{At}z^o + \int_0^t Ce^{A(t-\tau)}B\bar{v}(\tau)d\tau$$

This shows that the response of the system (1.1) under the feedback (4.10) has the desired form (4.3).

Necessity of (a). This part of the proof consists in a repeated use of the expressions which define $\gamma_i(x)$ and $\bar{\gamma}_i(x)$, in order to show that

$$L_{\hat{g}}L_{\hat{f}}^k\hat{h}(\xi,x) = CA^kB \tag{4.15}$$

for all $k \geq 0$, imply

$$\bar{C}_iB = 0$$

for all $i \geq 0$. One proves first that (4.15) implies

$$[L_gL_f\bar{\gamma}_i(x)]d(\xi,x) = \bar{C}_iAB \tag{4.16a}$$

and that this, in turn, implies

$$[L_g\gamma_i(x)]d(\xi,x) = C_iB \tag{4.16b}$$

for all $i \geq 0$. Then, (4.16) imply the desired result, because

$$\begin{aligned} 0 = (L_g\bar{\gamma}_i)d &= (L_g[K_1^i\gamma_1 + \ldots + K_{i-1}^i\gamma_{i-1} + K_i^iL_f\bar{\gamma}_{i-1}])d = \\ &= K_1^i(L_g\gamma_1)d + \ldots + K_{i-1}^i(L_g\gamma_{i-1})d + K_i^i(L_gL_f\bar{\gamma}_{i-1})d = \\ &= (K_1^iC_1 + \ldots + K_{i-1}^iC_{i-1} + K_i^iC_iA)B = \bar{C}_iB \end{aligned}$$

This completes the proof. □

(4.17) *Remark.* From the above statement, we see that a dynamic state-feedback which solves a Linear Model Matching Problem may easily be found in terms of data related to the solution of an Input-Output Linearization Problem. As a matter of fact, the availability of $P_i, K_1^i, \ldots, K_i^i$, $1 \leq i \leq q$ makes it possible to construct the matrices $C_i, \bar{C}_i$, $1 \leq i \leq q$ and, then, to check the existence condition (a). If this is satisfied, one takes any solution $\alpha(x)$ and $\beta(x)$ of (4.11) (i.e. any feedback solving the Linearization Problem) and constructs a solution of the Model Matching in the form (4.10).

(4.18) *Remark.* In the previous procedure, no special attention was paid to the properties of the zero-input term $Q(t,(\xi^o,x^o))$, which represents the effect of the initial states on the response of the closed loop system (4.2). If an asymptotically decreasing zero-input response is required, one should modify the outlined construction and use, instead of a solution $\alpha(x)$ and $\beta(x)$ of (4.11), a feedback which makes linear and *asymptotically stable* the input-output behavior of the extended system (4.5). This may be accomplished on the basis of the ideas discussed in the Remark (2.21). □

5. More on Linear Model Matching, Output Reproducibility and Noninteraction

In this section we will see that it is possible to solve a Linear Model Matching Problem even though the Input-Output Linearization Problem is *not* solvable. In particular, we will see that the condition (b) of Theorem (4.9) still implies the solvability of the Linear Model Matching problem, even in case the assumption of solvability of the Input-Output Linearization Problem does not hold.

To this end, note first of all that the so-called structure at the infinity can be associated with *any* system of the form (1.1) and not only with input-output-linearizable systems. This is because the string of integers $\{\delta_1,\delta_2,\ldots\}$, that we introduced by means of the Structure Algorithm, can also be independently defined in terms of dimensions of the codistributions generated by means of the Controlled Invariant Distribution Algorithm (see e.g. (3.9)). For this to be possible, it is only required that G, Ω_k and $\Omega_k \cap G^{\perp}$ have constant dimension, for all $k \geq 0$, around the point x^o, i.e. that x^o is a *regular point* for this algorithm (see chapter IV, section 3).

If this is the case, one may associate with the triplet (f,g,h) the sequence of integers

$$(5.1)\qquad r_k = \dim \frac{\Omega_k}{\Omega_k \cap G^{\perp}} \qquad k \geq 0$$

Given also a linear model (A,B,C), we may associate with the extended triplet (f^E, g^E, h^E) (see (4.5)), a similar sequence of integers

$$(5.2)\qquad r_k^E = \dim \frac{\Omega_k^E}{\Omega_k^E \cap G^{E\perp}} \qquad k \geq 0$$

where now the superscript "E" denotes objects pertinent to the extended system, namely

$$G^E = \mathrm{sp}\{g_1^E, \ldots, g_m^E, g_{m+1}^E, \ldots, g_{m+\mu}^E\}$$

$$\Omega_0^E = (\mathrm{sp}\{dh_1^E, \ldots, dh_\ell^E\})$$

$$\Omega_k^E = \Omega_{k-1}^E + L_{f^E}(\Omega_{k-1}^E \cap G^{E\perp}) + \sum_{i=1}^{m+\mu} L_{g_i^E}(\Omega_{k-1}^E \cap G^{E\perp})$$

The structure at the infinity $\{\delta_1, \delta_2, \ldots\}$ of a system of the form (1.1) is uniquely related to the sequence $\{r_1, r_2, \ldots\}$ and, therefore, the equality between the structure at the infinity of the system (1.1) and that of the extended system (4.5) (i.e. the condition (b) of Theorem (4.9)) is equivalent to the equality

$$r_k = r_k^E$$

for all $k \geq 0$.

We prove now that this is still sufficient for the solvability of the problem in question.

(5.3) *Theorem.* Suppose x^o is a regular point of the Algorithm IV.(3.17) for the triplet (f,g,h) and (x^o, z^o) is a regular point of the Algorithm IV.(3.17) for the triplet (f^E, g^E, h^E). Then the Linear Model Matching Problem is solvable at x^o if

$$(5.4)\qquad r_k = r_k^E$$

for all $k \geq 0$.

Proof. We first establish some notations. Let ν denote the dimension of the linear model (A,B,C). Throughout this proof we will be interested in some distributions and/or codistributions defined around the point (x^o, z^o) of $\mathbb{R}^n \times \mathbb{R}^\nu$ in the following way. We set

$$\bar{G}_u = sp\{g_1^E, \ldots, g_m^E\}$$

$$\bar{G}_v = sp\{g_{m+1}^E, \ldots, g_{m+\mu}^E\}$$

and we note that

$$G^E = \bar{G}_u \oplus \bar{G}_v \tag{5.5}$$

Moreover, we define a sequence of codistributions $\bar{\Omega}_k$, $k \geq 0$, as

$$\bar{\Omega}_k(x,z) = \Omega_k(x) \times \{0\} \subset T_x^* \mathbb{R}^n \times T_z^* \mathbb{R}^\nu$$

It is easy to verify that the sequence of codistributions thus defined is such that

$$\bar{\Omega}_{k+1} = \bar{\Omega}_k + \sum_{i=0}^{m+\mu} L_{g_i^E}(\bar{\Omega}_k \cap \bar{G}_u^\perp) \tag{5.6}$$

(with $g_0^E = f^E$) and also that

$$r_k = \dim \frac{\Omega_k}{\Omega_k \cap G^\perp} = \dim \frac{\bar{\Omega}_k}{\bar{\Omega}_k \cap \bar{G}_u^\perp} \tag{5.7}$$

Finally, we define another codistribution Γ as

$$\Gamma(x,z) = \{0\} \times T_z^* \mathbb{R}^\nu$$

We proceed now with the proof, which is divided into three steps.
(i) It will be shown that the assumption (5.4) implies

$$\bar{G}_v^\perp \supset \Omega_k^E \cap \bar{G}_u^\perp \tag{5.8}$$

for all $k \geq 0$. To this end, note first that the assumption (5.4), because of (5.7), may be rewritten as

$$\dim\left(\frac{\bar{\Omega}_k + \bar{G}_u^\perp}{\bar{G}_u^\perp}\right) = \dim\left(\frac{\Omega_k^E + G^{E\perp}}{G^{E\perp}}\right) \tag{5.9}$$

Suppose now that

$$(5.10) \qquad \bar{\Omega}_k + \Gamma = \Omega_k^E + \Gamma$$

for some k. Then, we may deduce the following implications

$$\bar{\Omega}_k + \Gamma = \Omega_k^E + \Gamma \Rightarrow \bar{\Omega}_k + \bar{G}_u^{\perp} = \Omega_k^E + \bar{G}_u^{\perp} \quad (\text{because } \Gamma \subset \bar{G}_u^{\perp})$$

$$\Rightarrow \dim \frac{\Omega_k^E + \bar{G}_u^{\perp}}{\bar{G}_u^{\perp}} = \dim \frac{\Omega_k^E + G^{E\perp}}{G^{E\perp}} \quad (\text{by } (5.9))$$

$$\Rightarrow \dim \Omega_k^E \cap \bar{G}_u^{\perp} = \dim \Omega_k^E \cap G^{E\perp}$$

$$\Rightarrow \Omega_k^E \cap \bar{G}_u^{\perp} = \Omega_k^E \cap G^{E\perp} \quad (\text{because } G^{E\perp} \subset \bar{G}_u^{\perp}).$$

The condition (5.10) also implies (because $\Gamma \subset \bar{G}_u^{\perp}$)

$$(5.11) \qquad \bar{G}_u^{\perp} \cap \bar{\Omega}_k + \Gamma = \bar{G}_u^{\perp} \cap (\bar{\Omega}_k + \Gamma) = \bar{G}_u^{\perp} \cap (\Omega_k^E + \Gamma) = \bar{G}_u^{\perp} \cap \Omega_k^E + \Gamma$$

Thus, we have

$$\bar{\Omega}_{k+1} + \Gamma = \bar{\Omega}_k + \sum_{i=0}^{m+\mu} L_{g_i^E}(\bar{G}_u^{\perp} \cap \bar{\Omega}_k) + \Gamma \qquad \text{by } (5.6)$$

$$= \bar{\Omega}_k + \sum_{i=0}^{m+\mu} L_{g_i^E}(\bar{G}_u^{\perp} \cap \bar{\Omega}_k + \Gamma) + \Gamma$$

$$= \bar{\Omega}_k + \sum_{i=0}^{m+\mu} L_{g_i^E}(\bar{G}_u^{\perp} \cap \Omega_k^E + \Gamma) + \Gamma \qquad \text{by } (5.11)$$

$$= \bar{\Omega}_k + \sum_{i=0}^{m+\mu} L_{g_i^E}(G^{E\perp} \cap \Omega_k^E + \Gamma) + \Gamma \qquad (\text{see above})$$

$$= \bar{\Omega}_k + \sum_{i=0}^{m+\mu} L_{g_i^E}(G^{E\perp} \cap \Omega_k^E) + \Gamma$$

$$= \Omega_k^E + \sum_{i=0}^{m+\mu} L_{g_i^E}(G^{E\perp} \cap \Omega_k^E) + \Gamma = \Omega_{k+1}^E + \Gamma \qquad \text{by } (5.10)$$

This shows that (5.10) holds also for k+1. Since (5.10) is true for $k = 0$, the previous argument shows that it is true for all $k \geq 0$. As a consequence, we have also

$$\Omega_k^E \cap \bar{G}_u^{\perp} = \Omega_k^E \cap G^{E\perp}$$

for all $k \geq 0$ and this, since $G^{E\perp} \subset \bar{G}_v^{\perp}$, implies (5.8). Note that (5.8) may be rewritten as

$$\bar{G}_v \subset \Omega_k^{E\perp} + \bar{G}_u . \tag{5.8'}$$

(ii) Since (x^o,z^o) is a regular point of the Algorithm IV.(3.17) for the triplet (f^E,g^E,h^E), there exists an integer k^* such that, in a neighborhood of (x^o,z^o) $\Omega_k^E = \Omega_{k^*}^E$ for all $k \geq k^*$. Moreover (see Lemma IV.(2.4)) the distribution

$$\Delta^{E*} = \Omega_{k^*}^{E\perp}$$

is such that

$$[f^E, \Delta^{E*}] \subset \Delta^{E*} + G^E$$

$$[g_i^E, \Delta^{E*}] \subset \Delta^{E*} + G^E \qquad 1 \leq i \leq m+\mu$$

From these, using (5.8'), we deduce that

$$[f^E, \Delta^{E*}] \subset \Delta^{E*} + \bar{G}_u \tag{5.12a}$$

$$[g_i^E, \Delta^{E*}] \subset \Delta^{E*} + \bar{G}_u \qquad 1 \leq i \leq m \tag{5.12b}$$

Since $\Omega_{k^*}^E \cap \bar{G}_u^{\perp} = \Omega_{k^*}^E \cap G^{E\perp}$ is nonsingular around (x^o,z^o) so is $\Delta^{E*} + \bar{G}_u$. Also Δ^{E*} and $\bar{G}_u$ are nonsingular and therefore one may use Lemma IV.(1.10) and deduce the existence of an $m\times 1$ vector $\alpha(x,z)$ of smooth functions, defined locally around (x^o,z^o), such that

$$[f^E + \sum_{i=1}^{m} g_i^E \alpha_i \, , \, \Delta^{E*}] \subset \Delta^{E*} \tag{5.13a}$$

Moreover, from the condition (5.8') one may deduce the existence of an $m\times\mu$ matrix $\gamma(x,z)$ of smooth functions, defined locally around (x^o,z^o), such that

$$g_{m+i}^E + \sum_{j=1}^{m} g_j^E \gamma_{ji} \in \Delta^{E*} \qquad 1 \leq i \leq \mu \tag{5.13b}$$

Finally, note that, because of the involutivity of Δ^{E*}, the above condition implies

$$[g^E_{m+i} + \sum_{j=1}^{m} g^E_j \gamma_{ji}, \Delta^{E*}] \subset \Delta^{E*} \qquad 1 \le i \le \mu \tag{5.13c}$$

and recall that

$$\Delta^{E*} \subset (\mathrm{sp}\{dh^E_1,...,dh^E_\ell\})^\perp . \tag{5.13d}$$

(iii) Consider the dynamical system

$$\dot{x}^E = f^E + \sum_{i=1}^{m} g^E_i \alpha_i + \sum_{i=1}^{\mu} (g^E_{m+i} + \sum_{j=1}^{m} g^E_j \gamma_{ji}) v_i$$

$$w = h^E(x^E)$$

This system is such that the conditions (5.13) hold. Thus, thanks to Theorem III.(3.12), we deduce that the inputs $v_1,...,v_\mu$ have no influence on the output w, i.e. that for all initial states (in the neighborhood where $\alpha(x,z)$ and $\gamma(x,z)$ are defined) the response of this system consists of a zero-input term $w_0(t,(x^o,z^o))$ alone. Thus, by means of the same arguments as the ones used at the end of the proof of Theorem (4.9), we conclude that the composition of the original system (1.1) with the dynamic feedback

$$\dot{z} = Az + Bv$$

$$u = \alpha(x,z) + \gamma(x,z)v$$

has a response of the form

$$y(t) = w_0(t,(x^o,z^o)) + Ce^{At}z^o + \int_0^t Ce^{A(t-\tau)}Bv(\tau)d\tau$$

This concludes the proof.

(5.16) *Remark*. The reader may easily check that the value of z^o is irrelevant in the previous discussions.

(5.17) *Remark*. We stress that the proof of the previous Theorem is constructive. The fulfillment of the conditions (5.4) makes it possible to find, locally around (x^o,z^o), a vector $\alpha(x,z)$ such that (5.13a) holds and a matrix $\gamma(x,z)$ such that (5.13b) holds. A dynamical state-feedback which solves the problem in question is the one described by the equations

(5.18a) $$\dot{\xi} = A\xi + Bv$$

(5.18b) $$u = \alpha(x,\xi) + \gamma(x,\xi)v \quad \square$$

As an application of this Theorem, we deduce now an interesting result which is rather useful in connection with problems of output reproducibility and noninteraction (via dynamic feedback).

(5.19) *Corollary*. Suppose $r_{k^*} = \ell$. Then there exists an integer $\delta > 0$ such that the Linear Model Matching Problem is solvable for a linear model (A,B,C) with transfer function

(5.20) $$W_M(s)=C(sI-A)^{-1}B = \begin{pmatrix} \frac{1}{s^\delta} & 0 & \cdots & 0 & 0 & \cdots & 0 \\ 0 & \frac{1}{s^\delta} & \cdots & 0 & 0 & \cdots & 0 \\ \cdot & \cdot & \cdots & \cdot & 0 & \cdots & 0 \\ 0 & 0 & \cdots & \frac{1}{s^\delta} & 0 & \cdots & 0 \end{pmatrix}$$

Proof. It is left as an exercice to the reader.

(5.21) *Remark*. Note that the transfer function (5.20) is right-invertible. Thus, given any smooth ℓ-vector-valued function $\bar{y}$, defined on $\mathbb{R}$, and such that

$$\bar{y}(0) = (\frac{d\bar{y}}{dt})_0 = \cdots = (\frac{d\bar{y}^{\delta-1}}{dt^{\delta-1}})_0 = 0$$

there exists a smooth μ-vector-valued function $\bar{v}$, defined on $\mathbb{R}$, such that

$$\bar{y}(t) = \int_0^t W_M(t-\tau)\bar{v}(\tau)d\tau. \quad \square$$

Now, suppose $r_{k^*} = \ell$ and suppose we have solved the problem of matching a linear model with transfer function (5.20). This means that we have found an appropriate dynamic state-feedback compensator (e.g. the one described in the proof of Theorem (5.3), which has the form (5.18)) under which the input-output behavior of the system (1.1) becomes

$$(5.22) \qquad y(t) = Q(t,(\xi^o,x^o)) + \int_0^t W_M(t-\tau)v(\tau)d\tau$$

Let y^* be any smooth ℓ-vector-valued function, defined on $\mathbb{R}$, such that

$$(5.23) \qquad \left(\frac{d^k y^*(t)}{dt^k}\right)_0 = \left(\frac{d^k Q(t,(\xi^o,x^o))}{dt^k}\right)_0$$

for $0 \leq k \leq \delta-1$.

From the Remark (5.21) we easily deduce that there exists an input v under which the right-hand-side of (5.22) becomes exactly y^*. Thus, the composition of (1.1) with the dynamic state-feedback compensator which solves the problem of matching the transfer function (5.20) is a system that, in the initial state (ξ^o,x^o), *can reproduce any output function* which satisfies the conditions (5.23).

Moreover, we note that in a linear system with transfer function (5.20) each output component is influenced only by the corresponding component of the input. Thus, we also see that if the condition $r_{k^*} = \ell$ holds, we can *achieve non-interaction via dynamic state-feedback.*

6. State-space linearization

In the first section of this chapter, we examined the problem of achieving, via feedback, a linear *input-output* response. The subsequent analysis developed in the second section showed that, from the point of view of a state-space description, in suitable local coordinates, the system thus linearized assumes (at least in the special case where $r_{q-1} = \ell$) the form

$$\dot{z} = Fz + Gv$$
$$\dot{\eta} = f(z,\eta) + g(z,\eta)v$$
$$y = Hz$$

In other words , the input-output-wise linear system one obtains by means of the techniques in question may be interpreted, at a state-space level, as the interconnection of a (possibly) nonlinear unobservable subsystem with a system that, in suitable local coordinates, is state-space-wise linear. Moreover, the latter subsystem was also shown being both reachable and observable (Remark (2.21)).

In other words again, we may say that the techniques developed at the beginning of this chapter modify the behavior of the original system in a way such as to make a part of it (i.e. the observable one) locally diffeomorphic to a reachable linear system.

Motivated by these considerations, we want to examine now the problem of modifying, via feedback, a given nonlinear system in a way such that not simply a part, but the whole of it, is locally diffeomorphic to a reachable linear system. In formal terms, the problem thus introduced may be characterized as follows.

State-Space Linearization Problem. Given a collection of vector fields $f, g_1, \ldots, g_m$ and an initial state x^o, find (if possible) a neighborhood U of x^o, a pair of feedback functions α and β (with invertible β) defined on U, a coordinates transformation $z = F(x)$ defined on U, a matrix $A \in \mathbb{R}^{n\times n}$ and a set of vectors $b_1 \in \mathbb{R}^n, \ldots, b_m \in \mathbb{R}^n$ such that

(6.1) $$F_*(f+g\alpha)\circ F^{-1}(z) = Az$$

(6.2) $$F_*(g\beta)_i\circ F^{-1}(z) = b_i \qquad 1 \le i \le m$$

for all $z \in F(U)$, and

(6.3) $$\sum_{k=0}^{n-1} \sum_{i=1}^{m} \mathrm{Im}(A^k b_i) = \mathbb{R}^n$$

(6.4) *Remark.* Let $x(t)$ denote a state trajectory of the system

$$\dot{x} = (f+g\alpha)(x) + \sum_{i=1}^{m} (g\beta_i)(x)u_i$$

and suppose $x(t) \in U$ for all $t \in [0,T]$ for some $T > 0$. If (6.1) and (6.2) hold, then for all $t \in [0,T]$

$$z(t) = F(x(t))$$

is a state trajectory of the linear system

$$\dot{z} = Az + \sum_{i=1}^{m} b_i u_i$$

Moreover, if (6.3) also holds, the latter is a reachable linear system. □

We shall describe first the solution of this problem in the special case of a system with a single input, which is rather easy. Then, we make some remarks about the usefulness of this linearization technique in problems of asymptotic stabilization. Finally, we conclude the section with the analysis of the (general) multi-input systems.

For the sake of simplicity, we state some intermediate results which may have their own independent interest.

(6.5) *Lemma.* Suppose $m = 1$ and let $g = g_1$. The State-Space Linearization Problem is solvable if and only if there exists a neighborhood V of x^o and a function $\varphi : V \to \mathbb{R}$ such that

$$L_g\varphi(x) = L_gL_f\varphi(x) = \ldots = L_gL_f^{n-2}\varphi(x) = 0 \tag{6.6}$$

for all $x \in V$, and

$$L_gL_f^{n-1}\varphi(x^o) \neq 0 \tag{6.7}$$

Proof. Necessity. Let (A,b) a reachable pair. Then, it is well known from the theory of linear system that there exist a nonsingular $n\times n$ matrix T and a $1\times n$ row vector k such that

$$T(A+bk)T^{-1} = \begin{pmatrix} 0 & 1 & 0 & \ldots & 0 \\ 0 & 0 & 1 & \ldots & 0 \\ \cdot & \cdot & \cdot & \ldots & \cdot \\ 0 & 0 & 0 & \ldots & 1 \\ 0 & 0 & 0 & \ldots & 0 \end{pmatrix} \qquad Tb = \begin{pmatrix} 0 \\ 0 \\ \cdot \\ 0 \\ 1 \end{pmatrix} \tag{6.8}$$

Suppose (6.1) and (6.2) hold, and set

$$\hat{z} = \hat{F}(x) = TF(x)$$

$$\hat{\alpha}(x) = \alpha(x) + \beta(x)kF(x)$$

$$\hat{\beta}(x) = \beta(x)$$

Then, it is easily seen that

$$\hat{F}_*(g\hat{\beta})\circ\hat{F}^{-1}(\hat{z}) = \begin{pmatrix} 0 \\ 0 \\ \cdot \\ 0 \\ 1 \end{pmatrix}$$

$$\hat{F}_*(f+g\hat{\alpha})\circ\hat{F}^{-1}(\hat{z}) = \begin{pmatrix} 0 & 1 & 0 & \dots & 0 \\ 0 & 0 & 1 & \dots & 0 \\ \cdot & \cdot & \cdot & \dots & \cdot \\ 0 & 0 & 0 & \dots & 1 \\ 0 & 0 & 0 & \dots & 0 \end{pmatrix} \hat{z}$$

From this, we deduce that there is no loss of generality in assuming that the pair A,b which makes (6.1) and (6.2) satisfied has directly the form specified in the right-hand-sides of (6.8).

Now, set

$$z = F(x) = \mathrm{col}(z_1(x),\dots,z_n(x))$$

If (6.1) holds (with A and b in the form of the right-hand-sides of (6.8)), we have for all $x \in U$,

$$F_*(f(x)+g(x)\alpha(x)) = AF(x)$$

that is

$$\frac{\partial z_1}{\partial x}(f(x)+g(x)\alpha(x)) = z_2(x)$$

$$\dots$$

$$\frac{\partial z_{n-1}}{\partial x}(f(x)+g(x)\alpha(x)) = z_n(x)$$

$$\frac{\partial z_n}{\partial x}(f(x)+g(x)\alpha(x)) = 0$$

If also (6.2) holds we have

$$F_* g(x)\beta(x) = b$$

that is

$$\frac{\partial z_1}{\partial x} g(x)\beta(x) = 0$$

$$\dots$$

$$\frac{\partial z_{n-1}}{\partial x} g(x)\beta(x) = 0$$

$$\frac{\partial z_n}{\partial x} g(x)\beta(x) = 1$$

Since $\beta(x)$ is nonzero for all $x \in U$, the second set of conditions imply

(6.9) $$\frac{\partial z_i}{\partial x} g(x) = L_g z_i(x) = 0 \qquad 1 \le i \le n-1$$

(6.10) $$\frac{\partial z_n}{\partial x} g(x) = L_g z_n(x) = \frac{1}{\beta(x)}$$

for all $x \in U$. These, in turn, together with the first set of conditions imply

(6.11) $$L_f z_i(x) = z_{i+1}(x) \qquad 1 \le i \le n-1$$

(6.12) $$L_f z_n(x) = -\frac{\alpha(x)}{\beta(x)}$$

for all $x \in U$.

If one sets

(6.13a) $$\varphi(x) = z_1(x)$$

the conditions (6.11) yield

(6.13b) $$z_{i+1}(x) = L_f^i \varphi(x) \qquad 0 \le i \le n-1$$

Thus, from (6.9) one obtains

$$L_g \varphi(x) = L_g L_f \varphi(x) = \dots = L_g L_f^{n-2} \varphi(x) = 0$$

for all $x \in U$ and, from (6.10),

$$L_g L_f^{n-1} \varphi(x) \neq 0$$

for all $x \in U$. This completes the proof of the necessity.

Sufficiency. Suppose (6.6) and (6.7) are true and let $U \subset V$ be a neighborhood of x^o such that $L_g L_f^{n-1} \varphi(x) \neq 0$ for all $x \in U$. Use (6.13) in order to define a set of functions $z_1, \dots, z_n$ on U. The functions thus defined are clearly such that (6.9) and (6.11) hold. Moreover, since $L_g z_n(x)$ is nonzero on U, one can define a nonzero function $\beta(x)$ and a function $\alpha(x)$ by means of (6.10) and (6.12). This pair of functions α and β and the mapping

$$F : x \longmapsto \mathrm{col}(z_1(x),\ldots,z_n(x))$$

are clearly such that

$$F_*(f(x) + g(x)\alpha(x)) = AF(z)$$

$$F_*(f(x)\beta(x)) = b$$

with A and b in the form of the right-hand-sides of (6.8). Thus, in order to complete the proof, we only have to show that F qualifies as a local coordinates transformation around x^o, i.e. that its differential F_* is nonsingular at x^o.

For, observe that the vector fields $\tilde{f} = f+g\alpha$ and $\tilde{g} = g\beta$ are such that

$$F_*\tilde{f}(x) = AF(x)$$

$$F_*\tilde{g}(x) = b$$

or, in other words, that $\tilde{f}$ is F-related to the vector field f' defined by

$$f'(z) = Az$$

and that $\tilde{g}$ is F-related to the vector field g' defined by

$$g'(z) = b$$

As a consequence, we have that the Lie bracket $[\tilde{f},\tilde{g}]$ is F-related to the Lie bracket $[f',g']$. Using this fact repeatedly, one may check that

$$F_*(ad^i_{\tilde{f}}\tilde{g})(x) = (ad^i_{f'}g')\circ F(x)$$

for all $0 \leq i \leq n-1$. The special form of f' and g' is such that

$$(ad^i_{f'}g') = (-1)^i A^i b$$

All together, these yield

$$F_*(g \quad ad_{\tilde{f}}\tilde{g} \ \ldots \ ad^{n-1}_{\tilde{f}}\tilde{g}) =$$

$$= (b \quad -Ab \ \ldots \ (-1)^{n-1}A^{n-1} \ b)$$

The matrix on the right-hand-side is nonsingular, because (A,b) is a reachable pair, and so is F_*. This completes the proof of the sufficiency. □

(6.14) *Lemma.* Let φ be a real-valued function defined on an open set V. Then the conditions (6.6) and (6.7) hold if and only if

$$L_g\varphi(x) = L_{[f,g]}\varphi(x) = \ldots = L_{(ad_f^{n-2}g)}\varphi(x) = 0 \tag{6.6'}$$

for all $x \in V$, and

$$L_{(ad_f^{n-1}g)}\varphi(x^o) \neq 0 \tag{6.7'}$$

Proof. We show, by induction, that the set of conditions

$$L_gL_f^0\varphi = \ldots = L_gL_f^k\varphi = 0 \tag{6.15a}$$

is equivalent to the set of conditions

$$L_{(ad_f^0g)}\varphi = \ldots = L_{(ad_f^kg)}\varphi = 0 \tag{6.15b}$$

and both imply

$$L_{(ad_f^ig)}L_f^j\varphi = (-1)^iL_gL_f^{i+j}\varphi \tag{6.16}$$

for all i,j such that $i+j = k+1$.

This is clearly true for $k = 0$. In this case (6.15a) and (6.15b) reduces to $L_g\varphi = 0$ and

$$L_{[f,g]}\varphi = L_fL_g\varphi - L_gL_f\varphi = -L_gL_f\varphi$$

Suppose (6.15a) and (6.15b) true for some k and (6.16) true for all i,j such that $i+j = k+1$. The latter yields, in particular,

$$L_{(ad_f^{k+1}g)}\varphi = (-1)^{k+1}L_gL_f^{k+1}\varphi$$

so that $L_gL_f^{k+1}\varphi = 0$ if and only if $L_{(ad_f^{k+1}g)}\varphi = 0$. Assume either one of these conditions holds. Then

$$L_{(ad_f^{k+2}g)}\varphi = L_f L_{(ad_f^{k+1}g)}\varphi - L_{(ad_f^{k+1}g)} L_f\varphi =$$

$$= (-1) L_f L_{(ad_f^k g)} L_f\varphi + (-1)^2 L_{(ad_f^k g)} L_f^2\varphi$$

$$= (-1)^{k+1} L_f L_g L_f^{k+1}\varphi + (-1)^2 L_{(ad_f^k g)} L_f^2\varphi$$

$$= (-1)^2 L_{(ad_f^k g)} L_f^2\varphi$$

$$= (-1)^2 L_f L_{(ad_f^{k-1}g)} L_f^2\varphi + (-1)^3 L_{(ad_f^{k-1}g)} L_f^3\varphi$$

$$= (-1)^{k+1} L_f L_g L_f^{k+1}\varphi + (-1)^3 L_{(ad_f^{k+1}g)} L_f^3\varphi$$

$$= (-1)^3 L_{(ad_f^{k-1}g)} L_f^3\varphi = \ldots$$

We see in this way that for all $0 \leq j \leq k+2$

$$L_{(ad_f^{k+2}g)}\varphi = (-1)^j L_{(ad_f^{k+2-j}g)} L_f^j\varphi = (-1)^{k+2} L_g L_f^{k+2}\varphi$$

and therefore that (6.16) is true for all i,j such that $i+j = k+2$.

From (6.15) and (6.16) the statement follows immediately. □

(6.17) *Remark.* We have proved, by the way, that either one of the two equivalent sets of conditions (6.15) imply

$$L_{(ad_f^i g)} L_f^j\varphi = (-1)^j L_{(ad_f^{i+j}g)}\varphi$$

for all i,j such that $i+j \leq k+1$. This fact will be used in the sequel.□

(6.18) *Theorem.* Suppose $m = 1$ and let $g = g_1$. The State-Space Linearization Problem is solvable if and only if:

(i) $\dim(\text{span}\{g(x^o), ad_f g(x^o), \ldots, ad_f^{n-1} g(x^o)\}) = n$

(ii) the distribution

$$\Delta = sp\{g, ad_f g, \ldots, ad_f^{n-2} g\} \tag{6.19}$$

is involutive in a neighborhood U of x^o.

(6.20) *Remark*. Note that the condition (i) implies that the tangent vectors $g(x), ad_f g(x), \ldots, ad_f^{n-1} g(x)$ are linearly independent for all x in a suitable neighborhood of x^o. Therefore the distribution (6.19) is nonsingular around x^o and has dimension (n-1).

Proof. We know from the previous Lemmas that the problem is solvable if and only if there exists a real-valued function φ defined in a neighborhood V of x^o such that the conditions (6.6') and (6.7') hold. These may be rewritten as

$$\langle d\varphi, ad_f^i g \rangle (x) = 0 \tag{6.6"}$$

for all $0 \leq i \leq n-2$ and all $x \in V$, and

$$\langle d\varphi, ad_f^{n-1} g \rangle (x^o) \neq 0 \tag{6.7"}$$

If both these conditions hold, then necessarily the tangent vectors $g(x^o), ad_f g(x^o), \ldots, ad_f^{n-1} g(x^o)$ are linearly independent. For, we see from Remark (6.17) that (6.6") implies

$$\langle dL_f^j \varphi, ad_f^i g \rangle = L_{(ad_f^i g)} L_f^j \varphi =$$

$$= (-1)^j L_{(ad_f^{i+j} g)} \varphi = (-1)^j \langle d\varphi, ad_f^{i+j} g \rangle$$

for all $i+j \leq n-1$. Therefore, using again (6.6") and (6.7") we have

$$\langle dL_f^j \varphi, ad_f^i g \rangle (x) = 0$$

for all i,j such that $i+j \leq n-2$ and all $x \in V$ and

$$\langle dL_f^j \varphi, ad_f^i g \rangle (x^o) \neq 0$$

for all i,j such that $i+j = n-1$.

The above conditions, all together, show that the matrix

$$\begin{pmatrix} d\varphi(x^o) \\ dL_f \varphi(x^o) \\ \vdots \\ dL_f^{n-1} \varphi(x^o) \end{pmatrix} (g(x^o) \; ad_f g(x^o) \ldots ad_f^{n-1} g(x^o)) = \tag{6.21}$$

$$= \begin{pmatrix} \langle d\varphi, g\rangle(x^o) & \langle d\varphi, ad_f g\rangle(x^o) & \dots & \langle d\varphi, ad_f^{n-1} g\rangle(x^o) \\ \langle dL_f\varphi, g\rangle(x^o) & \langle dL_f\varphi, ad_f g\rangle(x^o) & \dots & \langle dL_f\varphi, ad_f^{n-1} g\rangle(x^o) \\ \cdot & \cdot & \dots & \\ \langle dL_f^{n-1}\varphi, g\rangle(x^o) & \langle dL_f^{n-1}\varphi, ad_f g\rangle(x^o) & \dots & \langle dL_f^{n-1}\varphi, ad_f^{n-1} g\rangle(x^o) \end{pmatrix}$$

has rank n and, therefore, that the vectors $g(x^o), ad_f g(x^o), \dots, ad_f^{n-1} g(x^o)$ are linearly independent.

This proves the necessity of (i). If (i) holds then the distribution (6.19) has dimension n-1 around x^o and (6.6") tell us that the exact covector field $d\varphi$ spans $\Delta^\perp$ around x^o. So, because of Frobenius theorem (see Remark I.(3.7)) we conclude that Δ is completely integrable and thus involutive, i.e. the necessity of (ii).

Conversely, suppose (i) holds. Then the distribution (6.19) is nonsingular around x^o. If also (ii) holds, Δ is completely integrable around x^o and there exists a real-valued function φ, defined in a neighborhood V of x^o, such that $d\varphi$ spans $\Delta^\perp$ on V, i.e. such that (6.6") are satisfied. Moreover, the covector field $d\varphi$ is such that (6.7") also is satisfied, because otherwise $d\varphi$ would be annihilated by a set of n linearly independent vectors. This, in view of the previous Lemmas, completes the proof of the sufficiency. □

For the sake of convenience, we summarize now the *procedure* leading to the construction of the feedback α and β which solves the State-Space Linearization Problem in the case of a single input channel.

Suppose (i) and (ii) hold. Then, using Frobenius Theorem one constructs a function φ, defined in a neighborhood V of x^o, such that (6.6") and (6.7") hold. Then, one sets

$$\beta(x) = \frac{1}{L_g L_f^{n-1}\varphi(x)} \tag{6.22a}$$

and

$$\alpha(x) = \frac{-L_f^n\varphi(x)}{L_g L_f^{n-1}\varphi(x)} \tag{6.22b}$$

for all $x \in V$. This pair of feedback functions, together with the local coordinates transformation defined by

$$z_i(x) = L_f^{i-1}\varphi(x)$$

for $1 \leq i \leq n$, is such as to satisfy (6.1) and (6.2) with A and $b_1 = b$ in the form of the right-hand-sides of (6.8).

(6.23) *Remark.* There is a surprising affinity between some results described in this section and the ones described in the sections IV.3 and IV.4. For instance one may rephrase Lemma (6.5) by saying that the State-Space Linearization Problem is solvable if and only if one may define, for the system

$$\dot{x} = f(x) + g(x)u$$

a (dummy) output function

$$y = \varphi(x)$$

whose characteristic number is *exactly* n-1. Of course, this will be possible if and only if the conditions (i) and (ii) are satisfied.

Once such a dummy output function has been found, then the solution of a State-Space Linearization Problem proceeds like a solution of a (degenerate, because both ℓ and m are equal to 1) noninteracting control problem. As a matter of fact, we have from Lemma IV.(3.10) that the differentials $d\varphi, dL_f\varphi, \ldots, dL_f^{n-1}\varphi$ are linearly independent at x^o and thus that the mapping

$$F : x \longmapsto \text{col}(\varphi(x), L_f\varphi(x), \ldots, L_f^{n-1}\varphi(x))$$

qualifies as a local coordinates transformation. Then, from Corollary IV.(3.14) we learn that

$$\Delta^* = \bigcap_{i=0}^{n-1} (\text{sp}\{dL_f^i\varphi\})^{\perp} = 0$$

is the largest locally controlled invariant distribution contained in $(\text{sp}\{d\varphi\})^{\perp}$.

The feedback (6.22) coincides with a solution of IV.(3.15) (with $\gamma(x) = 0$ and $\delta(x) = 1$). Under this feedback the system becomes linear in the new coordinates, as it is seen from the constructions given in the section IV.4 (see Remark IV.(4.9)). □

We note that the formal statement of the State-Space Linearization Problem, given at the beginning of the section, does not incorporate any requirement about the image $F(U)$ of the coordinates transformation that makes it possible (6.1) and (6.2) to hold. However, one may wish to impose the additional requirement that the image $F(U)$ contains the origin of $\mathbb{R}^n$. In this case, the condition of Theorem (6.18) must be strenghtened a little bit.

Suppose the coordinates transformation $z = F(x)$ solving the State-Space Linearization Problem is such that

(6.24) $$z^o = F(x^o) = 0$$

Then, from (6.1) we deduce that necessarily

(6.25) $$f(x^o) + g(x^o)\alpha(x^o) = 0$$

If $f(x^o) = 0$, then the construction already proposed for the solution of the problem may be adapted to make (6.24) and (6.25) satisfied. As a matter of fact, one may always choose a function φ satisfying (6.6") and (6.7") in such a way that $\varphi(x^o) = 0$ (see, e.g., the construction proposed along the proof of Theorem I.(3.3)). If this is the case, then

$$z_1(x^o) = \varphi(x^o) = 0$$

and also, for $2 \leq i \leq n$,

$$z_i(x^o) = L_f^{i-1}\varphi(x^o) = \langle dL_f^{i-2}\varphi(x^o), f(x^o)\rangle = 0$$

because we have assumed $f(x^o) = 0$. Thus the proposed coordinates transformation satisfies (6.24). Moreover,

$$\alpha(x^o) = -\frac{L_f^n\varphi(x^o)}{L_gL_f^{n-1}\varphi(x^o)} = -\frac{\langle dL_f^{n-1}\varphi(x^o), f(x^o)\rangle}{L_gL_f^{n-1}\varphi(x^o)} = 0$$

and also (6.25) holds.

One may thus assert that if $f(x^o) = 0$, i.e. if the initial state x^o is an equilibrium state for the autonomous system

$$\dot{x} = f(x)$$

and if the State-Space Linearization Problem is solvable, one may always find a solution such that $\alpha(x^o) = 0$ and $F(x^o) = 0$.

If $f(x^o) \neq 0$, the condition (6.25) may be rewritten as

$$f(x^o) = cg(x^o) \tag{6.26}$$

where c is a nonzero real number. Again, if the State-Space Linearization Problem is solvable one may find a function φ such that $z_1(x^o) = \varphi(x^o) = 0$. But also, for $2 \leq i \leq n$, (6.26) ensures that

$$z_i(x^o) = \langle dL_f^{i-2}\varphi, f(x^o)\rangle = cL_gL_f^{i-2}\varphi(x^o) = 0$$

and thus (6.24) still holds. Moreover, the proposed α is such that

$$\alpha(x^o) = -\frac{\langle dL_f^{n-1}\varphi(x^o), f(x^o)\rangle}{L_gL_f^{n-1}\varphi(x^o)} = -c$$

as expected.

In this case, the initial state x^o is not an equilibrium state for the original system, but an α may be found such that x^o is an equilibrium state for the system

$$\dot{x} = f(x) + g(x)\alpha(x)$$

In summary, we have the following result.

(6.27) *Corollary*. Suppose $m = 1$ and let $g = g_1$. Suppose the State-Space Linearization Problem is solvable. Then, a solution with $F(x^o) = 0$ exists if and only if

$$f(x^o) \in sp\{g(x^o)\} \qquad \square$$

(6.28) *Remark*. When $F(x^o) = 0$, one may use the solution of the State-Space Linearization Problem for local stabilization purposes. Indeed, since (A,b) is a reachable pair, one may arbitrarily assign the eigenvalues to the matrix $(A + bk)$, via suitable choice of the $1\times n$ row vector k. If this is the case, the feedback control law

$$u = \alpha(x) + \beta(x)kF(x) + \beta(x)v$$

makes the system locally diffeomorphic, on U, to the asymptotically stable system

$$\dot{z} = (A + bk)z + bv \qquad \square$$

We now describe the extension of the previous discussion to the case of many inputs. This requires the introduction of some further notations, but the substance of the procedure is essentially the same as the one examined so far.

Given a set of vector fields $f, g_1, \ldots, g_m$ we define a sequence of distributions as follows

$$G_0 = \mathrm{sp}\{g_1, \ldots, g_m\}$$

$$G_i = G_{i-1} + [f, G_{i-1}]$$

The following Lemma describes the possibility of computing all G_i's in a simple way.

(6.29) *Lemma*. Suppose all G_i's are nonsingular. Then

(6.30) $$G_i = \mathrm{sp}\{\mathrm{ad}_f^k g_j : 0 \le k \le i,\ 1 \le j \le m\}$$

Proof. Suppose $G_i = \mathrm{sp}\{\mathrm{ad}_f^k g_j : 0 \le k \le i,\ 1 \le j \le m\}$. Suppose that at x^o some vectors $\mathrm{ad}_f^{k_1} g_{j_1}, \ldots, \mathrm{ad}_f^{k_r} g_{j_r}$ are linearly independent and span $G_i(x^o)$. Then on a neighborhood U of x^o any vector field τ in G_i may be written as $\tau = \sum_{\alpha=1}^{r} c_\alpha \mathrm{ad}_f^{k_\alpha} g_{j_\alpha}$, with $c_\alpha \in C^\infty(U)$. Then $[f, \tau] = \sum_{\alpha=1}^{r} (c_\alpha \mathrm{ad}_f^{k_\alpha+1} g_{j_\alpha} + (L_f c_\alpha) \mathrm{ad}_f^{k_\alpha} g_{j_\alpha})$. Therefore, on U, $G_{i+1} = \mathrm{sp}\{\mathrm{ad}_f^{k_\alpha+1} g_{j_\alpha}, \mathrm{ad}_f^{k_\alpha} g_{j_\alpha} : 1 \le \alpha \le r\}$. Since, by construction, all $\mathrm{ad}_f^k g_j$, $0 \le k \le i+1$ and $1 \le j \le m$ are in G_{i+1}, this proves that $G_{i+1} = \mathrm{sp}\{\mathrm{ad}_f^k g_j : 0 \le k \le i+1, 1 \le j \le m\}$. □

Since $G_i \subset G_{i+1}$ by definition, if the G_i's are nonsingular we have that

$$\dim \frac{G_{i+1}(x)}{G_i(x)} = \text{independent of } x$$

Thus we may define a sequence of integers $\nu_0, \nu_1, \ldots$ by setting

(6.31a) $$\nu_0 = \dim G_0$$

(6.31b) $$\nu_i = \dim \frac{G_i}{G_{i-1}} \qquad i \ge 1$$

The integers thus defined have the following property

(6.32) *Lemma*. The following condition holds

$$\nu_i \geq \nu_{i+1}$$

for all $i > 0$. Let ν_{i^*} denote the last nonzero element in the sequence $\{\nu_i : i \geq 0\}$. If

$$\dim G_{i^*} = n$$

then

$$\nu_0 + \nu_1 + \ldots + \nu_{i^*} = n$$

Proof. Consider G_i and G_{i-1}. By definition

$$\dim G_i(x) = \dim G_{i-1}(x) + \nu_i$$

From (6.30), we deduce that, given a point x^o, there will be ν_i vectors $ad_f^i g_{j_1}(x^o),\ldots,ad_f^i g_{j_{\nu_i}}(x^o)$ linearly independent and with the property that all vector fields in G_i may be written as linear combinations, with smooth coefficients, of vectors of G_{i-1} and of $sp\{ad_f^i g_{j_s} : 1 \leq s \leq \nu_i\}$. Thus

$$G_{i+1} = G_i + sp\{ad_f^{i+1} g_{j_s} : 1 \leq s \leq \nu_i\}$$

and

$$\nu_{i+1} \leq \nu_i \, . \quad \square$$

From the sequence $\{\nu_i : 0 \leq i \leq i^*\}$ we define another sequence of integers $m_0, m_1, \ldots, m_{i^*}$, setting

$$m_0 = \nu_{i^*}$$

$$m_0 + m_1 = \nu_{i^*-1}$$

$$m_0 + m_1 + m_2 = \nu_{i^*-2} \qquad (6.33)$$

$$\cdots$$

$$m_0 + m_1 + \ldots + m_{i^*} = \nu_0$$

(6.34) *Lemma*. The following conditions hold

$$m_0 > 0$$

$$m_i \geq 0 \qquad 1 \leq i \leq i^*$$

Moreover, if $\dim G_{i^*} = n$, then

$$\dim G^{\perp}_{i^*-i} = im_0 + \ldots + 2m_{i-2} + m_{i-1} \tag{6.35}$$

for $1 \leq i \leq i^*$. □

There is a need for a third sequence of integers $\{\kappa_i : 1 \leq i \leq \nu_0\}$ related to the previous ones by the following relations

$$(6.36)\qquad \begin{aligned}
&\kappa_i = i^*+1 && \text{if} && 1 \leq i \leq \nu_{i^*} \\
&\kappa_i = i^* && \text{if } m_1 > 0 && \text{and} \quad \nu_{i^*}+1 \leq i \leq \nu_{i^*-1} \\
&\kappa_i = i^*-1 && \text{if } m_2 > 0 && \text{and} \quad \nu_{i^*-1}+1 \leq i \leq \nu_{i^*-2} \\
&\ldots && \ldots && \\
&\kappa_i = 1 && \text{if } m_{i^*} > 0 && \text{and} \quad \nu_1+1 \leq i \leq \nu_0
\end{aligned}$$

With the help of these notations it is rather simple to state the necessary and sufficient conditions for the existence of a solution to the State-Space Linearization Problem in the general case where $m \geq 1$.

(6.37) *Theorem*. The State-Space Linearization Problem is solvable if and only if

(i) x^o is a regular point of the distribution G_i, for all $i \geq 0$
(ii) $\dim G_{i^*}(x^o) = n$
(iii) the distribution G_i is involutive, for all i such that $m_{i^*-i-1} \neq 0$.

Proof. We restrict ourselves to the proof of the sufficiency, which is constructive. Without loss of generality, we may assume that

$$\nu_0 = m$$

For, if this is not the case, since G_0 by assumption is nonsingular around x^o, we may always find a nonsingular $m \times m$ matrix $\bar{\beta}(x)$, defined in a neighborhood U of x^o, such that

$$G_0(x) = \mathrm{span}\{\bar{g}_1(x),\ldots,\bar{g}_{\nu_0}(x)\}$$

and

$$\bar{g}_{\nu_0+1}(x) = \ldots = \bar{g}_m(x) = 0$$

for all $x \in U$, where

$$\bar{g}_i(x) = (g(x)\bar{\beta}(x))_i$$

for $1 \leq i \leq m$. If a feedback (α,β) solves the State-Space Linearization Problem for the set $f,\bar{g}_1,\ldots,\bar{g}_{\nu_0}$, then it is easily seen that a feedback of the form

$$\alpha' = \bar{\beta}\begin{pmatrix}\alpha\\0\end{pmatrix} \qquad \beta' = \bar{\beta}\begin{pmatrix}\beta & 0\\0 & I\end{pmatrix}$$

solves the problem for the original set $f,g_1,\ldots,g_m$.

For the sake of simplicity, we break up the construction in two stages.

(i) Recursive construction of a coordinates transformation around the point x^o.

Step (1): By assumption

$$\dim G_{i^*} = n$$

and $\dim G^{\perp}_{i^*-1} = m_0 > 0$. Moreover, G_{i^*-1} is assumed to be involutive. Then, by Frobenius theorem, we know that there exist a neighborhood U_1 of x^o and m_0 functions $h_{01},\ldots,h_{0m_0}$ defined on U_1, whose differentials span $G^{\perp}_{i^*-1}(x)$ at all $x \in U_1$. In particular,

$$\langle dh_{0i}, ad^{\alpha}_f g_j\rangle(x) = 0 \tag{6.38}$$

for all $1 \leq j \leq m$, $1 \leq i \leq m_0$, $0 \leq \alpha \leq i^*-1$ and all $x \in U_1$. Moreover, the differentials $dh_{01}(x),\ldots,dh_{0m_0}(x)$ are linearly independent at all $x \in U_1$.

We claim that the $m_0\times m$ matrix

$$M_0 = \{m^{(0)}_{ij}(x)\} = \{\langle dh_{0i}, ad^{i^*}_f g_j\rangle(x)\}$$

has rank m_0 at all $x \in U_1$. For, suppose it is false at some $\bar{x} \in U_1$.

Then, there exist real numbers $c_1,\ldots,c_{m_0}$ such that

$$\langle \sum_{i=1}^{m_0} c_i dh_{0i}, ad_f^{i^*} g_j \rangle(\bar{x}) = 0$$

for all $1 \leq j \leq m$. This, together with (6.38), implies

$$\langle \sum_{i=1}^{m_0} c_i dh_{0i}, ad_f^{\alpha} g_j \rangle(\bar{x}) = 0 \tag{6.39}$$

for all $1 \leq j \leq m$, $0 \leq \alpha \leq i^*$, and this in turn implies

$$\langle \sum_{i=1}^{m_0} c_i dh_{0i}(\bar{x}), v \rangle = 0$$

for all $v \in G_{i^*}(\bar{x})$. Since $\dim G_{i^*} = n$, then $\sum_{i=1}^{m_0} c_i dh_{0i}(\bar{x})$ must be a zero covector, but since $dh_{01}(\bar{x}),\ldots,dh_{0m_0}(\bar{x})$ are independent, then $c_1 = \ldots = c_{m_0} = 0$.

Step (2): Consider the distribution G_{i^*-2}, which is such that

$$\dim G^{\perp}_{i^*-2} = 2m_0 + m_1$$

We claim that $dL_f h_{01},\ldots,dL_f h_{0m_0}$ are such that

$$\langle dL_f h_{0i}, ad_f^{\alpha} g_j \rangle(x) = 0$$

for all $1 \leq j \leq m$, $1 \leq i \leq m_0$, $0 \leq \alpha \leq i^*-2$ and all $x \in U_1$. This comes from the property

$$-\langle dL_f h_{0i}, ad_f^{\alpha} g_j \rangle = \langle dh_{0i}, ad_f^{\alpha+1} g_j \rangle - L_f \langle dh_{0i}, ad_f^{\alpha} g_j \rangle$$

in which both the terms are zero on U_1 because $\alpha \leq i^*-2$.

We claim also that the $2m_0$ differentials

$$\{dh_{01}(x),\ldots,dh_{0m_0}(x), dL_f h_{01}(x),\ldots,dL_f h_{0m_0}(x)\} \tag{6.40}$$

are linearly independent all $x \in U_1$. For, suppose this is false; then, for suitable reals c_{1i}, c_{2i}, we had

$$\sum_{i=1}^{m_0} c_{1i} dh_{0i}(\bar{x}) + \sum_{i=1}^{m_0} c_{2i} dL_f h_{0i}(\bar{x}) = 0 \tag{6.41}$$

at some $\bar{x} \in U$. This would imply

$$\langle \sum_{i=1}^{m_0} (c_{1i} dh_{0i} + c_{2i} dL_f h_{0i}), ad_f^{i^*-1} g_j \rangle (\bar{x}) = 0$$

for all $1 \leq j \leq m$. This in turn implies (because of (6.38))

$$\langle \sum_{i=1}^{m_0} c_{2i} dL_f h_{0i}, ad_f^{i^*-1} g_j \rangle (\bar{x}) = -\langle \sum_{i=1}^{m_0} c_{2i} dh_{0i}, ad_f^{i^*} g_j \rangle (\bar{x}) = 0$$

i.e. a contradiction, like in (6.39). Therefore $c_{21} = \ldots = c_{2m_0} = 0$ in (6.41), and also $c_{i1} = \ldots = c_{1m_0} = 0$ because $dh_{01}(\bar{x}),\ldots,dh_{0m_0}(\bar{x})$ are linearly independent.

If $m_1 = 0$, the $2m_0$ covectors (6.40) span $G^{\perp}_{i^*-2}$. If $m_1 > 0$, using again Frobenius theorem (because G_{i^*-2} is involutive), we may find m_1 more functions $h_{11}(x),\ldots,h_{1m_1}(x)$, defined in a neighborhood $U_2 \subset U_1$ of x^o, such that the $2m_0+m_1$ differentials

(6.43) $\{dh_{01}(x),\ldots,dh_{0m_0}(x), dL_f h_{01}(x),\ldots,dL_f h_{0m_0}(x), dh_{11}(x),\ldots,dh_{1m_1}(x)\}$

are linearly independent and

$$\langle dh_{1i}, ad_f^{\alpha} g_j \rangle (x) = 0$$

for all $1 \leq j \leq m$, $1 \leq i \leq m_1$, $0 \leq \alpha \leq i^*-2$ and all $x \in U_2$.

We claim that the $(m_0+m_1)\times m$ matrix

$$\begin{pmatrix} M_0 \\ M_1 \end{pmatrix}$$

where M_0 is as before and M_1 defined as

$$M_1 = \{m_{ij}^{(1)}(x)\} = \{\langle dh_{1i}, ad_f^{i^*-1} g_j \rangle (x)\}$$

has rank m_0+m_1 at all $x \in U_2$.

For suppose for some reals $c_{01},\ldots,c_{0m_0},c_{11},\ldots,c_{1m_1}$ we had

$$\langle \sum_{i=1}^{m_0} c_{0i} dh_{0i}(\bar{x}), ad_f^{i^*} g_j(\bar{x}) \rangle + \langle \sum_{i=1}^{m_1} c_{1i} dh_{1i}(\bar{x}), ad_f^{i^*-1} g_j(\bar{x}) \rangle = 0$$

at some $\bar{x} \in U_2$. Then (recall Remark (6.17))

$$(6.44) \quad \langle \sum_{i=1}^{m_0} c_{0i} dL_f h_{0i}(\bar{x}) + \sum_{i=1}^{m_1} c_{1i} dh_{1i}(\bar{x}), ad_f^{i^*-1} g_j(\bar{x}) \rangle = 0$$

The covector $\sum_{i=1}^{m_0} c_{0i} dL_f h_{0i}(\bar{x}) + \sum_{i=1}^{m_1} c_{1i} dh_{1i}(\bar{x})$ annihilates, as we have seen before, all $ad_f^{\alpha} g_j(\bar{x})$, $\alpha \leq i^*-2$, $1 \leq j \leq m$, but (6.44) tells us that it also annihilates $ad_f^{i^*-1} g_j(\bar{x})$. Thus, this covector annihilates all vectors in G_{i^*-1}.
From the previous discussion, we conclude that this covector must belong to span$\{dh_{01}(\bar{x}),\dots,dh_{0m_0}(\bar{x})\}$, but this is a contradiction, because the covectors (6.43) are linearly independent. Therefore, the c_{0i}'s and c_{1i}'s of (6.44) must be zero.

Eventually, with this procedure we end up with a set of functions

$$(6.45) \quad \begin{array}{l} h_{01},\dots,h_{0m_0}, L_f h_{01},\dots,L_f h_{0m_0},\dots,L_f^{i^*-1} h_{01},\dots,L_f^{i^*-1} h_{0m_0} \\ h_{11},\dots,h_{1m_1},\dots,L_f^{i^*-2} h_{11},\dots,L_f^{i^*-2} h_{1m_1} \\ \dots \\ h_{i^*-1,1},\dots,h_{i^*-1,m_{i^*-1}} \end{array}$$

(of course, some of these lines may be missing if some m_i is zero) with the following properties:

- the total number of functions is

$$i^* m_0 + (i^*-1)m_1 + \dots + 2m_{i^*-2} + m_{i^*-1} = n-m$$

- the n-m differentials of these functions are independent at all $x \in U$, a neighborhood of x^o,
- the $\nu_1 \times m$ matrix

$$\begin{pmatrix} M_0 \\ \vdots \\ M_{i^*-1} \end{pmatrix}$$

where M_i is $m_i \times m$ and

$$m_{\ell j}^{(i)}(x) = \langle dh_{i\ell}, ad_f^{i^*-i} g_j \rangle (x)$$

has rank ν_1 at all $x \in U$.

If $m_{i^*} > 0$, one may still find m_{i^*} more functions $h_{i^*1},\dots,h_{i^*m_{i^*}}$, that, together with the functions (6.45) and with the additional functions $L_f^{i^*}h_{01},\dots,L_f^{i^*}h_{0m_0},\dots,L_f h_{i^*-1,1},\dots,L_f h_{i^*-1,m_{i^*-1}}$, give rise to a set of n linearly independent differentials at x^o.

For convenience, let us relabel the functions h_{ij} and set

$$\varphi_i = h_{0i} \qquad \text{if} \qquad 1 \le i \le \nu_{i^*}$$

$$\varphi_i = h_{1,i-\nu_{i^*}} \qquad \text{if} \quad m_1 > 0 \quad \text{and} \quad \nu_{i^*}+1 \le i \le \nu_{i^*-1}$$

$$\dots \qquad \dots$$

$$\varphi_i = h_{i^*,i-\nu_1} \qquad \text{if} \quad m_{i^*} > 0 \quad \text{and} \quad \nu_1+1 \le i \le \nu_0 = m$$

The previous constructions tell us that the mapping $F : x \mapsto \text{col}(\xi_1(x),\dots,\xi_m(x))$, where

$$\xi_i(x) = \begin{pmatrix} \varphi_i(x) \\ L_f\varphi_i(x) \\ \vdots \\ L_f^{\kappa_i-1}\varphi_i(x) \end{pmatrix}$$

qualifies as a local diffeomorphism around x^o.

Moreover, by construction,

$$\langle dL_f^{\alpha}\varphi_i, g_j\rangle(x) = 0 \tag{6.46}$$

for all $0 \le \alpha \le \kappa_i-2$, $1 \le i,j \le m$, at all x around x^o, and the $m\times m$ matrix

$$A(x) = \{a_{ij}(x)\} = \{\langle dL_f^{\kappa_i-1}\varphi_i, g_j\rangle(x)\} \tag{6.47}$$

is nonsingular at $x = x^o$.

(ii) Construction of the linearizing feedback. From the conditions (6.46) and (6.47), we see that the control system

$$\dot{x} = f(x) + \sum_{i=1}^{m} g_i(x)u_i \tag{6.48a}$$

with (dummy) outputs

(6.48b) $$y_i = \varphi_i(x) \qquad 1 \le i \le m$$

is such that:

- the characteristic number ρ_i associated with the i-th output channel is exactly equal to $\kappa_i - 1$,
- the single-outputs noninteracting control problem is solvable around x^o.

Choose a feedback α and β as a solution of the equations

$$A(x)\alpha(x) = - \begin{pmatrix} L_f^{\kappa_1}\varphi_1(x) \\ \vdots \\ L_f^{\kappa_m}\varphi_m(x) \end{pmatrix}$$

$$A(x)\beta(x) = I$$

(they correspond to the equations IV.(4.4a) with $\gamma_i = 0$ and IV.(4.4b) with δ_i the i-th row of an $m \times m$ identity matrix). Under this feedback, the system (6.48) splits into m noninteracting single-input single-output channels. In particular, in the new coordinates defined at the previous stage, each subsystem is described by equations of the form (see IV.(4.8))

$$\dot{\xi}_i = \begin{pmatrix} 0 & 1 & 0 & \cdots & 0 & 0 \\ 0 & 0 & 1 & \cdots & 0 & 0 \\ \cdot & \cdot & \cdot & \cdots & \cdot & \cdot \\ 0 & 0 & 0 & \cdots & 0 & 1 \\ 0 & 0 & 0 & \cdots & 0 & 0 \end{pmatrix} \xi_i + \begin{pmatrix} 0 \\ 0 \\ \cdot \\ 0 \\ 1 \end{pmatrix} v_i$$

$$y_i = (1 \quad 0 \quad 0 \quad \cdots \quad 0 \quad 0)\xi_i$$

This completes the proof. □

7. Observer with Linear Error Dynamics

We consider in this section a problem which is in some sense *dual* of that considered in the previous one. We have seen that the solvability of the State-Space Linearization Problem enables us to design a feedback under which the system becames locally diffeomorphic to a linear system with prescribed eigenvalues. In the case of linear system, the dual notion of spectral assegnability via static state-fedback is the existence of state-obsevers with prescribed eigenvalues. Moreover, it is known that the dynamics of a state-observer and that of the observation error (i.e. the difference between the unknown state and the estimated state) are the same. In view of this, if we wish to dualize the results developed so far, we are led to the problem of the synthesis of (nonlinear) observers yielding an error dynamics that, possibly after some suitable coordinates transformation,becames linear and spectrally assignable.

For the sake of simplicity, we restrict ourselves to the consideration of systems without inputs and with scalar output, i.e. systems described by equations of the form

$$\dot{x} = f(x)$$

$$y = h(x)$$

with $y \in \mathbb{R}$.

Suppose there exists a coordinates transformation $z = F(x)$ under which the vector field f and the output map h become respectively

$$F_{*}f{\circ}F^{-1}(z) = Az + K(cz)$$

$$h{\circ}F^{-1}(z) = cz$$

where (A,c) is an observable pair and K is an n-vector valued function of a real variable.

If this is the case, then an observer of the form

$$\dot{\xi} = (A+kc)\xi - ky + K(y)$$

yields an observation error (in the z coordinates)

$$e = \xi - z = \xi - F(x)$$

governed by the differential equation

$$\dot{e} = (A + kc)e$$

which is linear and spectrally assignable (via the $n\times 1$ column vector k).

Motivated by these consideration, we examine the following problem.

Observer Linearization Problem. Given a vector field f, a real-valued function h and an initial state x^{o} find (if possible) a neighborhood U of x^{o}, a coordinates transformation $z = F(x)$ defined on U, a matrix $A \in \mathbb{R}^{n\times n}$ and a row vector $c \in \mathbb{R}^{1\times n}$, a mapping $K: h(U) \to \mathbb{R}^{n}$ such that

$$h \circ F^{-1}(z) = cz \tag{7.1}$$

$$F_{*} f \circ F^{-1}(z) - Az = K(cz) \tag{7.2}$$

for all $z \in F(U)$, and

$$\bigcap_{i=0}^{n-1} \ker(cA^{i}) = \{0\}. \quad \square \tag{7.3}$$

The conditions for the solvability of this problem can be described as follows

(7.4) *Lemma.* The Observer Linearization Problem is solvable only if

$$\dim(\operatorname{span}\{dh(x^{o}), dL_{f}h(x^{o}), \ldots, dL_{f}^{n-1}h(x^{o})\}) = n \tag{7.5}$$

Proof. The condition (7.3) says that the pair (A,c) is observable. Then, it is known from the theory of linear systems that there exist a nonsingular $n\times n$ matrix T and a $n\times 1$ column vector k such that

$$T(A+kc)T^{-1} = \begin{pmatrix} 0 & 0 & \cdots & 0 & 0 \\ 1 & 0 & \cdots & 0 & 0 \\ \cdot & \cdot & \cdots & \cdot & \cdot \\ 0 & 0 & \cdots & 1 & 0 \end{pmatrix} \qquad cT^{-1} = (0 \quad 0 \quad \cdots \quad 0 \quad 1) \tag{7.6}$$

Suppose (7.1) and (7.2) hold, and set

$$\hat{z} = \hat{F}(x) = TF(x)$$

$$\hat{K}(y) = T(K(y) - ky)$$

where $y \in h(U)$.

Then, it is easily seen that

$$h \circ \hat{F}^{-1}(\hat{z}) = (0 \quad 0 \ \dots \ 0 \quad 1)\hat{z}$$

$$\hat{F}_* f \circ \hat{F}^{-1}(\hat{z}) - \begin{pmatrix} 0 & 0 & \dots & 0 & 0 \\ 1 & 0 & \dots & 0 & 0 \\ 0 & 0 & \dots & 0 & 0 \\ \cdot & \cdot & \dots & \cdot & \cdot \\ 0 & 0 & \dots & 1 & 0 \end{pmatrix} \hat{z} = \hat{K}((0 \ 0 \ \dots \ 0 \ 1)\hat{z})$$

From this we deduce that there is no loss of generality in assuming that the pair (A,c) that makes (7.1) and (7.2) satisfied has directly the form specified in the right-hand-sides of (7.6).

Now, set

$$z = F(x) = \mathrm{col}(z_1(x), \dots, z_n(x))$$

If (7.1) and (7.2) hold, we have, for all $x \in U$

$$h(x) = z_n(x) \tag{7.7}$$

$$\begin{aligned} \frac{\partial z_1}{\partial x} f(x) &= k_1(z_n(x)) \\ \frac{\partial z_2}{\partial x} f(x) &= z_1(x) + k_2(z_n(x)) \\ &\dots \\ \frac{\partial z_n}{\partial x} f(x) &= z_{n-1}(x) + k_n(z_n(x)) \end{aligned} \tag{7.8}$$

where $k_1, \dots, k_n$ denote the n compoents of K.

Observe that

$$L_f h(x) = \frac{\partial z_n}{\partial x} f(x) = z_{n-1}(x) + k_n(z_n(x))$$

$$L_f^2 h(x) = \frac{\partial z_{n-1}}{\partial x} f(x) + ((\frac{\partial k_n}{\partial y})_{y=z_n})\frac{\partial z_n}{\partial x} f(x)$$

$$= z_{n-2}(x) + ((\frac{\partial k_n}{\partial y})_{y=z_n})\frac{\partial z_n}{\partial x} f(x) + k_{n-1}(z_n(x))$$

$$= z_{n-2}(x) + \hat{k}_{n-1}(z_n(x), z_{n-1}(x))$$

where

$$\hat{k}_{n-1}(z_n, z_{n-1}) = \frac{\partial k_n}{\partial z_n} z_{n-1} + \frac{\partial k_n}{\partial z_n} k_n(z_n) + k_{n-1}(z_n)$$

Proceeding in this way one obtains for each $L_f^i(x)$, for $2 \leq i \leq n-1$, an expression of the form

$$L_f^i h(x) = z_{n-i}(x) + \hat{k}_{n-i+1}(z_n(x), \ldots, z_{n-i+1}(x))$$

Differentianting with respect to x and arranging all these expressions together, one obtains

$$\begin{pmatrix} \frac{\partial h}{\partial x} \\ \frac{\partial L_f h}{\partial x} \\ \vdots \\ \frac{\partial L_f^{n-1} h}{\partial x} \end{pmatrix} = \begin{pmatrix} \frac{\partial h}{\partial z} \\ \frac{\partial L_f h}{\partial z} \\ \vdots \\ \frac{\partial L_f^{n-1} h}{\partial z} \end{pmatrix} \frac{\partial z}{\partial x} = \begin{pmatrix} 0 & 0 & \cdots & 0 & 1 \\ 0 & 0 & \cdots & 1 & * \\ \cdot & \cdot & \cdots & \cdot & \cdot \\ 1 & * & \cdots & * & * \end{pmatrix} F_*(x)$$

This, because of the nonsingularity of the matrix on the right-hand-side, proves the claim. □

If the condition (7.5) is satisfied, then it is possible to define, in a neighborhood U of x^o, a unique vector field τ which satisfies the conditions

$$L_\tau h(x) = L_\tau L_f h(x) = \ldots = L_\tau L_f^{n-2} h(x) = 0$$

$$L_\tau L_f^{n-1} h(x) = 1$$

for all $x \in U$.

As a matter of fact, one only needs to solve the set of equations

$$\begin{pmatrix} dh(x) \\ dL_f h(x) \\ \cdots \\ dL_f^{n-2} h(x) \\ dL_f^{n-1} h(x) \end{pmatrix} \tau(x) = \begin{pmatrix} 0 \\ 0 \\ \cdot \\ 0 \\ 1 \end{pmatrix} \tag{7.9}$$

The construction of this vector field τ is useful in order to find necessary and sufficient conditions for the solution of our problem.

(7.10) *Lemma*. The Observer Linearization Problem is solvable if and only if

(i) $\dim(\text{span}\{dh(x^o), dL_f h(x^o), \ldots, dL_f^{n-1} h(x^o)\}) = n$

(ii) There exists a mapping Φ of some open set V of $\mathbb{R}^n$ onto a neighborhood U of x^o that satisfies the equation

$$\frac{\partial\Phi}{\partial z} = (\tau \quad -ad_f\tau \quad \ldots \quad (-1)^{n-1} ad_f^{n-1}\tau) \circ \Phi(z) \tag{7.11}$$

for all $z \in V$, where τ is the unique vector field solution of (7.9).

Proof. Necessity. We already know that (i) is necessary. Suppose (7.1) and (7.2) are satisfied and set $\Phi(z) = F^{-1}(z)$ for all $z = F(U)$. Moreover, let θ be the (unique) vector field Φ-related to $\frac{\partial}{\partial z_1}$, i.e. let

$$\theta(x) = \Phi_*(\frac{\partial}{\partial z_1}) \circ \Phi^{-1}(x)$$

We claim that

$$ad_f^k\theta(x) = (-1)^k \Phi_*(\frac{\partial}{\partial z_{k+1}}) \circ \Phi^{-1}(x) \tag{7.12}$$

for all $0 \leq k \leq n-1$. To show this, we proceed by induction (because (7.12) is true by definition for $k = 0$), and we use the fact, deduced from (7.2), (see also (7.8)), that

$$f(x) = (F_*)^{-1}(Az + K(cz)) \circ F(x) =$$

$$= \Phi_*(k_1(z_n)\frac{\partial}{\partial z_1} + (z_1+k_2(z_n))\frac{\partial}{\partial z_2} + \ldots + (z_{n-1}+k_n(z_n))\frac{\partial}{\partial z_n}) \circ \Phi^{-1}(x)$$

Suppose (7.12) is true for some $k < n-1$. Then

$$ad_f^{k+1}\theta = [f, (-1)^k \Phi_*(\frac{\partial}{\partial z_{k+1}}) \circ \Phi^{-1}] =$$

$$= (-1)^k \Phi_*[k_1\frac{\partial}{\partial z_1} + (z_1+k_2)\frac{\partial}{\partial z_2} + \ldots + (z_{n-1}+k_n)\frac{\partial}{\partial z_n}, \frac{\partial}{\partial z_{k+1}}] \circ \Phi^{-1}$$

$$= (-1)^{k+1} \Phi_*(\frac{\partial}{\partial z_{k+1}}) \circ \Phi^{-1}$$

Collecting all (7.12) together one obtains

$$(\theta \quad -ad_f\theta \ \ldots \ (-1)^{n-1}ad_f^{n-1}\theta)\circ\Phi = \Phi_*(\frac{\partial}{\partial z_1} \quad \frac{\partial}{\partial z_2} \ \ldots \ \frac{\partial}{\partial z_n}) = \frac{\partial\Phi}{\partial z}$$

If we show that θ necessarily coincides with the unique solution of (7.9) the proof is completed, because the p.d.e. (7.11) will coincide with the one just found.

To this end, observe that

$$(-1)^k L_{ad_f^k\theta}h(x) = \frac{\partial h}{\partial x}\Phi_*(\frac{\partial}{\partial z_{k+1}})\circ\Phi^{-1}(x) =$$

$$= (\frac{\partial h\circ\Phi}{\partial z_{k+1}})\circ\Phi^{-1}(x)$$

but, since $h\circ\Phi(z) = z_n$, we have

$$L_{ad_f^k\theta}h(x) = 0$$

for all $0 \leq k \leq n-2$ and

$$(-1)^{n-1}L_{ad_f^{n-1}\theta}h(x) = 1$$

Using Lemma (6.14) we deduce that

$$L_\theta L_f^k h(x) = 0$$

for all $0 \leq k \leq n-2$ and (see also Remark (6.17))

$$L_\theta L_f^{n-1}h(x) = 1$$

Thus, the vector field θ necessarily coincides with the unique solution of (7.9).

Sufficiency. Suppose (i) holds and let τ denote the solution of (7.9). Using Remark (6.17) one may immediately note (see (6.21)) that the matrix

$$\begin{pmatrix} dh(x^o) \\ dL_f h(x^o) \\ \cdot \\ dL_f^{n-1}h(x^o) \end{pmatrix} \cdot (\tau(x^o) \quad ad_f\tau(x^o) \ \ldots \ ad_f^{n-1}\tau(x^o))$$

has rank n. Therefore, the vector fields $\tau(x), ad_f\tau(x), \ldots, ad_f^{n-1}\tau(x)$ are linearly independent at x^o.

Let Φ denote a solution of the p.d.e. (7.11) and let z^o be a point such that $\Phi(z^o) = x^o$. From the linear independence of the vector fields on the right-hand-side of (7.11) we deduce that Φ has rank n at z^o, i.e. that Φ is a diffeomorphism of a neighborhood of z^o onto a neighborhood of x^o.

Set $F = \Phi^{-1}$ and

$$F_* f \circ F^{-1}(z) = \hat{f}_1 \frac{\partial}{\partial z_1} + \hat{f}_2 \frac{\partial}{\partial z_2} + \ldots + \hat{f}_n \frac{\partial}{\partial z_n} \tag{7.13}$$

By definition, the mapping Φ is such that

$$\Phi_*\left(\frac{\partial}{\partial z_{k+1}}\right) \circ \Phi^{-1}(x) = (-1)^k ad_f^k \tau(x)$$

so that

$$F_* ad_f^k \tau \circ F^{-1}(z) = (-1)^k \frac{\partial}{\partial z_{k+1}} \tag{7.14}$$

for all $0 \leq k \leq n-1$.

Using (7.13) and (7.14), one obtains, for all $0 \leq k \leq n-2$

$$(-1)^{k+1} \frac{\partial}{\partial z_{k+2}} = F_* ad_f^{k+1} \tau \circ F^{-1}(z) = F_*[f, ad_f^k \tau] \circ F^{-1}(z) =$$

$$= (-1)^k [\hat{f}_1 \frac{\partial}{\partial z_1} + \ldots + \hat{f}_n \frac{\partial}{\partial z_n}, \frac{\partial}{\partial z_{k+1}}] =$$

$$= (-1)^{k+1} \left(\left(\frac{\partial \hat{f}_1}{\partial z_{k+1}}\right) \frac{\partial}{\partial z_1} + \ldots + \left(\frac{\partial \hat{f}_n}{\partial z_{k+1}}\right) \frac{\partial}{\partial z_n} \right)$$

that, because of the linear independence of $\frac{\partial}{\partial z_1}, \ldots, \frac{\partial}{\partial z_n}$, implies

$$\frac{\partial \hat{f}_i}{\partial z_{k+1}} = 0 \qquad \text{for} \quad i \neq k+2$$

$$\frac{\partial \hat{f}_{k+2}}{\partial z_{k+1}} = 1$$

From these, one deduces that $\hat{f}_1$ depends only on z_n and that $\hat{f}_i$, for $2 \leq i \leq n$, is such that $\hat{f}_i - z_{i-1}$ depends only on z_n. In other terms, one has

$$F_* f \circ F^{-1}(x) = \begin{pmatrix} k_1(z_n) \\ z_1 + k_2(z_n) \\ \vdots \\ z_{n-1} + k_n(z_n) \end{pmatrix}$$

where $k_1, \ldots, k_n$ are suitable functions of z_n alone, and this shows that the condition (7.2) holds.

Moreover, since

$$L_{ad_f^k \tau} h = 0$$

for all $0 \leq k < n-1$, and

$$L_{ad_f^{n-1}\tau} h = (-1)^{n-1}$$

we deduce that

$$\frac{\partial h \circ F^{-1}}{\partial z_1} = \ldots = \frac{\partial h \circ F^{-1}}{\partial z_{n-1}} = 0$$

and that

$$\frac{\partial h \circ F^{-1}}{\partial z_n} = 1$$

This shows that also (7.1) holds. □

The integrability of the p.d.e. (7.11) may be expressed in terms of a property of the vector fields $\tau, ad_f\tau, \ldots, ad_f^{n-1}\tau$. To this end, one may use the following consequence of Frobenius Theorem.

(7.15) *Theorem*. Let $\tau_1, \ldots, \tau_n$ be vector fields on $\mathbb{R}^n$. Consider the set of partial differential equations

$$\frac{\partial x}{\partial z_i} = \tau_i(x(z)) \tag{7.16}$$

where x denotes a mapping from an open set of $\mathbb{R}^n$ to an open set of $\mathbb{R}^n$. Let (z^o, x^o) be a point in $\mathbb{R}^n \times \mathbb{R}^n$ and suppose $\tau_1(x^o), \ldots, \tau_n(x^o)$ are linearly independent. There exist neighborhoods U of x^o and V of z^o and a diffeomorphism $x: V \to U$ solving the equation (7.16), and such that $x(z^o) = x^o$, if and only if

(7.17) $$[\tau_i, \tau_j] = 0$$

for all $1 \leq i,j \leq n$.

Proof. We limit ourselves to give a scketch of the proof of the sufficiency. To this end, set

$$\Delta_i = sp\{\tau_i\}$$

Then, the collection of distributions $\Delta_1, \ldots, \Delta_n$ is independent, spans the tangent space and is simultaneously integrable because of (7.17) (see Theorem I.(3.12)). As a consequence, we may find a coordinate chart (U, ξ), such that $z^o = \xi(x^o)$ and

$$\Delta_i(x) = sp\{(\frac{\partial}{\partial \xi_i})_x\}$$

for all $x \in U$. The above may be rewritten as

$$\xi_* \tau_i(x) = c_i(\xi)(\frac{\partial}{\partial \xi_i})_{\xi \circ \xi(x)}$$

for all $x \in U$, where c_i is a smooth real-valued function, and $c_i(z^o) \neq 0$. The condition (7.17) may be used again to show that c_i depends only on ξ_i. Thus, there exist functions $z_i = \varphi_i(\xi_i)$ such that $z_i^o = \varphi_i(z_i^o)$ and

$$\frac{\partial \varphi_i}{\partial \xi_i} c_i(\xi_i) = 1$$

The composed function

$$z = \Phi(x) = (\varphi_1, \ldots, \varphi_n) \circ \xi(x)$$

is clearly such that $\Phi(x^o) = z^o$ and

$$\Phi_* \tau_i(x) = (\frac{\partial}{\partial z_i}) \circ \Phi(x)$$

Thus $x = \Phi^{-1}(z)$ solves the p.d.e. (7.16). □

Merging Lemma (7.10) with Theorem (7.15) yields the desired result.

(7.18) *Theorem*. The Observer Linearization Problem is solvable if and only if

(i) $\dim(\mathrm{span}\{dh(x^o), dL_f h(x^o), \ldots, dL_f^{n-1} h(x^o)\}) = n$

(ii) the unique vector field τ solution of (7.9) is such that

$$[ad_f^i \tau, ad_f^j \tau] = 0 \qquad (7.19)$$

for all $0 \leq i,j \leq n-1$.

(7.20) *Remark*. Using the Jacobi identity repeatedly, one can easily show that the condition (7.19) may be replaced by the condition

$$[\tau, ad_f^k \tau] = 0$$

for all $k = 1,3,\ldots,2n-1$. □

In summary one may proceed as follows in order to obtain an observer with linear (and spectrally assignable) error dynamics. If condition (i) holds, one finds first the vector field τ solving the equation (7.9). If also condition (ii) holds, one solves the p.d.e. (7.11) and finds a function Φ, defined in a neighborhood V of z^o, such that $\Phi(z^o) = x^o$. Then one sets $F = \Phi^{-1}$. Eventually, one computes the mapping K as

$$K(z_n) = \begin{pmatrix} k_1(z_n) \\ k_2(z_n) \\ \vdots \\ k_n(z_n) \end{pmatrix} = F_* f \circ F^{-1}(z) - \begin{pmatrix} 0 \\ z_1 \\ \vdots \\ z_{n-1} \end{pmatrix}$$

At this point, the observer

$$\dot{\xi} = (A+kc)\xi - ky + K(y)$$

with (A,c) in the form of the right-hand-sides of (7.6) yields the desired result.

APPENDIX

BACKGROUND MATERIAL IN DIFFERENTIAL GEOMETRY

1. Some facts from advanced calculus

Let A be an open subset of $\mathbb{R}^n$ and f: A $\to \mathbb{R}$ a function. The value of f at $x = (x_1,\ldots,x_n)$ is denoted $f(x) = f(x_1,\ldots,x_n)$. The function f is said to be a function of class C^∞ (or, simply, C^∞ or also, a *smooth* function) if its partial derivatives of any order with respect to $x_1,\ldots,x_n$ exist and are continuous. A function f is said to be analytic (sometimes noted as C^ω) if it is C^∞ and for each point $x^o \in A$ there exists a neighborhood U of x^o, such that the Taylor series expansion of f at x^o converges to f(x) for all $x \in U$.

Example. A typical example of a function which is C^∞ but not analytic is the function f: $\mathbb{R} \to \mathbb{R}$ defined by

$$f(x) = 0 \quad \text{if} \quad x \leq 0$$

$$f(x) = \exp(-\frac{1}{x}) \quad \text{if} \quad x > 0 \qquad \square$$

A mapping F: A $\to \mathbb{R}^m$ is a collection $(f_1,\ldots,f_m)$ of functions f_i: A $\to \mathbb{R}$. The mapping F is C^∞ if all f_i's are C^∞.

Let $U \subset \mathbb{R}^n$ and $V \subset \mathbb{R}^n$ be open sets. A mapping F: U $\to$ V is a diffeomorphism if is bijective (i.e. one-to-one and onto) and both F and F^{-1} are of class C^∞. The *jacobian matrix* of F at a point x is the matrix

$$\frac{\partial F}{\partial x} = \begin{pmatrix} \frac{\partial f_1}{\partial x_1} & \cdots & \frac{\partial f_1}{\partial x_n} \\ \cdot & \cdots & \cdot \\ \frac{\partial f_n}{\partial x_1} & \cdots & \frac{\partial f_n}{\partial x_n} \end{pmatrix}$$

The value of $\frac{\partial F}{\partial x}$ at a point $x = x^o$ is sometimes denoted $(\frac{\partial F}{\partial x})_{x^o}$.

Theorem. (Inverse function theorem). Let A be an open set of $\mathbb{R}^n$ and F: A $\to \mathbb{R}^n$ a C^∞ mapping. If $(\frac{\partial F}{\partial x})_{x^o}$ is nonsingular at some $x^o \in A$, then there exists an open neighborhood U of x^o in A such that V = F(U) is open in $\mathbb{R}^n$ and the restriction of F to U is a diffeomorphism onto V.

Theorem. (Rank theorem). Let $A \subset \mathbb{R}^n$ and $B \subset \mathbb{R}^m$ be open sets, $F: A \to B$ a C^∞ mapping. Suppose $(\frac{\partial F}{\partial x})_x$ has rank k for all $x \in A$. For each point $x^o \in A$ there exist a neighborhood A_o of x^o in A and an open neighborhood B_o of $F(x^o)$ in B, two open sets $U \in \mathbb{R}^n$ and $V \subset \mathbb{R}^m$, and two diffeomorphisms $G: U \to A_o$ and $H: B_o \to V$ such that $H \circ F \circ G(U) \subset V$ and such that for all $(x_1,\ldots,x_n) \in U$

$$(H \circ F \circ G)(x_1,\ldots,x_n) = (x_1,\ldots,x_k,0,\ldots,0)$$

Remark. Let P_k denote the mapping $P_k: \mathbb{R}^n \to \mathbb{R}^m$ defined by

$$P_k(x_1,\ldots,x_n) = (x_1,\ldots,x_k,0,\ldots,0)$$

Then, since H and G are invertible, one may restate the previous expression as

$$F = H^{-1} \circ P_k \circ G^{-1}$$

which holds at all points of A_o.

Theorem. (Implicit function theorem). Let $A \subset \mathbb{R}^m$ and $B \subset \mathbb{R}^n$ be open sets. Let $F: A \times B \to \mathbb{R}^n$ be a C^∞ mapping. Let $(x,y) = (x_1,\ldots,x_m,y_1,\ldots,y_n)$ denote a point of $A \times B$. Suppose that for some $(x^o,y^o) \in A \times B$

$$F(x^o,y^o) = 0$$

and that the matrix

$$\frac{\partial F}{\partial y} = \begin{pmatrix} \frac{\partial f_1}{\partial y_1} & \cdots & \frac{\partial f_1}{\partial y_n} \\ \cdot & \cdots & \cdot \\ \frac{\partial f_n}{\partial y_1} & \cdots & \frac{\partial f_n}{\partial y_n} \end{pmatrix}$$

is nonsingular at (x^o,y^o). Then, there exists open neighborhoods A_o of x^o in A and B_o of y^o in B and a unique C^∞ mapping $G: A_o \to B_o$ such that

$$F(x,G(x)) = 0$$

for all $x \in A_o$.

Remark. As an application of the implicit function theorem, consider the following corollary. Let A be an open set in $\mathbb{R}^m$, let M be a $k\times n$ matrix whose entries are real-valued C^∞ functions defined on A and b a k-vector whose entries are also real-valued C^∞ functions defined on A. Suppose that for some $x^o \in A$

$$\text{rank } M(x^o) = k$$

Then, there exist an open neighborhood U of x^o and a C^∞ mapping $G : U \to \mathbb{R}^n$ such that

$$M(x)G(x) = b(x)$$

for all $x \in U$.

In other words, the equation

$$M(x)y = b(x)$$

has at least a solution which is a C^∞ function of x in a neighborhood of x^o. If $k = n$ this solution is unique.

2. Some elementary notions of topology

This section is a review of the most elementary topological concepts that will be encountered later on.

Let S be a set. A *topological structure*, or a *topology*, on S is a collection of subsets of S, called *open* sets, satisfying the axioms

(i) the union of any number of open sets is open
(ii) the intersection of any finite number of open sets is open
(iii) the set S and the empty set $\emptyset$ are open

A set S with a topology is called a *topological space*.

A *basis* for a topology is a collection of open sets, called *basic open sets*, with the following properties

(i) S is the union of basic open sets
(ii) a nonempty intersection of two basic open sets is a union of basic open sets.

A *neighborhood* of a point p of a topological space is any open set which contains p.

Let S_1 and S_2 be topological spaces and F a mapping $F: S_1 \to S_2$. The mapping F is *continuous* if the inverse image of every open set of S_2 is an open set of S_1. The mapping F is *open* if the image of an open set of S_1 is an open set of S_2. The mapping F is an *homeomorphism* if is a bijection and both continuous and open.

If F is an homeomorphism, the inverse mapping F^{-1} is also an homeomorphism.

Two topological spaces S_1,S_2 such that there is an homeomorphism $F:S_1 \to S_2$ are said to be *homeomorphic*.

A subset U of a topological space is said to be *closed* if its complement $\bar{U}$ in S is open. It is easy to see that the intersection of any number of closed sets is closed, the union of any finite number of closed sets is closed, and both S and $\emptyset$ are closed.

If S_o is a subset of a topological space S, there is a unique open set, noted $\mathrm{int}(S_o)$ and called the *interior* of S_o, which is contained in S_o and contains any other open set contained in S_o. As a matter of fact, $\mathrm{int}(S_o)$ is the union of all open sets contained in S_o. Likewise, there is a unique closed set, noted $\mathrm{cl}(S_o)$ and called the *closure* of S_o, which contains S_o and is contained in any other closed set which contains S_o. Actually, $\mathrm{cl}(S_o)$ is the intersection of all closed sets which contain S_o.

A subset of S is said to be *dense* in S if its closure coincides with S.

If S_1 and S_2 are topological spaces, then the cartesian product $S_1 \times S_2$ can be given a topology taking as a basis the collection of all subsets of the form $U_1 \times U_2$, with U_1 a basic open set of S_1 and U_2 a basic open set of S_2. This topology on $S_1 \times S_2$ is sometimes called the *product topology*.

If S is a topological space and S_1 a subset of S, then S_1 can be given a topology taking as open sets the subsets of the form $S_1 \cap U$, with U any open set in S. This topology on S_1 is sometimes called the *subset topology*.

Let $F: S_1 \to S_2$ be a continuous mapping of topological spaces, and let $F(S_1)$ denote the image of F. Clearly, $F(S_1)$ with the subset topology is a topological space . Since F is continuous, the inverse image of any open set of $F(S_1)$ is an open set of S_1. However, not all open sets of S_1 are taken onto open sets of $F(S_1)$. In other words, the mapping $F': S_1 \to F(S_1)$ defined by $F'(p) = F(p)$ is continuous but not necessarily open. The set $F(S_1)$ can be given another topology, taking as open sets in $F(S_1)$ the images of open sets in S_1. It is easily seen that this new topology, sometimes called the *induced topology*, contains the subset topology (i.e. any set which is open in the subset topology is open also in the induced topology), and that the mapping F' is now open. If F is an injection, then S_1 and $F(S_1)$ endowed with the induced topology are homeomorphic.

A topological space S is said to satisfy the *Hausdorff separation axiom* (or, briefly, to be an Hausdorff space) if any two different points p_1 and p_2 have disjoint neighborhoods.

3. Smooth manifolds

Definition. A *locally Euclidean space* X of dimension n is a topological space such that, for each $p \in X$, there exists a homeomorphism φ mapping some open neighborhood of p onto an open set in $\mathbb{R}^n$. □

Definition. A *Manifold* N of dimension n is a topological space which is locally Euclidean of dimension n, is Hausdorff and has a countable basis. □

It is not possible that an open subset U of $\mathbb{R}^n$ be homeomorphic to an open subset V of $\mathbb{R}^m$, if $n \neq m$ (Brouwer's theorem on invariance of domain). Therefore, the *dimension* of a locally Euclidean space is a well-defined object.

A *coordinate chart* on a manifold N is a pair (U,φ), where U is an open set of N and φ a homeomorphism of U onto an open set of $\mathbb{R}^n$. Sometimes φ is represented as a set $(\varphi_1,\ldots,\varphi_n)$, and $\varphi_i: U \to \mathbb{R}$ is called the i-th *coordinate function*. If $p \in U$, the n-tuple of real numbers $(\varphi_1(p),\ldots,\varphi_n(p))$ is called the set of *local coordinates* of p in the coordinate chart (U,φ). A coordinate chart (U,φ) is called a *cubic* coordinate chart if $\varphi(U)$ is an open cube about the origin in $\mathbb{R}^n$. If $p \in U$ and $\varphi(p) = 0$, then the coordinate chart is said to be *centered* at p.

Let (U,φ) and (V,ψ) be two coordinate charts on a manifold N, with $U \cap V \neq \emptyset$. Let $(\psi_1,\ldots,\psi_n)$ be the set of coordinate functions associated with the mapping ψ. The homeomorphism

$$\psi\circ\varphi^{-1} : \varphi(U \cap V) \to \psi(U \cap V)$$

taking, for each $p \in U \cap V$, the set of local coordinates $(\varphi_1(p),\ldots,\varphi_n(p))$ into the set of local coordinates $(\psi_1(p),\ldots,\psi_n(p))$, is called a *coordinates transformation* on $U \cap V$. Clearly, $\varphi\circ\psi^{-1}$ gives the inverse mapping, which expresses $(\varphi_1(p),\ldots,\varphi_n(p))$ in terms of $(\psi_1(p),\ldots,\psi_n(p))$.

Frequently, the set $(\varphi_1(p),\ldots,\varphi_n(p))$ is represented as an n-vector $x = \mathrm{col}(x_1,\ldots,x_n)$, and the set $(\psi_1(p),\ldots,\psi_n(p))$ as an n-vector $y = \mathrm{col}(y_1,\ldots,y_n)$. Consistently, the coordinate transformation $\psi \circ \varphi^{-1}$ can be represented in the form

$$y = \begin{pmatrix} y_1 \\ \vdots \\ y_n \end{pmatrix} = \begin{pmatrix} y_1(x_1,\ldots,x_n) \\ \vdots \\ y_n(x_1,\ldots,x_n) \end{pmatrix} = y(x)$$

and the inverse transformation $\varphi \circ \psi^{-1}$ in the form

$$x = x(y)$$

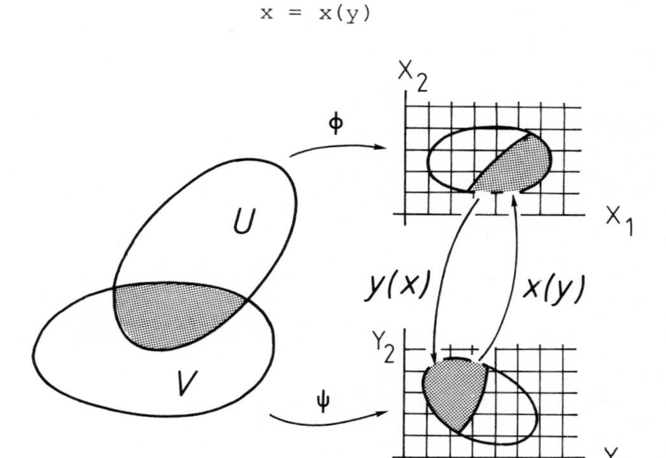

Two coordinate charts (U,φ) and (V,ψ) are C^∞-*Compatible* if, whenever $U \cap V \neq \emptyset$, the coordinate transformation $\psi \circ \varphi^{-1}$ is a diffeomorphism, i.e. if $y(x)$ and $x(y)$ are both C^∞ maps.

A C^∞ *atlas* on a manifold N is a collection $A = \{(U_i,\varphi_i)\}_{i\in I}$ of pairwise C^∞-compatible coordinate charts, with the property that $\bigcup_{i\in I} U_i = N$. An atlas is *complete* if not properly contained in any other atlas.

Definition. A smooth or C^∞ *manifold* is a manifold equipped with a complete C^∞ atlas. □

Remark. If A is any C^∞ atlas on a manifold N, there exists a *unique* complete C^∞ atlas A^* containing A. The latter is defined as the set of all coordinate charts (U,φ) which are compatible with every coordinate chart (U_i,φ_i) of A. This set contains A, is a C^∞ atlas, and is complete by construction. □

Some elementary examples of smooth manifolds are the ones described below.

Example. Any open set U of $\mathbb{R}^n$ is a smooth manifold, of dimension n. For, consider the atlas A consisting of the (single) coordinate chart (U, identity map on U) and let A^* denote the unique complete atlas containing A. In particular, $\mathbb{R}^n$ is a smooth manifold.

Remark. One may define different complete C^∞ atlases on the same manifold, as the following example shows. Let $N = \mathbb{R}$, and consider the coordinate charts $(\mathbb{R},\varphi)$ and $(\mathbb{R},\psi)$, with

$$\varphi(x) = x$$

$$\psi(x) = x^3$$

Since $\varphi^{-1}(x) = x$ and $\psi^{-1}(x) = x^{1/3}$,

$$\varphi \circ \psi^{-1}(x) = x^{1/3}$$

and the two charts are not compatible. Therefore the unique complete atlas A^*_φ which includes $(\mathbb{R},\varphi)$ and the unique complete atlas A^*_ψ which includes $(\mathbb{R},\psi)$ are different. This means that the same manifold N may be considered as a substrate of two different objects (two *smooth* manifolds), one arising with the atlas A^*_φ and the other with the atlas A^*_ψ . □

Example. Let U be an open set of $\mathbb{R}^m$ and let $\lambda_1,\ldots,\lambda_{m-n}$ be real-valued C^∞ functions defined on U. Let N denote the (closed) subset of U on which all functions $\lambda_1,\ldots,\lambda_{m-n}$ vanish, i.e. let

$$N = \{x \in U : \lambda_i(x) = 0,\ 1 \leq i \leq m-n\}$$

Suppose the rank of the jacobian matrix

$$\begin{pmatrix} \frac{\partial\lambda_1}{\partial x_1} & \cdots & \frac{\partial\lambda_1}{\partial x_m} \\ \cdot & \cdots & \cdot \\ \frac{\partial\lambda_{m-n}}{\partial x_1} & \cdots & \frac{\partial\lambda_{m-n}}{\partial x_m} \end{pmatrix}$$

is m-n at all $x \in N$. Then N is a smooth manifold of dimension n.

The proof of this essentially depends on the Implicit Function Theorem, and uses the following arguments. Let $x^o=(x_1^o,\ldots,x_n^o,x_{n+1}^o,\ldots,x_m^o)$ be a point of N and assume, without loss of generality, that the matrix

$$\begin{pmatrix} \frac{\partial\lambda_1}{\partial x_{n+1}} & \cdots & \frac{\partial\lambda_1}{\partial x_m} \\ \cdot & \cdots & \cdot \\ \frac{\partial\lambda_{m-n}}{\partial x_{n+1}} & \cdots & \frac{\partial\lambda_{m-n}}{\partial x_m} \end{pmatrix}$$

is nonsingular at x^o. Then, there exist neighborhoods A_o of $(x_1^o,\ldots,x_n^o)$ in $\mathbb{R}^n$ and B_o of $(x_{n+1}^o,\ldots,x_m^o)$ in $\mathbb{R}^{m-n}$ and a C^∞ mapping $G: A_o \to B_o$ such that

$$\lambda_i(x_1,\ldots,x_n,g_1(x_1,\ldots,x_n),\ldots,g_{m-n}(x_1,\ldots,x_n)) = 0$$

for all $1 \leq i \leq m-n$. This makes it possible to describe points of N around x^o as m-tuples $(x_1,\ldots,x_m)$ such that $x_{n+i} = g_i(x_1,\ldots,x_n)$ for $1 \leq i \leq m-n$. In this way one can construct a coordinate chart around each point x^o of N and the coordinate charts thus defined form a C^∞ atlas.

A manifold of this type is sometimes called a smooth *hypersurface* in $\mathbb{R}^m$. An important example of hypersurface is the *sphere* S^{m-1}, defined by taking $n = m-1$ and

$$\lambda_1 = x_1^2 + x_2^2 + \ldots + x_m^2 - 1$$

The set of points of $\mathbb{R}^m$ on which $f_1(x) = 0$ consists of all the points on a sphere of radius 1 centered at the origin. Since

$$\left(\frac{\partial\lambda_1}{\partial x_1} \cdots \frac{\partial\lambda_1}{\partial x_m}\right)$$

never vanishes on this set, the required conditions are satisfied and the set is a smooth manifold, of dimension m-1.

Example. An open subset N' of a smooth manifold N is itself a smooth manifold. The topology of N' is the subset topology. If (U,φ) is a coordinate chart of a complete C^∞ atlas of N, such that $U \cap N' \neq \emptyset$, then the pair (U',φ') defined as

$$U' = U \cap N'$$

$$\varphi' = \text{restriction of } \varphi \text{ to } U'$$

is a coordinate chart of N'. In this way, one may define a complete C^∞ atlas of N'. The dimension of N' is the same as that of N.

Example. Let M and N be smooth manifolds, of dimension m and n. Then the cartesian product M×N is a smooth manifold. The topology of M×N is the product topology. If (U,φ) and (V,ψ) are coordinate charts of M and N, the pair $(U\times V,(\varphi,\psi))$ is a coordinate chart of M×N. The dimension of M×N is clearly m+n.

An important example of this type of manifold is the *torus* $T^2 = S^1\times S^1$, the cartesian product of two circles. □

Let λ be a real-valued function defined on a manifold N. If (U,φ) is a coordinate chart on N, the composed function

$$\hat{\lambda} = \lambda\circ\varphi^{-1} : \varphi(U) \to \mathbb{R}$$

taking, for each $p \in U$, the set of local coordinates $(x_1,\ldots,x_n)$ of p into the real number $\lambda(p)$, is called an *expression of λ in local coordinates*.

In practice, whenever no confusion arises, one often uses the same symbol λ to denote $\lambda\circ\varphi^{-1}$, and write $\lambda(x_1,\ldots,x_n)$ to denote the value of λ at a point p of local coordinates $(x_1,\ldots,x_n)$.

If N and M are manifolds, of dimension n and m, $F : N \to M$ is a mapping, (U,φ) a coordinate chart on N and (V,ψ) a coordinate chart on M, the composed mapping

$$\hat{F} = \psi\circ F\circ\varphi^{-1}$$

is called an expression of F in local coordinates. Note that this definition make sense only if $F(U) \cap V \neq \emptyset$. If this is the case, then $\hat{F}$ is well defined for all n-tuples $(x_1,\ldots,x_n)$ whose image under $F\circ\varphi^{-1}$ is a point in V.

Here again, one often uses F to denote $\psi\circ F\circ\varphi^{-1}$, writes $y_i = f_i(x_1,\ldots,x_n)$ to denote the value of the i-th coordinate of F(p), p being a point of local coordinates $(x_1,\ldots,x_n)$, and also

$$y = \begin{pmatrix} y_1 \\ \vdots \\ y_m \end{pmatrix} = \begin{pmatrix} f_1(x_1,\ldots,x_n) \\ \vdots \\ f_m(x_1,\ldots,x_n) \end{pmatrix} = F(x)$$

Definition. Let N and M be smooth manifolds. A mapping $F: N \to M$ is a *smooth* mapping if for each $p \in N$ there exists coordinate charts (U,φ)

of N and (V,ψ) of M, with $p \in U$ and $F(p) \in V$, such that the expression of F in local coordinates is C^∞.

Remark. Note that the property of being smooth is independent of the choice of the coordinate charts on N and M. Different coordinate charts (U',φ') and (V',ψ') are by definition C^∞ compatible with the former and

$$\hat{F}' = \psi' \circ F \circ \varphi'^{-1} =$$

$$= \psi' \circ \psi^{-1} \circ \psi \circ F \circ \varphi^{-1} \circ \varphi \circ \varphi'^{-1} =$$

$$= (\psi' \circ \psi^{-1}) \circ \hat{F} \circ (\varphi' \circ \varphi^{-1})^{-1}$$

being a composition of C^∞ functions is still C^∞. □

Definition. Let N and M be smooth manifolds, both of dimension n. A mapping $F : N \to M$ is a *diffeomorphsim* if F is bijective and both F and F^{-1} are smooth mappings. Two manifolds N and M are *diffeomorphic* if there exists a diffeomorphism $F : N \to M$. □

The *rank* of a mapping $F : N \to M$ at a point $p \in N$ is the rank of the jacobian matrix

$$\begin{pmatrix} \frac{\partial f_1}{\partial x_1} & \cdots & \frac{\partial f_1}{\partial x_n} \\ \cdot & \cdots & \cdot \\ \frac{\partial f_m}{\partial x_1} & \cdots & \frac{\partial f_m}{\partial x_n} \end{pmatrix}$$

at $x = \varphi(p)$. It must be stressed that, although apparently dependent on the choice of local coordinates, the notion of rank thus defined is actually coordinate-independent. The reader may easily verify that the ranks of the jacobian matrices of two different expressions of F in local coordinates are equal.

Theorem. Let N and M be smooth manifolds both of dimension n. A mapping $F : N \to M$ is a diffeomorphism if and only if F is bijective, F is smooth and $\text{rank}(F) = n$ at all points of N.

Remark. In some cases, the assumption that functions, mappings, etc. are C^∞, may be replaced by the stronger assumption that functions, mappings, etc. are analytic. In this way one may define the notion of analytic manifold, analytic mappings of manifolds, and so on. We shall make this assumption explicitly whenever needed.

4. Submanifolds

Definitions. Let $F : N \to M$ be a smooth mapping of manifolds.

(i) F is an *immersion* if $\mathrm{rank}(F) = \dim(N)$ for all $p \in N$.

(ii) F is a *univalent immersion* if F is an immersion and is injective.

(iii) F is an *embedding* if F is a univalent immersion and the topology induced on F(N) by the one of N coincides with the topology of F(N) as a subset of M. □

Remark. The mapping F, being smooth, is in particular a continuous mapping of topological spaces. Therefore (see section 2) the topology induced on F(N) by the one of N may properly contain the topology of F(N) as a subset of M. This motivates the definition (iii). □

The difference between (i),(ii) and (iii) is clarified in the following examples.

Examples. Let $N = \mathbb{R}$ and $M = \mathbb{R}^2$. Let t denote a point in N and (x_1,x_2) a point in M. The mapping F is defined by

$$x_1(t) = at-\sin t$$
$$x_2(t) = \cos t$$

and, then,

$$\mathrm{rank}(F) = \mathrm{rank}\begin{pmatrix} a - \cos t \\ \\ - \sin t \end{pmatrix}$$

If $a = 1$ this mapping is *not* an immersion because $\mathrm{rank}(F) = 0$ at $t = 2k\pi$ (for any integer k).

If $0 < a < 1$ the mapping is an immersion, because $\mathrm{rank}(F) = 1$ for all t, but *not* a univalent immersion, because $F(t_1) = F(t_2)$ for all t_1,t_2 such that $t_1 = 2k\pi-\tau$, $t_2 = 2k\pi+\tau$ and $\sin \tau = a\tau$.

As a second example we consider the so-called "figure-eight".Let N be the open interval $(0,2\pi)$ of the real line and $M = \mathbb{R}^2$. Let t denote a point in N and (x_1,x_2) a point in M. The mapping F is defined by

$$x_1(t) = \sin 2t$$

$$x_2(t) = \sin t$$

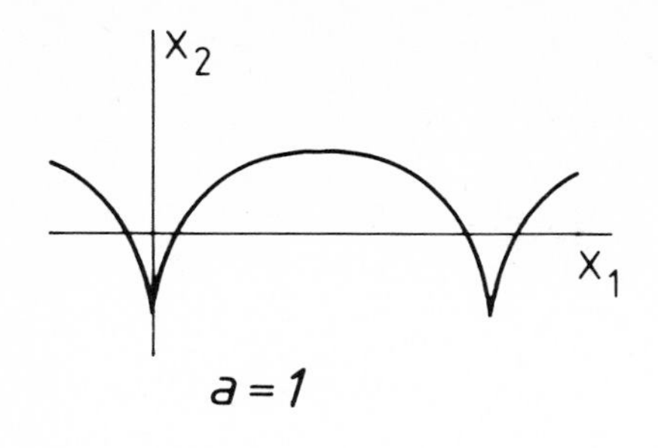

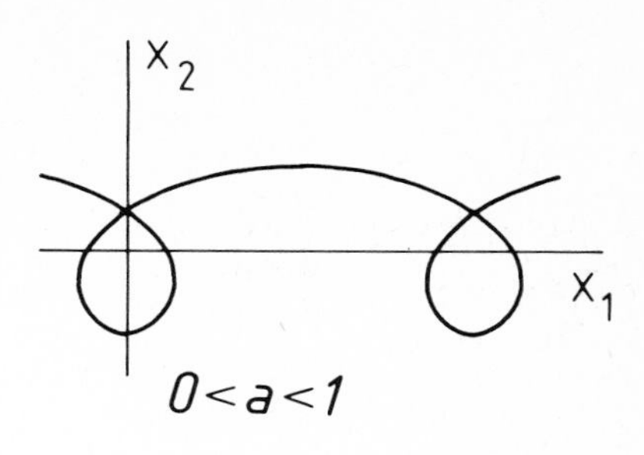

This mapping is an immersion because

$$\text{rank}(F) = \text{rank}\begin{pmatrix} \frac{dx_1}{dt} \\ \\ \frac{dx_2}{dt} \end{pmatrix} = \begin{pmatrix} 2\cos 2t \\ \\ \cos t \end{pmatrix} = 1$$

for all $0 < t < 2\pi$. It is also univalent because

$$F(t_1) = F(t_2) \Rightarrow t_1 = t_2$$

However, the mapping is *not* an embedding. For, consider the image of F.

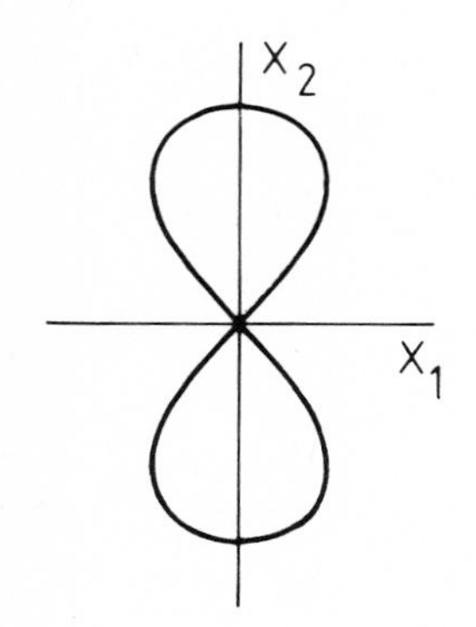

The mapping F takes the open set $(\pi-\varepsilon, \pi+\varepsilon)$ of N onto a subset U' of F(N) which is open *by definition* in the topology induced by the one of N, but is not an open set in the topology of F(N) as a subset of M. This is because U' cannot be seen as the intersection of F(N) with an open set of $\mathbb{R}^2$.
As a third example one may consider the mapping $F : \mathbb{R} \to \mathbb{R}^3$ given by

$$x_1(t) = \cos 2\pi t$$
$$x_2(t) = \sin 2\pi t$$
$$x_3(t) = t$$

whose image is an "helix" winding on an infinite cylinder whose axis is the x_3 axis. The reader may easily check that is an embedding. □

The following theorem shows that the every immersion *locally* is an embedding.

Theorem. Let $F : N \to M$ be an immersion. For each $p \in N$ there exists a neighborhood U of p with the property that the restriction of F to U is an embedding.

Example. Consider again the "figure eight" discussed above. If U is any interval of the type $(\delta, 2\pi-\delta)$, then the critical situation we had before disappears and the image U' of $(\pi-\varepsilon, \pi+\varepsilon)$ is now open also in the topology of F(N) as a subset of $\mathbb{R}^2$. □

The notions of univalent immersion and of embedding are used in the following way.

Definition. The image F(N) of a univalent immersion is called an *immersed submanifold* of M. The image F(N) of an embedding is called an *embedded submanifold* of M.

Remark. Conversely, one may say that a subset M' of M is an immersed (respectively, embedded) submanifold of M if there is another manifold N and a univalent immersion (respectively, embedding) $F: N \to M$ such that $F(N) = M'$. □

The use of the word "submanifold" in the above definition clearly indicates the possibility of giving F(N) the structure of a smooth manifold, and this may actually be done in the following way. Let $M'=F(N)$ and $F': N \to M'$ denote the mapping defined by

$$F'(p) = F(p)$$

for all $p \in N$. Clearly, F' is a bijection. If the topology of M' is the one induced by that of N (i.e. open sets of M' are the images under F' of open sets of N), F' is a homeomorphism. Consequently, any coordinate chart (U, φ) of N induces a coordinate chart (V, ψ) of M', defined as

$$V = F'(U) \quad , \quad \psi = \varphi \circ (F')^{-1}$$

C^∞-compatible charts of N induce C^∞-compatible charts of M' and so complete C^∞-atlases induce complete C^∞-atlases. This gives M' the structure of a smooth manifold.

The smooth manifold M' thus defined is *diffeomorphic* to the smooth manifold N. A diffeomorphism between M' and N is indeed F' itself, which is bijective, smooth, and has rank equal to the dimension of N at each $p \in N$.

Embedded submanifolds can also be characterized in a different way, based on the following considerations.

Let M be a smooth manifold of dimension m and (U,φ) a cubic coordinate chart. Let n be an integer, $0 \leq n < m$, and p a point of U. The subset of U

$$S_p = \{q \in U: x_i(q) = x_i(p),\ i = n+1,\dots,m\}$$

is called an n-dimensional *slice* of U passing through p. In other words a slice of U is the locus of all points of U for which some coordinates (e.g. the last m-n) are constant.

Theorem. Let M be a smooth manifold of dimension m. A subset M' of M is an embedded submanifold of dimension $n < m$ if and only if for each $p \in M'$ there exists a cubic coordinate chart (U,φ) of M, with $p \in U$, such that $U \cap M'$ coincides with an n-dimensional slice of U passing through p. □

This theorem provides a more "intrinsic" characterization of the notion of an embedded submanifold (of a manifold M), directly related to the existence of special coordinate charts (of M). Note that, if (U,φ) is a coordinate chart of M such that $U \cap M'$ is an n-dimensional slice of U, the pair (U',φ') defined as

$$U' = U \cap M'$$

$$\varphi'(p) = (x_1(p),\dots,x_n(p))$$

is a coordinate chart of M'. This is illustrated in the following figure (where $M = \mathbb{R}^3$ and $n = 2$).

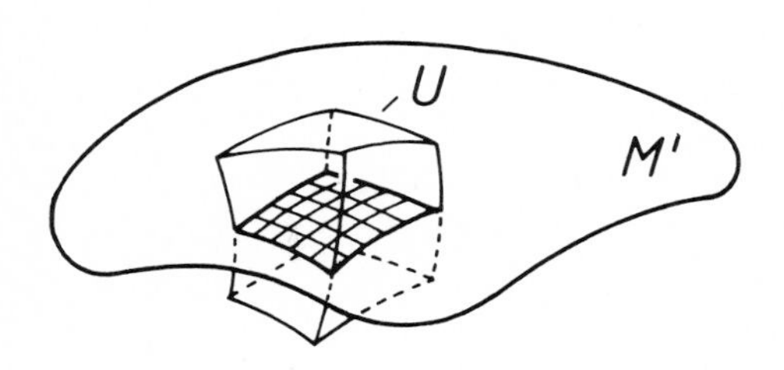

Remark. Note that an open subset M' of M is indeed an embedded submanifold of M, of the same dimension m. Thus, a submanifold M' of M may be a *proper* subset of M, although being a manifold of the same dimension.

Remark. It can be proven that any smooth hypersurface in $\mathbb{R}^m$ is an embedded submanifold of $\mathbb{R}^m$. Moreover it has also been shown that if N is an n-dimensional smooth manifold, there exist an integer $m \geq n$ and a mapping $F : N \rightarrow \mathbb{R}^m$ which is an embedding (Whitney's embedding theorem). In other words, any manifold is diffeomorphic to an embedded submanifold of $\mathbb{R}^m$, for a suitably large m.

Remark. Let V be an n-dimensional subspace of $\mathbb{R}^m$. Any subset of $\mathbb{R}^m$ of the form

$$x^o+V = \{x \in \mathbb{R}^m : x = x'+x^o;\ x' \in V\},$$

where x^o is some fixed point of $\mathbb{R}^m$, is indeed a smooth hypersurface and so an embedded submanifold of $\mathbb{R}^m$, of dimension n. This is sometimes called a *flat* submanifold of $\mathbb{R}^n$.

5. Tangent vectors

Let N be a smooth manifold of dimension n. A real-valued function λ is said to be *smooth in a neighborhood* of p, if the domain of λ includes an open set U of N containing p and the restriction of λ to U is a smooth function. The set of all smooth functions in a neighborhood of p is denoted $C^\infty(p)$. Note that $C^\infty(p)$ forms a vector space over the field $\mathbb{R}$. For, if λ,γ are functions in $C^\infty(p)$ and a,b are real numbers, the function $a\lambda+b\gamma$ defined as

$$(a\lambda+b\gamma)(q) = a\lambda(q)+b\gamma(q)$$

for all q in a neighborhood of p, is again a function in $C^\infty(p)$. Note also that two functions $\lambda,\gamma \in C^\infty(p)$ may be multiplied to give another element of $C^\infty(p)$, written $\lambda\gamma$ and defined as

$$(\lambda\gamma)(q) = \lambda(q)\cdot\gamma(q)$$

for all q in a neighborhood of p.

Definition. A *tangent vector* v at p is a map $v\colon C^\infty(p) \to \mathbb{R}$ with the following properties:

(i) (linearity) : $v(a\lambda+b\gamma) = av(\lambda)+bv(\gamma)$ for all $\lambda,\gamma \in C^\infty(p)$ and $a,b \in \mathbb{R}$

(ii) (Leibnitz rule): $v(\lambda\gamma) = \gamma(p)v(\lambda)+\lambda(p)v(\gamma)$ for all $\lambda,\gamma \in C^\infty(p)$.

Definition. Let N be a smooth manifold. The *tangent space* to N at p, written T_pN, is the set of all tangent vectors at p.

Remark. A map which satisfies the properties (i) and (ii) is also called a *derivation*.

Remark. The set T_pN forms a vector space over the field $\mathbb{R}$ under the rules of scalar multiplication and addition defined in the following way. If v_1, v_2 are tangent vectors and c_1, c_2 real numbers, $c_1v_1+c_2v_2$ is a new tangent vector which takes the function $\lambda \in C^\infty(p)$ into the real number

$$(c_1v_1+c_2v_2)(\lambda) = c_1v_1(\lambda) + c_2v_2(\lambda)$$

Remark. We shall see later on that, if the manifold N is a smooth hypersurface in $\mathbb{R}^m$, the object previously defined may be naturally identified with the intuitive notion of "tangent hyperplane" at a point. □

Let (U,φ) be a (fixed) coordinate chart around p. With this coordinate chart one may associate n tangent vectors at p, denoted

$$\left(\frac{\partial}{\partial\varphi_1}\right)_p, \ldots, \left(\frac{\partial}{\partial\varphi_n}\right)_p$$

defined in the following way

$$\left(\frac{\partial}{\partial\varphi_i}\right)_p(\lambda) = \left(\frac{\partial(\lambda\circ\varphi^{-1})}{\partial x_i}\right)_{x=\varphi(p)}$$

for $1 \leq i \leq n$. The right-hand-side is the value taken at $x = (x_1,\ldots,x_n) = \varphi(p)$ of the partial derivative of the function $\lambda\circ\varphi^{-1}(x_1,\ldots,x_n)$ with respect to x_i (recall that the function $\lambda\circ\varphi^{-1}$ is an expression of λ in local coordinates).

Theorem. Let N be a smooth manifold of dimension n. Let p be any point of N. The tangent space T_pN to N at p is an n-dimensional vector space over the field $\mathbb{R}$. If (U,φ) is a coordinate chart around p, then the tangent vectors $\left(\frac{\partial}{\partial\varphi_1}\right)_p, \ldots, \left(\frac{\partial}{\partial\varphi_n}\right)_p$ form a basis of T_pN. □

The basis $\{\left(\frac{\partial}{\partial\varphi_1}\right)_p, \ldots, \left(\frac{\partial}{\partial\varphi_n}\right)_p\}$ of T_pN is sometimes called the *natural basis* induced by the coordinate chart (U,φ).

Let v be a tangent vector at p. From the above theorem it is seen that

$$v = \sum_{i=1}^{n} v_i \left(\frac{\partial}{\partial\varphi_i}\right)_p$$

where $v_1,\ldots,v_n$ are real numbers. One may compute the v_i's explicitly in the following way. Let φ_i be the i-th coordinate function. Clearly

$\varphi_i \in C^\infty(p)$, and then

$$v(\varphi_i) = \sum_{j=1}^{n} v_j \left(\frac{\partial}{\partial \varphi_j}\right)_p (\varphi_i) = \sum_{j=1}^{n} v_j \left(\frac{\partial(\varphi_i \circ \varphi^{-1})}{\partial x_j}\right)_{x=\varphi(p)} = v_i$$

because $\varphi_i \circ \varphi^{-1}(x_1,\ldots,x_n) = x_i$. Thus the real number v_i coincides with the value of v at φ_i , the i-th coordinate function.

A change of coordinates around p clearly induces a *change of basis* in T_pN. The computations involved are the following ones. Let (U,φ) and (V,ψ) be coordinate charts around p. Let $\{(\frac{\partial}{\partial \psi_1})_p,\ldots,(\frac{\partial}{\partial \psi_n})_p\}$ denote the natural basis of T_pN induced by the coordinate chart (V,ψ). Then

$$\left(\frac{\partial}{\partial \psi_i}\right)_p (\lambda) = \left(\frac{\partial(\lambda \circ \psi^{-1})}{\partial y_i}\right)_{y=\psi(p)} = \left(\frac{\partial(\lambda \circ \varphi^{-1} \circ \varphi \circ \psi^{-1})}{\partial y_i}\right)_{y=\psi(p)} =$$

$$= \sum_{j=1}^{n} \left(\frac{\partial(\lambda \circ \varphi^{-1})}{\partial x_j}\right)_{x=\varphi(p)} \cdot \left(\frac{\partial(\varphi_j \circ \psi^{-1})}{\partial y_i}\right)_{y=\psi(p)} =$$

$$= \sum_{j=1}^{n} \left(\left(\frac{\partial}{\partial \varphi_j}\right)_p (\lambda)\right) \left(\frac{\partial(\varphi_j \circ \psi^{-1})}{\partial y_i}\right)_{y=\psi(p)}$$

In other words

$$\left(\frac{\partial}{\partial \psi_i}\right)_p = \sum_{j=1}^{n} \left(\frac{\partial(\varphi_j \circ \psi^{-1})}{\partial y_i}\right)_{y=\psi(p)} \left(\frac{\partial}{\partial \varphi_j}\right)_p$$

Note that the quantity

$$\frac{\partial(\varphi_j \circ \psi^{-1})}{\partial y_i}$$

is the element on the j-th row and i-th column of the *jacobian matrix* of the coordinate transformation

$$x = x(y)$$

So the elements of the columns of the jacobian matrix of $x = x(y)$ are the coefficients which express the vectors of the "new" basis as linear combinations of the vectors of the "old" basis.

If v is a tangent vector, and $(v_1,\dots,v_n),(w_1,\dots,w_n)$ the n-tuples of real numbers which express v in the form

$$v = \sum_{i=1}^{n} v_i \left(\frac{\partial}{\partial\varphi_i}\right)_p = \sum_{i=1}^{n} w_i \left(\frac{\partial}{\partial\psi_i}\right)_p$$

then

$$\begin{pmatrix} v_1 \\ \vdots \\ v_n \end{pmatrix} = \begin{pmatrix} \frac{\partial x_1}{\partial y_1} & \cdots & \frac{\partial x_1}{\partial y_n} \\ \cdot & \cdots & \cdot \\ \frac{\partial x_n}{\partial y_1} & \cdots & \frac{\partial x_n}{\partial y_n} \end{pmatrix} \begin{pmatrix} w_1 \\ \vdots \\ w_n \end{pmatrix}$$

Definition. Let N and M be smooth manifolds. Let $F : N \to M$ be a smooth mapping. The *differential* of F at $p \in N$ is the map

$$F_* : T_pN \to T_{F(p)}M$$

defined as follows. For $v \in T_pN$ and $\lambda \in C^\infty(F(p))$,

$$(F_*(v))(\lambda) = v(\lambda \circ F)$$

Remark. F_* is a map of the tangent space of N at a point p into the tangent space of M at the point $F(p)$. If $v \in T_pN$, the value $F_*(v)$ of F_* at v is a tangent vector in $T_{F(p)}M$. So one has to express the way in which $F_*(v)$ maps the set $C^\infty(F(p))$, of all functions which are smooth a neighborhood of $F(p)$, into $\mathbb{R}$. This is actually what the definition specifies. Note that there is one of such maps for *each* point p of N.

Theorem. The differential F_* is a linear map. □

Since F_* is a linear map, given a basis for T_pN and a basis for $T_{F(p)}M$ one may wish to find its matrix representation. Let (U,φ) be a coordinate chart around p, (V,ψ) a coordinate chart around $q = F(p)$, $\{(\frac{\partial}{\partial\varphi_1})_p,\dots,(\frac{\partial}{\partial\varphi_n})_p\}$ the natural basis of T_pN and $\{(\frac{\partial}{\partial\psi_1})_q,\dots,(\frac{\partial}{\partial\psi_m})_q\}$ the natural basis of T_qM . In order to find a matrix representation of F_* , one has simply to see how F_* maps $(\frac{\partial}{\partial\varphi_i})_p$ for each $1 \leq i \leq n$.

$$\left(F_*\left(\frac{\partial}{\partial\varphi_i}\right)_p\right)(\lambda) = \left(\frac{\partial}{\partial\varphi_i}\right)_p(\lambda\circ F) = \left(\frac{\partial(\lambda\circ F\circ\varphi^{-1})}{\partial x_i}\right)_{x=\varphi(p)} =$$

$$= \left(\frac{\partial(\lambda\circ\psi^{-1}\circ\psi\circ F\circ\varphi^{-1})}{\partial x_i}\right)_{x=\varphi(p)} = \sum_{j=1}^{m}\left(\frac{\partial(\lambda\circ\psi^{-1})}{\partial y_j}\right)_{y=\psi(q)}\left(\frac{\partial(\psi_j\circ F\circ\varphi^{-1})}{\partial x_i}\right)_{x=\varphi(p)}$$

$$= \sum_{j=1}^{m}\left(\left(\frac{\partial}{\partial\psi_j}\right)_q(\lambda)\right)\left(\frac{\partial(\psi_j\circ F\circ\varphi^{-1})}{\partial x_i}\right)_{x=\varphi(p)}$$

In other words

$$F_*\left(\frac{\partial}{\partial\varphi_i}\right)_p = \sum_{j=1}^{m}\left(\frac{\partial(\psi_j\circ F\circ\varphi^{-1})}{\partial x_i}\right)_{x=\varphi(p)}\left(\frac{\partial}{\partial\psi_j}\right)_q$$

Now, recall that $\psi\circ F\circ\varphi^{-1}$ is an expression of F in local coordinates. Then, the quantity

$$\frac{\partial(\psi_j\circ F\circ\varphi^{-1})}{\partial x_i}$$

is the element on the j-th row and i-th column of the *jacobian matrix* of the mapping expressing F in local coordinates. Using again

$$F(x) = F(x_1,\ldots,x_n) = \begin{pmatrix} F_1(x_1,\ldots,x_n) \\ \vdots \\ F_m(x_1,\ldots,x_n) \end{pmatrix}$$

to denote $\psi\circ F\circ\varphi^{-1}$, one has simply

$$F_*\left(\frac{\partial}{\partial\varphi_i}\right)_p = \sum_{j=1}^{m}\left(\frac{\partial F_j}{\partial x_i}\right)\left(\frac{\partial}{\partial\psi_j}\right)_q$$

If $v \in T_pN$ and $w = F_*(v) \in T_{F(p)}M$ are expressed as

$$v = \sum_{i=1}^{n} v_i\left(\frac{\partial}{\partial\varphi_i}\right)_p \qquad w = \sum_{i=1}^{m} w_i\left(\frac{\partial}{\partial\psi_i}\right)_q$$

then

$$\begin{pmatrix} w_1 \\ \vdots \\ w_m \end{pmatrix} = \begin{pmatrix} \frac{\partial F_1}{\partial x_1} & \cdots & \frac{\partial F_1}{\partial x_n} \\ \cdot & \cdots & \cdot \\ \frac{\partial F_m}{\partial x_1} & \cdots & \frac{\partial F_m}{\partial x_n} \end{pmatrix}\begin{pmatrix} v_1 \\ \vdots \\ v_n \end{pmatrix}$$

Remark. The matrix representation of F_* is exactly the jacobian of its expression in local coordinates. From this, it is seen that the rank of a mapping coincides with the rank of the corresponding differential.

Remark (Chain rule). It is easily seen that, if F and G are smooth mappings, then

$$(G \circ F)_* = G_* F_*$$

The following examples may clarify the notion of tangent space and the one of differential.

Example. The tangent vectors on $\mathbb{R}^n$. Let $\mathbb{R}^n$ be equipped with the "natural" complete atlas already considered in previous examples (i.e. the one including the chart ($\mathbb{R}^n$, identity map on $\mathbb{R}^n$)). Then, if v is a tangent vector at a point x and λ a smooth function

$$v(\lambda) = \sum_{i=1}^{n} v_i \left(\frac{\partial}{\partial x_i}\right)_x (\lambda) = \sum_{i=1}^{n} \left(\frac{\partial \lambda}{\partial x_i}\right)_x v_i$$

So, $v(\lambda)$ is just the value of the *derivative* of λ *along the direction* of the vector

$$\mathrm{col}(v_1, \ldots, v_n)$$

at the point x.

Remark. Let $F : N \to M$ be a univalent immersion. Let $n = \dim(N)$ and $m = \dim(M)$. By definition, F_* has rank n at each point. Therefore the image $F_*(T_pN)$ of F_*, at each point p, is a subspace of $T_{F(p)}M$ isomorphic to T_pN. The subspace $F_*(T_pN)$ can actually be *identified* with the tangent space at F(p) to the submanifold $M' = F(N)$. In order to understand this point, let F' denote the function $F' : N \to M'$ defined as

$$F'(p) = F(p)$$

for all $p \in N$. F' is a diffeomorphism and so F'_* is an isomorphism. Therefore the image $F'_*(T_pN)$ is exactly the tangent space at F'(p) to M'. Any tangent vector in $T_{F(p)}M'$ is the image $F'_*(v)$ of a (unique) vector $v \in T_pN$ and can be identified with the (unique) vector $F_*(v)$ of $F_*(T_pN)$.

In other words, the tangent space at p to a *submanifold* M' of M can be identified with a *subspace* of the tangent space at p to M.

The same considerations can be repeated in local coordinates. It is known that an immersion is locally an embedding. Therefore, around every point $p \in M'$ it is possible to find a coordinate chart (U,φ) of M, with the property that the pair (U',φ') defined by

$$U' = \{q \in U : \varphi_i(q) = \varphi_i(p),\ i = n+1,\dots,m\}$$

$$\varphi' = (\varphi_1,\dots,\varphi_n)$$

is a coordinate chart of M'. According to this choice, the tangent space to M' at p is identified with the n-dimensional subspace of T_pM spanned by the tangent vectors $\{(\frac{\partial}{\partial\varphi_1})_p,\dots,(\frac{\partial}{\partial\varphi_n})_p\}$. □

Example. The tangent vector to a smooth curve in $\mathbb{R}^n$. We define first the notion of a smooth curve in $\mathbb{R}^n$. Let $N = (t_1,t_2)$ be an open interval on the real line. A *smooth curve* in $\mathbb{R}^n$ is the image of a univalent immersion $\sigma : N \to \mathbb{R}^n$. Thus, a smooth curve is an immersed submanifold of $\mathbb{R}^n$. In N and $\mathbb{R}^n$ one may choose natural local coordinates as usual and, letting t denote an element of N, express σ by means of an n-tuple of scalar-valued functions $\sigma_1,\dots,\sigma_n$ of t.

A smooth curve is a 1-dimensional immersed submanifold of $\mathbb{R}^n$. At a point $\sigma(t_o)$, the tangent space to the curve is a 1-dimensional vector space which, as we have seen, may be identified with a subspace of the tangent space to $\mathbb{R}^n$ at this point. A basis of the tangent space to the curve at $\sigma(t_o)$ is given by the image under σ_* of $(\frac{d}{dt})_{t_o}$, a tangent vector at t_o to N. This image is computed as follows

$$\sigma_*(\frac{d}{dt})_{t_o} = \sum_{i=1}^{n} (\frac{d\sigma_i}{dt})_{t_o} (\frac{\partial}{\partial x_i})_{\sigma(t_o)}$$

Thinking of $t \in N$ as time and $\sigma(t)$ as a point moving in $\mathbb{R}^n$, we may interpret the vector

$$\text{col}((\frac{d\sigma_1}{dt})_{t_o},\dots,(\frac{d\sigma_n}{dt})_{t_o})$$

as the velocity along the curve, evaluated at the point $\sigma(t^o)$. So, we have that the velocity vector at a point of the curve spans the tangent space to the curve at this point. From this point of view, we see that the notion of tangent space to a 1-dimensional manifold may be identified with the geometric notion of tangent line to a curve in a Euclidean space.

Example. Let h be a smooth function $h : \mathbb{R}^2 \to \mathbb{R}$ and $F : \mathbb{R}^2 \to \mathbb{R}^3$ a mapping defined by

$$F(x_1,x_2) = (x_1,x_2,h(x_1,x_2))$$

This mapping is an embedding and therefore $F(\mathbb{R}^2)$, a surface in $\mathbb{R}^2$, is an embedded submanifold of $\mathbb{R}^3$. At each point $F(x)$ of this surface, the tangent space, identified as a subspace of the tangent space to $\mathbb{R}^3$ at this point, may be computed as

$$\text{span}\{F_*(\frac{\partial}{\partial x_1})_x \ , \ F_*(\frac{\partial}{\partial x_2})_x\}$$

Now,

$$F_*(\frac{\partial}{\partial x_1})_x = \sum_{i=1}^{3} (\frac{\partial F_1}{\partial x_i})(\frac{\partial}{\partial x_i})_{F(x)} = (\frac{\partial}{\partial x_1})_{F(x)} + (\frac{\partial h}{\partial x_1})(\frac{\partial}{\partial x_3})_{F(x)}$$

$$F_*(\frac{\partial}{\partial x_2})_x = (\frac{\partial}{\partial x_2})_{F(x)} + (\frac{\partial h}{\partial x_2})(\frac{\partial}{\partial x_3})_{F(x)}$$

This tangent space to $F(\mathbb{R}^2)$ at some point $(x_1^o,x_2^o,h(x_1^o,x_2^o))$ is the set of tangent vectors whose expressions in local coordinates are of the form

$$v = \begin{pmatrix} \alpha \\ \beta \\ (\frac{\partial h}{\partial x_1})\alpha+(\frac{\partial h}{\partial x_2})\beta \end{pmatrix}$$

α,β being real numbers and $\frac{\partial h}{\partial x_1}$, $\frac{\partial h}{\partial x_2}$ being evaluated at $x_1 = x_1^o$ and $x_2 = x_2^o$. From this point of view, we see that the notion of tangent space to a 2-dimensional manifold may be identified with the geometric notion of tangent plane to a surface in a Euclidean space. □

One may define objects dual to the ones considered so far.

Definition. Let N be a smooth manifold. The *cotangent space* to N at p, written T_p^*N, is the dual space of T_pN. Elements of the cotagent space are called *tangent covectors*.

Remark. Recall that a dual space V^* of a vector space V is the space of all linear functions from V to $\mathbb{R}$. If $v^* \in V^*$, then $v^*: V \to \mathbb{R}$ and the value of v^* at $v \in V$ is written as $\langle v^*,v\rangle$. V^* forms a vector space over the field $\mathbb{R}$, with rules of scalar multiplication and addition which define $c_1v_1^*+c_2v_2^*$ in the following terms

$$\langle c_1 v_1^* + c_2 v_2^*, v \rangle = c_1 \langle v_1^*, v \rangle + c_2 \langle v_2^*, v \rangle$$

If $e_1,\ldots,e_n$ is a basis of V, the unique basis $e_1^*,\ldots,e_n^*$ of V^* which satisfies

$$\langle e_i^*, e_j \rangle = \delta_{ij}$$

is called a *dual basis*.

If V and W are vector spaces, $F : V \to W$ a linear mapping and $w \in W$, $v \in V$, the mapping $F^* : W^* \to V^*$ defined by

$$\langle F^*(w^*), v \rangle = \langle w^*, F(v) \rangle$$

is called the *dual* mapping (of F). □

Let λ be a smooth function $\lambda : N \to \mathbb{R}$. There is a natural way of identifying the differential λ_* of λ at p with an element of T_p^*N. For, observe that λ_* is a linear mapping

$$\lambda_* : T_pN \to T_{\lambda(p)}\mathbb{R}$$

and that $T_{\lambda(p)}\mathbb{R}$ is isomorphic to $\mathbb{R}$. The natural isomorphism between $\mathbb{R}$ and $T_{\lambda(p)}\mathbb{R}$ is the one in which the element c of $\mathbb{R}$ corresponds to the tangent vector $c(\frac{d}{dt})_t$. If $c(\frac{d}{dt})_t$ is the value at v of the differential λ_* at p, then c must depend linearly on v, i.e. there must exist a co-vector, denoted $(d\lambda)_p$, such that

$$\lambda_*(v) = \langle (d\lambda)_p, v \rangle (\frac{d}{dt})_t$$

Given a basis of T_pN, the covector $(d\lambda)_p$ (like any other covector), may be represented in matrix form. Let $\{(\frac{\partial}{\partial\varphi_1})_p,\ldots,(\frac{\partial}{\partial\varphi_n})_p\}$ be the natural basis of T_pN induced by the coordinate chart (U,φ). The image under λ_* of a vector

$$v = \sum_{i=1}^{n} v_i (\frac{\partial}{\partial\varphi_i})_p$$

is the vector

$$\lambda_*(v) = (\sum_{i=1}^{n} \frac{\partial\lambda}{\partial x_i} v_i)(\frac{d}{dt})_t$$

and this shows that

$$\langle (d\lambda)_p, v \rangle = \left(\frac{\partial\lambda}{\partial x_1} \quad \cdots \quad \frac{\partial\lambda}{\partial x_n} \right) \begin{pmatrix} v_1 \\ \vdots \\ v_n \end{pmatrix}$$

Remark. Note also that the value at λ of a tangent vector v is equal to the value at v of the tangent covector $(d\lambda)_p$, i.e.

$$v(\lambda) = \langle (d\lambda)_p, v \rangle \qquad \square$$

The dual basis of $\{(\frac{\partial}{\partial\varphi_1})_p, \ldots, (\frac{\partial}{\partial\varphi_n})_p\}$ is computed as follows. From the equality $v(\lambda) = \langle (d\lambda)_p, v \rangle$ we deduce that

$$\langle (d\varphi_i)_p, (\frac{\partial}{\partial\varphi_j})_p \rangle = (\frac{\partial}{\partial\varphi_j})_p(\varphi_i) = \frac{\partial(\varphi_i \circ \varphi^{-1})}{\partial x_j} = \frac{\partial x_i}{\partial x_j} = \delta_{ij}$$

so that the desired dual basis is exactly provided by the set of tangent covectors $\{(d\varphi_1)_p, \ldots, (d\varphi_n)_p\}$.

If v^* is any tangent covector, expressed as

$$v^* = \sum_{i=1}^{n} v_i^* (d\varphi_i)_p ,$$

the real numbers $v_1^*, \ldots, v_n^*$ are such that

$$v_i^* = \langle v^*, (\frac{\partial}{\partial\varphi_i})_p \rangle$$

Note also that, if v is any tangent vector expressed as

$$v = \sum_{i=1}^{n} v_i (\frac{\partial}{\partial\varphi_i})_p$$

the real numbers $v_1, \ldots, v_n$ are such that

$$v_i = \langle (d\varphi_i)_p, v \rangle .$$

6. Vector fields

Definition. Let N be a smooth manifold, of dimension n. A *vector field* f on N is a mapping assigning to each point $p \in N$ a tangent vector f(p) in T_pN. A vector field f is *smooth* if for each $p \in N$ there exists a coordinate chart (U,φ) about p and n real-valued smooth functions $f_1,\dots,f_n$ defined on U, such that, for all $q \in U$,

$$f(q) = \sum_{i=1}^{n} f_i(q)\left(\frac{\partial}{\partial\varphi_i}\right)_q$$

Remark. Because of C^∞-compatibility of coordinate charts, given any coordinate chart (V,ψ) about p other than (U,φ), one may find a neighborhood $V' \subset V$ of p and n real-valued smooth functions $f_1',\dots,f_n'$ defined on V', such that, for all $q \in V'$.

$$f(q) = \sum_{i=1}^{n} f_i'(q)\left(\frac{\partial}{\partial\psi_i}\right)_q$$

Thus, the notion of smooth vector field is independent of the coordinates used.

Remark. If (U,φ) is a coordinate chart of N, on the *submanifold* U of N one may define a special set of smooth vector fields, denoted $\frac{\partial}{\partial\varphi_1},\dots,\frac{\partial}{\partial\varphi_n}$, in the following way

$$\left(\frac{\partial}{\partial\varphi_i}\right):\ p \longmapsto \left(\frac{\partial}{\partial\varphi_i}\right)_p$$

It must be stressed, however, that such a set of vector fields is an object defined only on U. □

For any fixed coordinate chart (U,φ), the set of tangent vectors $\{(\frac{\partial}{\partial\varphi_1})_q,\dots,(\frac{\partial}{\partial\varphi_n})_q\}$ is a basis of T_qN at each $q \in U$, and therefore there is a *unique* set of smooth functions $\{f_1,\dots,f_n\}$ that makes it possible to express the value of a vector field f at q in the form

$$f(q) = \sum_{i=1}^{n} f_i(q)\left(\frac{\partial}{\partial\varphi_i}\right)_q$$

Expressing each f_i in local coordinates, as

$$\hat{f}_i = f_i \circ \varphi^{-1}$$

provides an expression in local coordinates of the vector field f itself. So, if p is a point of coordinates $(x_1,\dots,x_n)$ in the chart (U,φ), $f(p)$ is a tangent vector of coefficients $(\hat{f}_1(x_1,\dots,x_n),\dots$ $\dots,\hat{f}_n(x_1,\dots,x_1))$ in the natural basis $\{(\frac{\partial}{\partial\varphi_1})_p,\dots,(\frac{\partial}{\partial\varphi_n})_p\}$ of T_pN

induced by (U,φ). Most of the times, whenever possible, the symbol f_i replaces $f_i \circ \varphi^{-1}$ and the expression of f in local coordinates is given a form of an n-vector $f = \mathrm{col}(f_1,\ldots,f_n)$.

Remark. Let f be a smooth vector field, (U,φ) and (V,ψ) two coordinate charts about p and $f(x) = f(x_1,\ldots,x_n)$, $f'(y) = f'(y_1,\ldots,y_n)$ the corresponding expressions of f in local coordinates. Then

$$f'(y) = \left(\frac{\partial y}{\partial x} f(x)\right)_{x=x(y)} \qquad \square$$

The notion of vector field makes it possible to introduce the concept of a *differential equation on a manifold* N. For, let f be a smooth vector field. A smooth curve $\sigma:(t_1,t_2) \to N$ is an *integral curve* of f if

$$\sigma_*\left(\frac{d}{dt}\right)_t = f(\sigma(t))$$

for all $t \in (t_1,t_2)$. The left-hand-side is a tangent vector to the submanifold $\sigma((t_1,t_2))$ at the point $\sigma(t)$; the right-hand-side is a tangent vector to N at $\sigma(t)$. As usual, we identify the tangent space to a submanifold of N at a point with a subspace of the tangent space to N at this point.

In local coordinates, $\sigma(t)$ is expressed as an n-tuple $(\sigma_1(t),\ldots,\sigma_n(t))$, and $f(\sigma(t))$ as

$$f(\sigma(t)) = \sum_{i=1}^{n} f_i(\sigma_1(t),\ldots,\sigma_n(t))\left(\frac{\partial}{\partial \varphi_i}\right)_{\sigma(t)}$$

Moreover

$$\sigma_*\left(\frac{d}{dt}\right)_t = \sum_{i=1}^{n} \frac{d\sigma_i}{dt}\left(\frac{\partial}{\partial \varphi_i}\right)_{\sigma(t)}$$

Therefore, the expression of σ in local coordinates is such that

$$\frac{d\sigma_i}{dt} = f_i(\sigma_1(t),\ldots,\sigma_n(t))$$

for all $1 \leq i \leq n$. This shows that the notion of integral curve of a vector field corresponds to the notion of solution of a set of n ordinary differential equations of the first order.

For this reason, one often uses the notation

$$\dot{\sigma}(t) = \sigma_*(\frac{d}{dt})_t$$

to indicate the image of $(\frac{d}{dt})_t$ under the differential σ_* at t.

The following theorem contains all relevant informations about the properties of integral curves of vector fields.

Theorem. Let f be a smooth vector field on a manifold N. For each $p \in N$, there exist an open interval - depending on p and written I_p - of $\mathbb{R}$ such that $0 \in I_p$ and a smooth mapping

$$\Phi : W \to N$$

defined on the subset W of $\mathbb{R}\times N$

$$W = \{(t,p) \in \mathbb{R}\times N: t \in I_p\}$$

with the following properties:

(i) $\Phi(0,p) = p$,

(ii) for each p the mapping $\sigma_p : I_p \to N$ defined by

$$\sigma_p(t) = \Phi(t,p)$$

is an integral curve of f,

(iii) if $\mu : (t_1,t_2) \to N$ is another integral curve of f satisfying the condition $\mu(0) = p$, then $(t_1,t_2) \subset I_p$ and the restriction of σ_p to (t_1,t_2) coincides with μ,

(iv) $\Phi(s,\Phi(t,p)) = \Phi(s+t,p)$ whenever both sides are defined,

(v) whenever $\Phi(t,p)$ is defined, there exists an open neighborhood U of p such that the mapping $\Phi_t : U \to N$ defined by

$$\Phi_t(q) = \Phi(t,q)$$

is a diffeomorphism onto its image, and

$$\Phi_t^{-1} = \Phi_{-t}$$

Remark. Properties (i) and (ii) say that σ_p is an integral curve of f passing through p at t = 0. Property (iii) says that this curve is unique and that the domain I_p on which σ_p defined is maximal. Property (iv) and (v) say that the family of mappings $\{\Phi_t\}$ is a one-parameter (namely, the parameter t) group of local diffeomorphisms, under the operation of composition. □

Example. Let $N = \mathbb{R}$ and use x to denote a point in N. Consider the vector field

$$f(x) = (x^2+1)\left(\frac{\partial}{\partial x}\right)_x$$

An integral curve σ of f must be such that

$$\dot{\sigma}(t) = \left(\frac{d\sigma}{dt}\right)\left(\frac{\partial}{\partial x}\right)_x = (\sigma^2(t)+1)\left(\frac{\partial}{\partial x}\right)_x$$

so

$$\frac{d\sigma}{dt} = \sigma^2+1$$

A solution of this equation has the form

$$\sigma(t) = tg(t+tg^{-1}(x^o))$$

with x^o being indeed the value of σ at $t = 0$. Clearly, for each x^o the solution is defined for

$$-\frac{\pi}{2} < t+tg^{-1}(x^o) < \frac{\pi}{2}$$

Thus W is the set

$$W = \{(t,x^o) : t \in (-\frac{\pi}{2}-tg^{-1}(x^o), \frac{\pi}{2}-tg^{-1}(x^o))\}$$

which has the form indicated below. □

The mapping Φ is called the *flow* of f. Often, for practical purposes, the notation Φ_t replaces Φ, with the understanding that t is a variable. To stress the dependence on f, sometimes Φ_t is written as Φ_t^f.

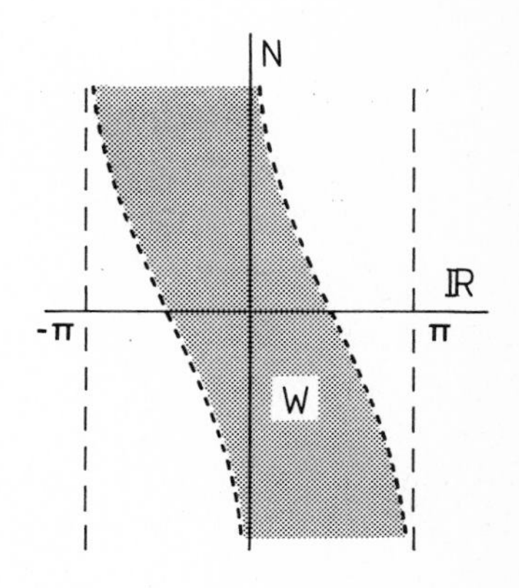

Definition. A vector field f is *complete* if, for all $p \in N$, the interval I_p coincides with $\mathbb{R}$, i.e. - in other words - if the flow Φ of f is defined on the whole cartesian product $\mathbb{R}\times N$. □

The integral curves of a complete vector field are thus defined, whatever the initial point p is, for all $t \in \mathbb{R}$.

Definition. Let f be a smooth vector field on N and λ a smooth real-valued function on N. The *Lie derivative of* λ *along* f is a function $N \to \mathbb{R}$, written $L_f\lambda$ and defined by

$$(L_f\lambda)(p) = (f(p))(\lambda)$$

(i.e. $(L_f\lambda)(p)$ is the value at λ of the tangent *vector* f(p) at p). □

The function $L_f\lambda$ is a smooth function. In local coordinates, $L_f\lambda$ is represented by

$$(L_f\lambda)(x_1,\ldots,x_n) = \left(\frac{\partial\lambda}{\partial x_1} \cdots \frac{\partial\lambda}{\partial x_n}\right)\begin{pmatrix} f_1 \\ \vdots \\ f_n \end{pmatrix}$$

If f_1, f_2 are vector fields and λ a real-valued function, we denote

$$L_{f_1}L_{f_2}\lambda = L_{f_1}(L_{f_2}\lambda)$$

The set of all smooth vector fields on a manifold N is denoted by the symbol V(N). This set is a *vector space* over $\mathbb{R}$ since if f, g are vector fields and a,b are real numbers, their linear combination af+bg is a vector field defined by

$$(af+bg)(p) = af(p)+bg(p)$$

If a,b are smooth real-valued functions on N, one may still define a linear combination af+bg by

$$(af+bg)(p) = a(p)f(p)+b(p)f(p)$$

and this gives V(N) the structure of a *module* over the ring, denoted $C^\infty(N)$, of all smooth real-valued functions defined on N. The set V(N) can be given, however, a more interesting algebraic structure in this way.

Definition. A vector space V over $\mathbb{R}$ is a *Lie algebra* if in addition to its vector space structure it is possible to define a binary operation $V \times V \to V$, called a product and written $[\cdot,\cdot]$, which has the following properties

(i) it is skew commutative, i.e.

$$[v,w] = -[w,v]$$

(ii) it is bilinear over $\mathbb{R}$, i.e.

$$[\alpha_1 v_1+\alpha_2 v_2, w] = \alpha_1[v_1,w]+\alpha_2[v_2,w]$$

(where α_1,α_2 are real numbers)

(iii) it satisfies the so called *Jacobi identity*, i.e.

$$[v,[w,z]]+[w,[z,v]]+[z,[v,w]] = 0. \quad \square$$

The set V(N) forms a Lie algebra with the vector space structure already discussed and a product $[\cdot,\cdot]$ defined in the following way. If f and g are vector fields, $[f,g]$ is a new vector field whose value at p, a tangent vector in T_pN, maps $C^\infty(p)$ into $\mathbb{R}$ according to the rule

$$([f,g](p))(\lambda) = (L_f L_g \lambda)(p)-(L_g L_f \lambda)(p)$$

In other words, $[f,g](p)$ takes λ into the real number $(L_f L_g\lambda)(p)-$ $+(L_g L_f\lambda)(p)$. Note that one may write more simply

$$L_{[f,g]}\lambda = L_f L_g\lambda - L_g L_f \lambda$$

Theorem. V(M) with the product $[f,g]$ thus defined is a Lie algebra. $\square$

The product $[f,g]$ is called the *Lie bracket* of the two vector fields f and g.

Remark. If f,g are smooth vector fields and λ,γ smooth real-valued functions, then

$$[\lambda f,\gamma g] = \lambda\cdot\gamma\cdot[f,g] + \lambda\cdot L_f\gamma\cdot g - \gamma\cdot L_g\lambda\cdot f$$

Note that $\lambda,\gamma,L_f\gamma,L_g\lambda$ are elements of $C^\infty(N)$, and $g,f,[f,g]$ elements of V(N). $\square$

The reader may easily find that the expression of $[f,g]$ in local coordinates is given by the n-vector

$$\begin{pmatrix} \frac{\partial g_1}{\partial x_1} & \cdots & \frac{\partial g_1}{\partial x_n} \\ \cdot & \cdots & \cdot \\ \frac{\partial g_n}{\partial x_1} & \cdots & \frac{\partial g_n}{\partial x_n} \end{pmatrix} \begin{pmatrix} f_1 \\ \vdots \\ f_n \end{pmatrix} - \begin{pmatrix} \frac{\partial f_1}{\partial x_1} & \cdots & \frac{\partial f_1}{\partial x_n} \\ \cdot & \cdots & \cdot \\ \frac{\partial f_n}{\partial x_1} & \cdots & \frac{\partial f_n}{\partial x_n} \end{pmatrix} \begin{pmatrix} g_1 \\ \vdots \\ g_n \end{pmatrix} = \frac{\partial g}{\partial x} f - \frac{\partial f}{\partial x} g$$

If, in particular, $N = \mathbb{R}^n$ and

$$f(x) = Ax \ , \ g(x) = Bx$$

then

$$[f,g](x) = (BA-AB)x$$

The matrix $[A,B] = (BA-AB)$ is called the *commutator* of A,B.

The importance of the notion of Lie bracket of vector fields is very much related to its applications in the study of nonlinear control systems. For the moment, we give hereafter two interesting properties.

Theorem. Let N' be an embedded submanifold of N. Let U' be an open set of N' and f,g two smooth vector fields of N such that for all $p \in U'$

$$f(p) \in T_pN' \quad \text{and} \quad g(p) \in T_pN'.$$

Then also

$$[f,g](p) \in T_pN'$$

for all $p \in U'$. □

In other words, the Lie bracket of two vector fields "tangent" to a fixed submanifold is still tangent to that submanifold.

Theorem. Let f,g be two smooth vector fields on N. Let Φ_t^f denote the flow of f. For each $p \in N$.

$$\lim_{t \to 0} \frac{1}{t}[(\Phi_{-t}^f)_* g(\Phi_t^f(p)) - g(p)] = [f,g](p)$$

Remark. The first term of the expression under bracket is a tangent vector at p, obtained in the following way. With p, the mapping Φ_t^f (always defined for sufficiently small t) associates a point $q = \Phi_t^f(p)$. The vector field g is evaluated at q, and the value $g(q) \in T_qN$ is taken back to T_pN via the differential $(\Phi_{-t}^f)_*$ (which maps the tangent space at q onto the tangent space at $p = \Phi_{-t}^f(q)$). Thus, the mapping $p \mapsto (\Phi_{-t}^f)_* g(\Phi_t^f(p))$ defines a vector field, on the domain of Φ_t^f.

Remark. Let f be a smooth vector field on N, g a smooth vector field on M and $F : N \to M$ a smooth function. The vector fields f,g are said to be F-*related* if

$$F_* f = g \circ F$$

Note that the vector field $(\Phi^f_{-t})_* g(\Phi^f_t(p))$ considered in the above Remark is Φ^f_{-t}-related to g.

Remark. If $\bar{f}$ is F-related to f and $\bar{g}$ is F-related to g, then $[\bar{f},\bar{g}]$ is F-related to $[f,g]$.

Definition. Let f,g be two smooth vector fields on N. The *Lie derivative of* g *along* f is a vector field on N, written $L_f g$ and defined by

$$(L_f g)(p) = \lim_{t\to 0} \frac{1}{t}[(\Phi^f_{-t})_* g(\Phi^f_t(p)) - g(p)]. \square$$

Thus, by definition, the Lie derivative $L_f g$ of g along f coincides with the Lie bracket $[f,g]$. There is also a third notation often used, which expreses the Lie derivative of q along f as

$$L_f g = ad_f g$$

Both notations may be used recurrently, taking

$$L^0_f g = g \quad \text{and} \quad L^k_f g = L_f(L^{k-1}_f g)$$

or

$$ad^0_f g = g \quad \text{and} \quad ad^k_f g = ad_f(ad^{k-1}_f g)$$

Remark. The Lie derivative of g along f may be interpreted as the value at t = 0 of the derivative with respect to t of a function defined as

$$W(t) = (\Phi^f_{-t})_* g(\Phi^f_t(p))$$

Moreover, it is easily seen that for any $k \geq 0$

$$\left(\frac{d^k W(t)}{dt^k}\right)_{t=0} = L^k_f g(p) = ad^k_f\, g(p)$$

If W(t) is analytic in a neighborhood of t = 0, then W(t) can be expanded in the form

$$W(t) = \sum_{k=0}^{\infty} ad^k_f g(p) \frac{t^k}{k!}$$

known as *Campell-Baker-Hausdorff formula.* □

One may define an object which dualizes the notion of a vector field.

Definition. Let N be a smooth manifold of dimension n. A *covector field* (also called *one-form*) ω on N is a mapping assigning to each point $p \in N$ a tangent covector $\omega(p)$ in T_p^*N. A covector field f is *smooth* if for each $p \in N$ there exists a coordinate chart (U,φ) about p and n real-valued smooth functions $\omega_1,\dots,\omega_n$ defined on U, such that, for all $q \in U$

$$\omega(q) = \sum_{i=1}^{n} \omega_i(q)(d\varphi_i)_q \qquad \square$$

The notion of smooth covector field is clearly independent of the coordinate used. The expression of a covector field in local coordinates is often given the form of a row vector $\omega = \text{row}(\omega_1,\dots,\omega_n)$ in which the ω_i's are real-valued functions of $x_1,\dots,x_n$.

If ω is a covector field and f is a vector field, $\langle \omega,f \rangle$ denotes the smooth real-valued function defined by

$$\langle \omega,f \rangle(p) = \langle \omega(p),f(p)\rangle$$

With any smooth function $\lambda:N \to \mathbb{R}$ one may associate a covector field by taking at each p the cotangent vector $(d\lambda)_p$. The covector field thus defined is usually still represented by the symbol dλ. However, the converse is *not always true.*

Definition. A covector field ω is *exact* if there exists a smooth real valued function $\lambda:N \to \mathbb{R}$ such that

$$\omega = d\lambda \qquad \square$$

The set of all smooth covector fields on a manifold N is denoted by the symbol $V^*(N)$.

One may also define the notion of Lie derivative of a covector field ω along a vector field f. In order to do this, one has to introduce first the notion of a covector field Φ_t^f-*related* to a given covector field ω. Let p be a point of the domain of Φ_t^f. Recall that $(\Phi_t^f)_* : T_pN \to T_{\Phi_t^f(p)}N$ is a linear mapping and let $(\Phi_t^f)^*:T^*_{\Phi_t^f(p)}N \to T_p^*N$ denote the dual mapping. With ω and Φ_t^f we associate a new covector field, whose value at a point p in the domain of Φ_t^f is defined by

$$(\Phi_t^f)^* \omega(\Phi_t^f(p))$$

The covector field thus defined is said to be Φ_t^f-related to ω.

Theorem. Let f be a smooth vector field and ω a smooth covector field on N. For each $p \in N$ the limit

$$\lim_{t\to 0} \frac{1}{t}[(\Phi_t^f)^* \omega(\Phi_t^f(p)) - \omega(p)]$$

exists.

Definition. The *Lie derivative of* ω *along* f is a covector field on N, written $L_f\omega$, whose value at p is set equal to the value of the limit

$$\lim_{t\to 0} \frac{1}{t}[(\Phi_t^f)^* \omega(\Phi_t^f(p)) - \omega(p)] \qquad \square$$

The expression of $L_f\omega$ in local coordinates is given by the (row) n-vector

$$(f_1 \dots f_n)\begin{pmatrix} \frac{\partial \omega_1}{\partial x_1} & \dots & \frac{\partial \omega_n}{\partial x_1} \\ \cdot & \dots & \cdot \\ \frac{\partial \omega_1}{\partial x_n} & \dots & \frac{\partial \omega_n}{\partial x_n} \end{pmatrix} + (\omega_1 \dots \omega_n)\begin{pmatrix} \frac{\partial f_1}{\partial x_1} & \dots & \frac{\partial f_1}{\partial x_n} \\ \cdot & \dots & \cdot \\ \frac{\partial f_n}{\partial x_1} & \dots & \frac{\partial f_n}{\partial x_n} \end{pmatrix} = (\frac{\partial \omega^T}{\partial x} f)^T + \omega \frac{\partial f}{\partial x}$$

where the superscript "T" denotes "transpose".

Remark. The three types of Lie derivatives $L_f\lambda, L_fg, L_f\omega$ defined above are related by the following Leibnitz-type relation

$$L_f\langle \omega, g\rangle = \langle L_f\omega, g\rangle + \langle \omega, L_fg\rangle$$

Remark. If ω is an exact covector field, i.e. if $\omega = d\lambda$ for some λ,

$$\langle d\lambda, f\rangle = L_f\lambda$$

and

$$L_f d\lambda = d(L_f\lambda)$$

Remark. If ω is a covector field, f a vector field, λ and γ real-valued functions, then

$$L_{\lambda f}\gamma\omega = \lambda\cdot\gamma\cdot L_f\omega + \lambda\cdot L_f\gamma\cdot\omega + \gamma\langle\omega, f\rangle d\lambda$$

Note that $\lambda, \gamma, L_f\gamma, \langle\omega, f\rangle$ are elements of $C^\infty(N)$ and $\omega, L_f\omega, d\lambda$ elements of $V^*(N)$.

BIBLIOGRAPHICAL NOTES

Chapter I

The definition of distribution used here is taken from Sussmann (1973); in most of the references in Differential Geometry quoted in the Appendix, the term "distribution" without any further specification is used to denote what we mean here for "nonsingular distribution". Different proofs of Frobenius' Theorem are available. The one used here is mutuated from Lobry (1970) and Sussmann (1973). Theorems on simultaneous integrability of distributions are due to Jakubczyk-Respondek (1980) and Respondek (1982).

The importance in control theory of the notion of invariance of a distribution under a vector field was pointed out independently by Hirschorn (1981) and Isidori et al. (1981a). A more general notion of invariance, under a group of local diffeomorphisms, was given by Sussmann (1973). The local decompositions described in section 5 are consequences of ideas of Krener (1977).

Theorem (6.15) and (6.20) were first proved by Sussmann-Jurdjevic (1972). The proof described here is due to Krener (1974). An earlier version of Theorem (6.15), dealing with "reachability" along trajectories traversed in either time direction, was given by Chow (1939). Controllability of systems evolving on Lie groups was studied by Brockett (1972a). Theorem (7.8), although in a slightly different version, is due to Hermann-Krener (1977).

Chapter II

The proof of Theorems (1.4) and (1.7) may be found in Sussmann (1973). An independent proof of Theorem (1.11) was given earlier by Hermann (1962) and an independent proof of Corollary (1.13) by Nagano (1966).

The relevance of the so-called "control Lie algebra" in the analysis of global reachability derives from the work of Chow (1939) and was subsequently elucidated by Lobry (1970), Haynes-Hermes (1970), Elliott (1971) and Sussmann-Jurdjevic (1972), among the others. The properties of the "observation space" were studied by Hermann-Krener (1977), and, in the case of discrete-time systems, by Sontag (1979).

Reachability, observability and decompositions of bilinear systems were studied by Brockett (1972b), Goka et al. (1973) and d'Alessandro et al. (1974).

Chapter III

The functional expansions illustrated in the first section were introduced by Fliess since 1973. A comprehensive exposition of the subject, together with several additional results, may be found in Fliess (1981). The expressions of the Kernels of the Volterra series expansion were discovered by Lesjak-Krener (1978); the expansions (2.12) are due to Fliess et al. (1983). The structure of the Volterra kernels was earlier analyzed by Brockett (1976), who proved that any individual kernel can always be interpreted as a kernel of a suitable bilinear system, and related results may also be found in Gilbert (1977). The expressions of the kernels of a bilinear system were first calculated by Bruni et al. (1971). Multivariable Laplace transforms of Volterra kernels and their properties are extensively studied by Rugh (1981).

Functional expansions for nonlinear discrete-time systems have been studied by Sontag (1979) and Normand Cyrot (1983).

The way the invariance analysis is dealt with reflects jointly ideas developed by Isidori et al. (1981a) and Claude (1982); the former contains, in particular, a different proof of Theorem (3.9). Left invertibility of nonlinear systems was mostly studied by Hirschorn (1979a) (1979b); our presentation follows an idea of Nijmeijer (1982b).

Definitions and properties of generalized Hankel matrices were developed by Fliess (1974). Theorem (5.8) was proved independently by Isidori (1973) and Fliess. The notion of Lie rank and Theorem (5.11) are due to Fliess (1983). Equivalence of minimal realizations was extensively studied by Sussmann (1977); the version given here of the uniqueness Theorem essentially develops an idea of Hermann-Krener (1977); related results may also be found in Fliess (1983).

An independent approach to realization theory was followed by Jakubczyk (1980). Realization of finite Volterra series was studied by Crouch (1981). Constructive realization methods from the Laplace transform of a Volterra kernel may be found in the work of Rugh (1983). Realization theory of discrete-time response maps was extensively studied by Sontag (1979).

Chapter IV

Controlled invariant distribution is the nonlinear version of the notion of controlled invariant subspace, introduced independently by Basile-Marro (1969) and Wonham-Morse (1970). For a comprehensive presentation of the theory of multivaribale linear control systems, the reader is referred to the classical treatise of Wonham (1979). Controlled invariant distributions were introduced independently by Hirschorn (1981) and, in the more general form described here, by Isidori et al. (1981a). The proof of Lemma (1.10) may be found in Hirschorn (1981), Isidori et al. (1981b) and Nijmeijer (1981).

Lemma (3.6) is due to Claude (1982); the special case where the matrix A(x) has rank ℓ was dealt with in Isidori et al. (1981a). The algorithm (3.17) has been suggested by Krener (1985).

Early results on nonlinear decoupling and noninteracting control were given by Singh-Rugh (1972) and Freund (1975). The possibility of solving decoupling problems in a differential-geometric setting was described by Hirschorn (1981) and Isidori et al. (1981a). The notion of controllability distribution, the nonlinear version of the one of controllability subspace, was introduced by Isidori-Krener (1982) and Nijmeijer (1982a). The solution of noninteracting control problems via controllability distributions is described in Nijmeijer (1983).

Controlled invariance for general nonlinear control systems (i.e. systems where the control does not enter linearly) is studied in Nijmeijer-Van der Schaft (1983). Controlled invariance for discrete-time nonlinear system is studied in Grizzle (1985) and Monaco-Normand Cyrot (1985).

Chapter V

The input-output linearization problem was treated by Isidori-Ruberti (1984). A slightly different version of this problem was earlier studied by Claude-Fliess-Isidori (1983) and, in the case of discrete-time systems, by Monaco-Normand Cyrot (1983). The so-called "structure algorithm" was introduced by Silverman (1969) and its importance in connection with the computation of the "zero structure at infinity" was outlined by Van Dooren et al. (1979). The possibility of computing a "zero structure at infinity" on the coefficients of the formal power series associated with the external behavior of a nonlinear system was pointed out by Isidori (1983); a geometric approach to the definition of a "zero structure at infinity" is followed by Nijmeijer-Schumacher (1985). The problem of matching a linear model via dynamic state feedback was studied by Isidori (1985) and Di Benedetto-Isidori (1985). The proof of Theorem (5.9) is the nonlinear version of a proof of Malabre (1984).

The state-space linearization problem was proposed and solved for single-input systems by Brockett (1978). Complete solution for multi-input systems was found by Jakubczyk-Respondek (1980). Independent work of Su (1982) and Hunt et al. (1983a) lead to a slightly weaker formulation, together with a constructive algorithm for the solution. The possibility of using noninteracting control techniques for the solution of such a problem was pointed out in Isidori et al. (1981a); the construction suggested here essentially recaptures an idea of Marino (1982). Additional results on this subject may be found in Sommer (1980), Hunt et al. (1983b), Boothby et al. (1985), and Cheng et al. (1985).

The observer linearization problem was studied independently by Bestle-Zeitz (1983) and Krener-Isidori (1983), for single-output systems, and by Krener-Respondek (1985) for multi-output systems.

Appendix

For a comprehensive introduction to the subjects dealt with in this appendix, the reader is referred to Boothby (1975), Warner (1970), Singer-Thorpe (1967).

REFERENCES

G. BASILE, G. MARRO

(1969) Controlled and conditioned invariant subspaces in linear systems theory, J. Optimiz. Th. & Appl. 3, pp. 306-315.

D. BESTLE, M. ZEITZ

(1983) Canonical form observer design for non-linear time-variable system, Int. J. Contr. 38, pp. 419-431.

W.A. BOOTHBY

(1975) "An Introduction to Differentiable Manifolds and Riemaniann Geometry", Academic Press: New York.

F.R. BRICKELL, R.S. CLARK

(1970) "Differentiable Manifolds", Van Nostrand: New York.

R.W. BROCKETT

(1972a) System theory on group manifolds and coset spaces, SIAM J. Contr. 10, pp. 265-284.

(1972b) On the algebraic structure of bilinear systems, in "Theory and Applications of Variable Structure Systems", R. Mohler and A. Ruberti Eds., Academic Press: New York, pp. 153-168.

(1976) Volterra series and geometric control theory, Automatica 12, pp. 167-176.

(1978) Feedback invariants for nonlinear systems, Preprints 6th IFAC Congress, Helsinki, pp. 1115-1120.

C. BRUNI, G. DI PILLO, G. KOCH

(1971) On the mathematical models of bilinear systems, Ricerche di Automatica 2, pp. 11-26.

P. BRUNOVSKY

(1970) A classification of linear controllable systems, Kybernetika 6, pp. 173-188.

D. CHENG, T.J. TARN, A. ISIDORI

(1985) Global external linearization of nonlinear systems via feedback, IEEE Trans. Aut. Contr. AC-30, to appear.

W.L. CHOW

(1939) Uber systeme von linearen partiellen differentialgleichungen ester ordnung, Math. Ann. 117, pp. 98-105.

D. CLAUDE

(1982) Decoupling of nonlinear systems, Syst. & Contr. Lett. 1, pp. 242-248.

D. CLAUDE, M. FLIESS, A. ISIDORI

(1983) Immersion, directe et par bouclage, d'un systeme non lineaire dans un lineaire, C.R. Acad. Sci. Paris 296, pp. 237-240.

P. CROUCH

(1981) Dynamical realizations of finite Volterra Series, SIAM J. Contr. Optimiz. 19, pp. 177-202.

P. d'ALESSANDRO, A. ISIDORI, A. RUBERTI

(1974) Realization and structure theory of bilinear dynamical systems, SIAM J. Contr. 12, pp. 517-535.

M.D. DI BENEDETTO, A. ISIDORI

(1985) The matching of nonlinear models via dynamic state-feedback, to be published.

D.L. ELLIOTT

(1970) A consequence of controllability, J. Diff. Eqs. 10, pp. 364-370.

M. FLIESS

(1974) Matrices de Hankel, J. Math. pures et appl. 53, pp. 197-224.

(1981) Fonctionnelles causales non lineaires et indeterminées non commutatives, Bull. Soc. math. France 109, pp. 3-40.

(1983) Realisation locale des systemes non lineaires, algebres de Lie filtrees transitives et series generatrices non commutatives, Invent. math. 71, pp. 521-537.

M. FLIESS, M. LAMNABHI, F. LAMNABHI-LAGARRIGUE

(1983) An algebraic approach to nonlinear functional expansions, IEEE Trans. Circ. Syst. CAS-30, pp. 554-570.

E. FREUND

(1975) The structure of decoupled nonlinear systems, Int. J. Contr. 21, pp. 651-659.

E.G. GILBERT

(1977) Functional expansion for the response of nonlinear differential systems, IEEE Trans. Aut. Contr. AC-22, pp. 909-921.

T. GOKA, T.J. TARN, J. ZABORSZKY

(1973) On the controllability of a class of discrete bilinear systems, Automatica 9, pp. 615-622.

J.W. GRIZZLE

(1985) Controlled invariance for discrete time nonlinear systems with an application to the disturbance decoupling problem, to be published.

G.W. HAYNES, H. HERMES

(1970) Nonlinear controllability via Lie theory, SIAM J. Contr. 8, pp. 450-460.

R. HERMANN

(1962) The differential geometry of foliations, J. Math. and Mech. 11, pp. 302-316.

R. HERMANN, A.J. KRENER

(1977) Nonlinear controllability and observability, IEEE Trans. Aut. Contr. AC-22, pp. 728-740.

R.M. HIRSCHORN

(1979a) Invertibility of nonlinear control systems, SIAM J. Contr. & Optimiz. 17, pp. 289-297.

(1979b) Invertibility of multivariable nonlinear control systems, IEEE Trans. Aut. Contr. AC-24, pp. 855-865.

(1981) (A,B)-invariant distributions and disturbance decoupling of nonlinear systems, SIAM J. Contr. & Optimiz. 19, pp. 1-19.

L.R. HUNT, R. SU, G. MEYER

(1983a) Design for multi-input nonlinear systems, in "Differential geometric control theory", R.W. Brockett, R.S. Millman and H. Sussmann Eds., Birkhauser: Boston, pp. 268-298.

(1983b) Global transformations of nonlinear systems, IEEE Trans. Aut. Contr. AC-28, pp. 24-31.

A. ISIDORI

(1973) Direct construction of minimal bilinear realizations from non-linear input-output maps, IEEE Trans. Aut. Contr. AC-18, pp. 626-631.

(1983) Nonlinear feedback, structure at infinity and the input-output linearization problem, in "Mathematical theory of networks and systems", P.A. Fuhrmann Ed., Springer Verlag: Berlin, pp. 473-493.

(1985) The matching of a prescribed linear input-output behavior in a nonlinear system, IEEE Trans. Aut. Contr. AC-30, to appear.

A. ISIDORI, A.J. KRENER, C. GORI GIORGI, S. MONACO

(1981a) Nonlinear decoupling via feedback: a differential geometric approach, IEEE Trans. Aut. Contr. AC-26, pp. 331-345.

(1981b) Locally (f,g)-invariant distributions, Syst. & Contr. Lett. 1, pp. 12-15.

A. ISIDORI, A. RUBERTI

(1984) On the synthesis of linear input-output responses for nonlinear systems, Syst. & Contr. Lett. 4, pp. 17-22.

B. JAKUBCZYK

(1980) Existence and uniqueness of realizations of nonlinear systems, SIAM J. Contr. & Optimiz. 18, pp. 455-471.

B. JAKUBCZYK, W. RESPONDEK

(1980) On linearization of control systems, Bull. Acad. Polonaise Sci. Ser. Sci. Math. 28, pp. 517-522.

R.E. KALMAN

(1972) Kronecker invariants and feedback, in "Ordinary differential equations", C. Weiss Ed., Academic Press: New York, pp. 459-471.

S.R. KOU, D.L. ELLIOTT, T.J. TARN

(1973) Observability of nonlinear systems, Inform. Contr. 22, pp. 89-99.

A.J. KRENER

(1974) A generalization of Chow's Theorem and the bang-bang Theorem to nonlinear control systems, SIAM J. Contr. 12, pp. 43-52.

(1977) A decomposition theory for differentiable systems, SIAM J. Contr. & Optimiz. 15, pp. 289-297.

(1985) (Adf,g), (adf,g) and locally (adf,g) invariant and controllability distributions, SIAM J. Contr. & Optimiz., to appear.

A.J. KRENER, A. ISIDORI

(1982) (Adf,G) invariant and controllability distributions, in "Feedback Control of Linear and Nonlinear Systems", D. Hinrichsen and A. Isidori, Eds. Springer Verlag: Berlin, pp. 157-164.

(1983) Linearization by output injection and nonlinear observers, Syst. & Contr. Lett. 3, pp. 47-52.

A.J. KRENER, W. RESPONDEK

(1985) Nonlinear observers with linearizable error dynamics, SIAM J. Contr. & Optimiz., to appear.

C. LESJAK, A.J. KRENER

(1978) The existence and uniqueness of Volterra series for nonlinear systems, IEEE Trans. Aut. Contr. AC-23, pp. 1090-1095.

C. LOBRY

(1979) Contrôlabilité des systèmes non lineaires, SIAM J. Contr. 8, pp. 573-605.

M. MALABRE

(1984) Structure a l'infini des triplets invariants: application a la pursuite parfaite de modele, in "Analysis and Optimization of Systems", A Bensoussan and J.L. Lions Eds., Springer-Verlag: Berlin, pp. 43-53.

R. MARINO

(1982) Feedback equivalence of nonlinear systems with applications to power systems equations, D.Sc. Dissertation, Washington University, St. Louis.

R. MARINO, W.M. BOOTHBY, D.L. ELLIOTT

(1985) Geometric properties of linearizable control systems, Math. Syst. Theory, to appear.

S.H. MIKHAIL, M.H. WONHAM

(1978) Local decomposability and the disturbance decoupling problem in nonlinear autonomous systems, Proc. 16th Allerton Conf., pp. 664-669.

S. MONACO, D. NORMAND-CYROT

(1983) The immersion under feedback of a multidimensional discrete-time nonlinear system into a linear system, Int. J. Contr. 38, pp. 245-261.

(1985) Invariant distributions for discrete-time nonlinear systems, Syst. & Contr. Lett. 5, pp. 191-196.

T. NAGANO

(1966) Linear differential systems with singularities and applications to transitive Lie algebras, J. Math. Soc. Japan 18, pp. 398-404.

H. NIJMEIJER

(1981) Controlled invariance for affine control systems, Int. J. Contr. 34, pp. 824-833.

(1982a) Controllability distributions for nonlinear systems. Syst. & Contr. Lett. 2, pp. 122-129.

(1982b) Invertibility of affine nonlinear control systems: a geometric approach, Syst. & Contr. Lett. 2, pp. 163-168.

(1983) Feedback decomposition of nonlinear control systems, IEEE Trans. Aut. Contr. AC-28, pp. 861-862.

(1985) Right invertibility of nonlinear control systems: a geometric approach, to be published.

H. NIJMEIJER, J.M. SCHUMACHER

(1985) Zeros at infinity for affine nonlinear control systems, IEEE Trans. Aut. Contr. AC-30, to appear.

H. NIJMEIJER, A.J. VAN DER SCHAFT

(1982) Controlled invariance for nonlinear systems, IEEE Trans. Aut. Contr. AC-27, pp. 904-914.

D. NORMAND-CYROT

(1983) Theorie et pratique des systèmes nonlineaires en temps discret, Thèse d'Etat, Université de Paris Sud.

W. RESPONDEK

(1982) On decomposition of nonlinear control systems, Syst. & Contr. Lett. 1, pp. 301-308.

W.J. RUGH

(1981) "Nonlinear System Theory: the Volterra-Wiener Approach", Johns Hopkins Press: Baltimore.

(1983) A method for constructing minimal linear-analytic realizations for polynomial systems, IEEE Trans. Aut. Contr. AC-28, pp. 1036-1043.

L.M. SILVERMAN

(1969) Inversion of multivariable linear systems, IEEE Trans. Aut. Contr. AC-14, pp. 270-276.

L.M. SINGER, J.A. THORPE

(1967) "Lecture Notes on Elementary Topology and Geometry", Scott, Foresman: Glenview.

S.N. SINGH, W.J. RUGH

(1972) Decoupling in a class of nonlinear systems by state variable feedback, Trans ASME J. Dyn. Syst. Meas. Contr. 94, pp. 323-329.

R. SOMMER

(1980) Control design for multivariable nonlinear time-varying systems, Int. J. Contr. 31, pp. 883-891.

E. SONTAG

(1979) "Polynomial Response Maps", Springer Verlag: Berlin.

R. SU

(1982) On the linear equivalents of nonlinear systems, Syst. & Contr. Lett. 2, pp. 48-52.

H. SUSSMANN

(1973) Orbits of families of vector fields and integrability of distributions, Trans. American Math. Soc. 180, pp. 171-188.

(1977) Existence and uniqueness of minimal realizations of nonlinear systems, Math. Syst. Theory 10, pp. 263-284.

H. SUSSMANN, V. JURDJEVIC

(1972) Controllability of nonlinear systems, J. Diff. Eqs. 12,pp. 95-116.

P.M. VAN DOOREN, P. DEWILDE, J. WANDEWALLE

(1979) On the determination of the Smith-MacMillan form of a rational matrix from its Laurent expansion, IEEE Trans. Circ. Syst. CT-26, pp. 180-189.

F.W. WARNER

(1970) "Foundations of differentiable manifolds and Lie groups",Scott, Foresman: Glenview.

M.H. WONHAM

(1979) "Linear Multivariable Control: a Geometric Approach", Berlin: Springer Verlag.

M.H. WONHAM, A.S. MORSE

(1970) Decoupling and pole assignment in linear multivariable systems: a geometric approach, SIAM J. Contr. 8, pp. 1-18.

Lecture Notes in Control and Information Sciences

Edited by M. Thoma

Vol. 43: Stochastic Differential Systems
Proceedings of the 2nd Bad Honnef Conference
of the SFB 72 of the DFG at the University of Bonn
June 28 – July 2, 1982
Edited by M. Kohlmann and N. Christopeit
XII, 377 pages. 1982.

Vol. 44: Analysis and Optimization of Systems
Proceedings of the Fifth International
Conference on Analysis and Optimization of Systems
Versailles. December 14–17, 1982
Edited by A. Bensoussan and J. L. Lions
XV, 987 pages, 1982

Vol. 45: M. Arató
Linear Stochastic Systems
with Constant Coefficients
A Statistical Approach
IX, 309 pages. 1982

Vol. 46: Time-Scale Modeling of Dynamic Networks
with Applications to Power Systems
Edited by J. H. Chow
X, 218 pages. 1982

Vol. 47: P. A. Ioannou, P. V. Kokotovic
Adaptive Systems with Reduced Models
V, 162 pages. 1983

Vol. 48: Yaakov Yavin
Feedback Strategies for Partially
Observable Stochastic Systems
VI, 233 pages, 1983

Vol. 49: Theory and Application of Random Fields
Proceedings of the IFIP-WG 7/1
Working Conference
held under the joint auspices of the
Indian Statistical Institute
Bangalore, India, January 1982
Edited by G. Kallianpur
VI. 290 pages. 1983

Vol. 50: M. Papageorgiou
Applications of Automatic Control Concepts
to Traffic Flow Modeling and Control
IX, 186 pages. 1983

Vol. 51: Z. Nahorski, H.F. Ravn, R.V.V. Vidal
Optimization of Discrete Time Systems
The Upper Boundary Approach
V, 137 pages 1983

Vol. 52: A. L. Dontchev
Perturbations, Approximations and Sensitivity Analysis
of Optimal Control Systems
IV, 158 pages. 1983

Vol. 53: Liu Chen Hui
General Decoupling Theory of Multivariable
Process Control Systems
XI, 474 pages. 1983

Vol. 54: Control Theory for Distributed
Parameter Systems and Applications
Edited by F. Kappel, K. Kunisch,
W. Schappacher
VII, 245 pages. 1983.

Vol. 55: Ganti Prasada Rao
Piecewise Constant Orthogonal Functions
and Their Application to Systems and Contro
VII, 254 pages. 1983.

Vol. 56: Dines Chandra Saha, Ganti Prasada
Identification of Continuous
Dynamical Systems
The Poisson Moment Functional
(PMF) Approach
IX, 158 pages. 1983.

Vol. 57: T. Söderström, P. G. Stoica
Instrumental Variable Methods
for System Identification
VII, 243 pages. 1983.

Vol. 58: Mathematical Theory of
Networks and Systems
Proceedings of the MTNS-83 International
Symposium
Beer Sheva, Israel, June 20–24, 1983
Edited by P. A. Fuhrmann
X, 906 pages. 1984

Vol. 59: System Modelling and Optimization
Proceedings of the 11th IFIP Conference
Copenhagen, Denmark, July 25-29, 1983
Edited by P. Thoft-Christensen
IX, 892 pages. 1984

Vol. 60: Modelling and Performance
Evaluation Methodology
Proceedings of the International Seminar
Paris, France, January 24–26, 1983
Edited by F. Bacelli and G. Fayolle
VII, 655 pages. 1984

Vol. 61: Filtering and Control of Random
Processes
Proceedings of the E.N.S.T.-C.N.E.T. Colloquium
Paris, France, February 23–24, 1983
Edited by H. Korezlioglu, G. Mazziotto, and
J. Szpirglas
V, 325 pages. 1984